Very High Energy Gamma Ray Astronomy

NATO ASI Series

Advanced Science Institutes Series

A series presenting the results of activities sponsored by the NATO Science Committee, which aims at the dissemination of advanced scientific and technological knowledge, with a view to strengthening links between scientific communities.

The series is published by an international board of publishers in conjunction with the NATO Scientific Affairs Division

A Life Sciences	Plenum Publishing Corporation
B Physics	London and New York
C Mathematical	D. Reidel Publishing Company
and Physical Sciences	Dordrecht, Boston, Lancaster and Tokyo
D Behavioural and Social Sciences	Martinus Nijhoff Publishers
E Engineering and	Dordrecht, Boston and Lancaster
Materials Sciences	
F Computer and Systems Sciences	Springer-Verlag
G Ecological Sciences	Berlin, Heidelberg, New York, London,
	Paris, and Tokyo

Series C: Mathematical and Physical Sciences Vol. 199

Very High Energy Gamma Ray Astronomy

edited by

K. E. Turver

Department of Physics, University of Durham,
Durham, U.K.

D. Reidel Publishing Company

Dordrecht / Boston / Lancaster / Tokyo

Published in cooperation with NATO Scientific Affairs Division

Proceedings of the NATO Advanced Research Workshop on
Very High Energy Gamma Ray Astronomy
Durham, U.K.
August 11-15, 1986

Library of Congress Cataloging in Publication Data

NATO Advanced Research Workshop on Very High Energy Gamma Ray Astronomy (1986:
    Durham)
    Very high energy gamma ray astronomy.

    (NATO ASI series. Series C, Mathematical and physical sciences; vol. 199)
    "Proceedings of the NATO Advanced Research Workshop on Very High Energy Gamma
Ray Astronomy, Durham, U.K., August 11–15, 1986"—T.p. verso.
    "Sponsored by the Scientific Affairs Division of NATO and the University of Durham"—
Pref.
    "Published in cooperation with NATO Scientific Affairs Division."
    Includes Index.
    1. Gamma ray astronomy—Congresses. 2. Radio sources (Astronomy)—Congres-
ses. 3. Nuclear astrophysics—Congresses. I. Turver, K. E. II. North Atlantic Treaty
Organization. Scientific Affairs Division. III. NATO ASI series. Series C, Mathematical and
physical sciences; vol. 199.
    QB471.N38   1986   523.01'97222   87–4641
    ISBN 978-94-010-8205-1   ISBN 978-94-009-3831-1   (eBook)
    DOI 10.1007/978-94-009-3831-1

Published by D. Reidel Publishing Company
P.O. Box 17, 3300 AA Dordrecht, Holland

Sold and distributed in the U.S.A. and Canada
by Kluwer Academic Publishers,
101 Philip Drive, Assinippi Park, Norwell, MA 02061, U.S.A.

In all other countries, sold and distributed
by Kluwer Academic Publishers Group,
P.O. Box 322, 3300 AH Dordrecht, Holland

D. Reidel Publishing Company is a member of the Kluwer Academic Publishers Group

CONTENTS

CONTRIBUTED PAPERS.

CONTENTS

PREFACE

An Advanced Research Workshop on Very High Energy Gamma Ray Astronomy and Related Topics was held at Durham, England during August 11-15 1986. The meeting was sponsored by the Scientific Affairs Division of NATO and the University of Durham.

It is four years since the first Workshop dedicated to High Energy Gamma Ray Astronomy was held at Ootacamund, India. At that meeting the developments in Very High Energy Gamma Ray Astronomy over a period of more than 20 years were reported and the methodology, limitations , improvements and prospects for further progess were discussed. The possible requirement for a follow-up meeting was clear if the optimistic future foreseen for the field at the Ooty meeting was correct. The Durham meeting was suggested to fill this role. Although the arrangements for the Durham meeting were discussed as long ago as 1983 with possible dates in 1984 or 1986, the eventual date in 1986 has proved admirable and has coincided with a time when further advances have been reported. An important feature of the proposal for the Durham meeting was the emphasis on a series of Workshop sessions, the conclusions of each to be summarized by a Rapporteur. The purpose of these sessions was to provide a consensus view of many of the important areas in the field at a time of increasing interest by the rest of the astrophysics community. Although such sessions are difficult to arrange the view of many participants was that they had fulfilled their role and thanks are due especially to the participants who acted as Discussion leaders and Rapporteurs

In these Proceedings we have grouped the Invited papers together and have followed them with the seven Rapporteur papers. The short Contributed papers conclude the report. No attempt to note verbatim the discussions following the Invited or Contributed papers was made because it was hoped that relevant points would be discussed in the Workshop seesions.

The Workshop was made possible by the support of the NATO Science Committee which is gratefully acknowledged. The success of the meeting was due in part to the efforts of the two Organizing Committees. The Organizing Committee reponsible for obtaining the financial support and for the outline of the scientific programme comprised Dr J J Quenby, Dr T C Weekes, Dr K E Turver and Professor A W Wolfendale FRS. The Local Organizing Committee which was responsible for the day-to-day

arrangements for the meeting was drawn from the staff of the Durham High
Energy Gamma Ray Astronomy group and included Dr K E Turver, Dr K J
Orford, Dr T J L McComb, Dr N A Dipper, Miss P M Chadwick and Mr E W
Lincoln. Mrs Mary Bradley was the Workshop Secretary and has also
contributed greatly to the production of these Proceedings.

Durham,   K E Turver
October 1986.

HIGH ENERGY RADIATION FROM THE YOUNG PULSARS

F. Graham Smith
University of Manchester
Nuffield Radio Astronomy Laboratories,
Jodrell Bank, Macclesfield, Cheshire SK11 9DL

The young pulsars, typified by the Crab and Vela Pulsars, are distinguished by their short periods and large rates of period increase. The category does not include the "millisecond" pulsars, which have very small rates of period increase: these are old pulsars, whose high spin rates are the result of a spin-up process in a binary system.

We make this distinction because of the observed pulse structure in the young pulsars. Old pulsars usually produce a single radio pulse and nothing at higher photon energies. Some produce a double radio pulse, usually separated by about half a period, i.e. by 180° in rotation. The high energy photons from the young pulsars occur in a double pulse separated by 0.4 periods, i.e. by about 145°.

The Crab pulsar shows both kinds of pulse together at low radio frequencies. A pair of pulses coincides with the pair seen at all energies up to the highest gamma-ray energies, and a single pulse with a different spectrum is seen immediately before the first of the pair. This single pulse, the "precursor" corresponds to the radio pulse seen in the older pulsars. The pair seen from infra-red upwards in energy covers more than ten decades in spectrum, with almost the same shape. This paper interprets the pair of pulses as curvature radiation from restricted locations in the outer magnetosphere. The radio pulses which coincide with the high-energy pulses must originate in the same location, but the radiation mechanism is different; notably they are so bright that the emission must be coherent.

There are several other complications in the observed pattern of pulses: for example, the radio pulse in the Vela Pulsar is a typical radio pulse, and the gamma-rays occur in the pair, but there is no X-ray pulse and the optical pulse is in a closer pair. Here we concentrate on the high-energy pair.

Why can we not assign the same type of source to all pulsar radiation? There is good evidence that the normal radio pulse originates near a magnetic pole, or from both poles when the orientation and the position of the observer are suitable. The radio emission is in a beam whose edges are defined by the magnetic field lines of the polar cap. The same location cannot apply for gamma rays, since these would convert into positron-electron pairs on their passage through the strong magnetic

1

K. E. Turver (ed.), Very High Energy Gamma Ray Astronomy, 1–5.
© *1987 by D. Reidel Publishing Company.*

field of the magnetosphere. We must look for a location more than half-
way out to the velocity of light cylinder.

All the high energy radiation from the Crab Pulsar, and the gamma
rays from Vela, occur in the pulse pair separated by 0.4 rotations.
This spacing is very stable: it has not changed in the Crab by 1° of
rotation in 16 years (Jones et al. 1980, and data from 1985). The whole
pulse profile has changed by less than 1% of the peak in that time. The
beam forming the pulse is anchored firmly to the neutron star by the
magnetic field; the timing of the pulse is used to follow the rotation
phase of the pulsar within an accuracy of about 1° rotation, corresponding
to about 200 m on the surface.

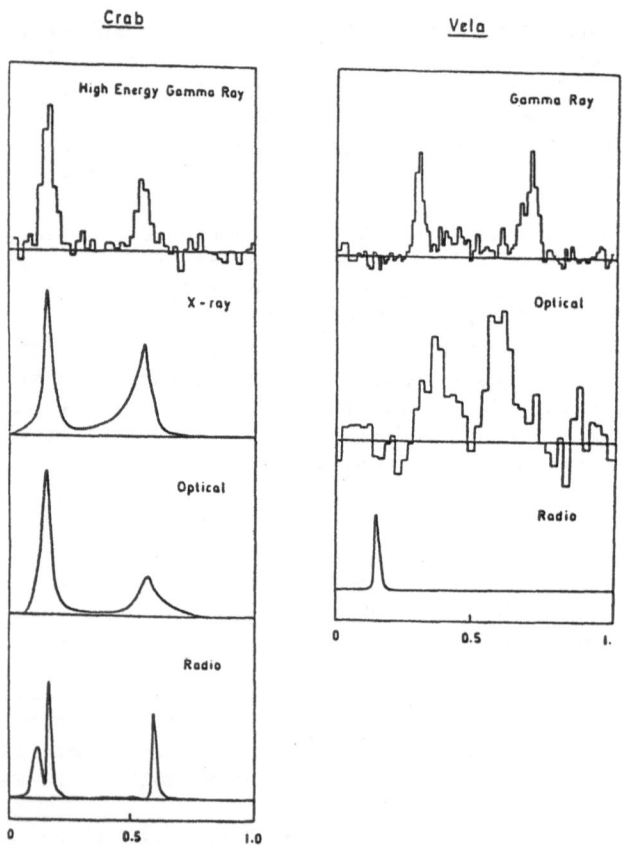

Fig. 1.   Observed pulse profiles for the Crab
          and Vela Pulsars.

Incidentally the timing measurements at Jodrell Bank now extend
continuously over four years. No glitch of more than 100 microseconds
has occurred during this time, and the rotation phase variations are
contained within one rotation.

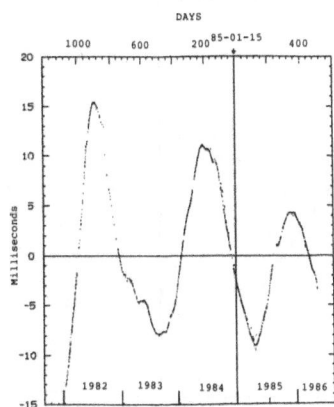

Fig. 2.   Timing Residuals for the Crab Pulsar.

The structure of the beam is evidently determined by the dipole
magnetic field of the neutron star. I have used a simple model (Smith
1986) in which the magnetic dipole is orthogonal to the rotation axis,
and the observer's line of sight is also orthogonal to the rotation axis.
There is a relativistic drag on the field configuration, but we can
assume that the plasma itself does not force a further distortion. In
the model I have looked at the geometry of emission from the field lines
which just close at the velocity of light cylinder: these are the field
lines which close in are at the edge of the polar cap. The emission is
due to the streaming of high energy particles along these field lines.
The mechanism is curvature radiation.

This location was pointed out recently by Cheng, Ho and Ruderman
(1986) as the only region where particles can be accelerated to very
high energies. The charge density is depleted in a slot-like gap, the
outer magnetosphere gap, and a large electric field develops along the
magnetic field lines. Electron-positron pairs are created here by high
energy gamma rays: the particles are accelerated and radiate gamma rays
which in turn create more pairs. This cascade process occurs throughout
the gap region, which extends over the outer half of the magnetosphere.

My analysis was purely geometrical. Allowing for the relativistic
distortion of the field lines, and for aberration, what would be the
arrival times of pulses due to beamed radiation from the limiting field
lines? Further, how would these times relate to the arrival times of
radio pulses from the polar caps? I found that the simple model fitted
the observed times if the double pulse was emitted at about 0.9 of the
radial distance to the light cylinder.

Analysis of the spectrum and intensity of the observed radiation
from the Crab Pulsar now allows an estimate of the numbers and energy
of the high-energy particles in the outer magnetosphere gaps. Curvature
radiation from a particle with relativistic factor Γ extends to
frequencies above Γ^3 times the fundamental $(2\pi R)^{-1}$, where R is the
radius of curvature of the field line. The critical frequency ν_c is

$$\nu_c = 7.2 \times 10^9 \ R^{-1} \ \Gamma^3$$

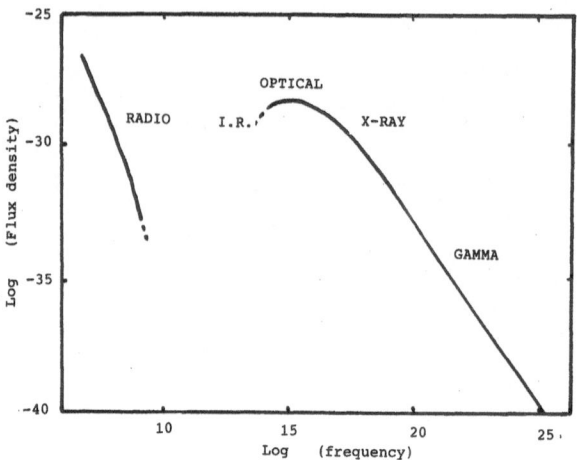

Fig. 3.   Spectrum of the Crab Pulsar.

We observe radiation up to about 10 ν_c. The highest energy gamma-rays
are at 10^{27} Hz, and R is 10^8 cm in the outer gap. Hence we find $\Gamma = 10^8$.
At the lowest energy, in the infrared, we observe $\nu = 10^{14}$ Hz, requiring
$\nu_c \approx 10^{16}$ Hz and $\Gamma = 10^4$.

    The spectral index γ for particle energy is related to the radiation
spectral index α by

$$\gamma = 3\alpha + 1 \quad .$$

At high energy $\gamma = 4$ and at low energy $\gamma = 1$. Most of the particle
energy is contained in particles with energy $\Gamma = 10^4$ to 10^4, i.e. 10^{10}
to 10^{11} eV. The highest energy particles lose half their energy in
one microsecond. Equilibrium between the accelerating field and loss
by curvature radiation occurs at 5×10^{13} eV, but most particles have
lower energies.

    The total radiated power is 10^{35} erg sec^{-1}. The power radiated
by a single electron is

$$10^{-8.3} \Gamma^4 R^{-2} \text{ erg sec}^{-1} \quad .$$

Hence the observed power is effectively the radiation from 10^{37} electrons
with $\Gamma = 10^4$. Since these traverse the radiating zone in about 1 millisecond, the outward flow of energy is calculated to be 10^{38} erg sec^{-1};
this is a flow leaving the magnetosphere and providing the energy for the
surrounding Crab Nebula. It is in fact the energy which is known to be
required for the maintenance of the Nebula. In comparison the rate of
loss of rotational energy is 10^{40} ergs.

    Less detail is available for the Vela Pulsar. The energetic
particles radiating the gamma-rays must have relativistic factor $\Gamma > 10^7$,
but the spectrum does not extend to lower energies. The optical pulses
may originate closer to the pulsar surface, where the model allows for a
closer pair of pulses.

A useful test of the model is provided by observations of the polarisation of the optical pulses. (If the model is correct the same polarisation would be observed in X-rays and gamma rays). In the simple model
the linear polarisation would reach a maximum of about 65%, and the
position angle would be constant through the period. Radiation can be
observed through the whole period, the minimum level being about 1% of
the peak. Observations reported by Jones et al. (1980) have now been
repeated with improved accuracy using the Isaac Newton Telescope on La
Palma. Preliminary results show that the plane of polarisation lies
within a range of only 60°, which is encouragingly close to zero; the
polarisation reaches 50%, but only well away from the pulse peaks. We
require a geometric model that depolarises the radiation at the pulse
peaks: this would not be surprising if the source is considerably
extended.

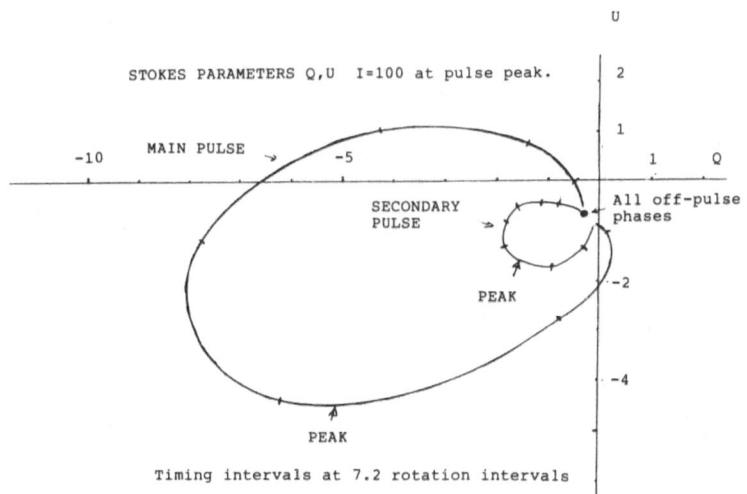

Fig. 4.   Optical Polarisation of the Crab Pulsar.
The Stokes Parameters Q,U are plotted
through the main and secondary pulses.

We may conclude that particles with energies from 10^{10} to 5×10^{13}eV
are generated within the pulsar magnetosphere, and stream out beyond the
velocity of light cylinder. Their fate in the surrounding nebula depends,
of course, on the density of that nebula, but it is certainly reasonable
to regard pulsars as the origin of at least part of the cosmic rays.

References

Cheng, K.S., Ho, C. & Ruderman, M.A., 1986. Astrophys.J. 300, 500
Jones, D.H.P., Smith, F.G. & Nelson, J.E., 1980. Nature, 283, 50.
Smith, F.G., 1986. Mon.Not.R.astr.Soc., 219, 729.

X-RAY EMISSION FROM MASS ACCRETING NEUTRON STARS

J. Truemper
Max-Planck-Institut fuer Physik und Astrophysik
Institut fuer Extraterrestrische Physik
8046 Garching b. Muenchen, FRG

ABSTRACT. After a brief review of the general situation concerning mass accreting compact stellar objects in binary systems we concentrate on a single source - Hercules X-1 - which is the best-known among all accreting neutron stars. In particular we shall discuss recent progress which has been made in understanding (1) the geometrical situation of the neutron star and its immediate surroundings and (2) the 35-day-cycle of the source.

1. INTRODUCTION

Most of the approximately 100 known bright X-ray sources in our galaxy are compact objects in binary systems. They exhibit a wide range of phenomena primarily observed in X-rays and optical light - due to the rather large number of key parameters describing their physical situation:
- mass and evolutionary stage of the stellar companion
- mass and nature of the compact object
- nature of the mass transfer (wind, disk)
- mass transfer rate
- magnetic field of the stars
- magnitude and orientation of the angular momenta (stellar, orbital)

As far as black holes are concerned there are now two good candidates; Cyg X-1 and 0620-00 (McClintock 1986), for which the black hole argument rests on the mass derived for the compact object. Unfortunately, some other black hole indicators (high time variability, or the X-ray spectral appearance) have proven to be unreliable: Circinus X-1 which had been on the candidate list because of its timing and spectral characteristics resembling those of Cyg X-1 has been identified as a burster (Tennant et al. 1986).

Systems containing a degenerate dwarf (cataclysmic variables) can be classified in three main types depending on the magnetic moment of the degenerate dwarf: (1) In the case of the dwarf novae the magnetic moment is not large enough to influence the matter flow; (2) in the DQ

7

K. E. Turver (ed.), Very High Energy Gamma Ray Astronomy, 7–21.

Her system or intermediate polars the accretion flow is stopped at the
boundary of the white dwarf's magnetosphere and channeled onto the
magnetic poles leading to X-ray pulsations. (3) In the AM Her system the
magnetic moment of the DD is so strong that the interaction with the
magnetic field of the companion leads to a situation of bound rotation.
For recent reviews of the cataclysmic variables see Cordova and Mason
(1983) and Mason (1985).

The binary systems containing a neutron star can be crudely divided
into two groups: (1) the high mass binary systems which are young and
often show X-ray pulsations indicative of a strong magnetic field
($\sim 10^{12}$ G) of the neutron star. (2) the low mass X-ray binaries (LMXB)
which form an old (galactic bulge) population containing neutron stars
with rather low ($< 10^{10}$ G) magnetic fields. Specific X-ray phenomena
exhibited by members of this group are X-ray bursts and quasiperiodic
oscillations (QPO). Recent reviews on high mass X-ray binaries (HMXB)
can be found in Rappaport and Joss (1983) and on low mass X-ray binaries
in Lewin and Joss (1983) and van der Klis (1986). The optical properties
of theses sources have been reviewed by van Paradijs (1983).

We conclude this very crude introductory overview by pointing out
the analogies between degenerate dwarf and neutron star systems in
table I.

TABLE I

Deg. Dwarfs	Neutron stars	remark
dwarf novae	LMXB	weak magnetic momenta QPO's, bursts
DQ Her systems (intermediate polars) $P_{rot} \neq P_{orbit}$	HMXB	medium magnetic momenta X-ray pulsations
Am Her system $P_{rot} = P_{orbit}$	-	high magnetic momenta

The neutron star analogy to the AM Her system does not exist because
the maximum magnetic momenta of neutron stars are comparatively small
($\mu \sim B_s R^3$).

2. HER X-1, A KEY OBJECT

In the following I'd like to discuss an object for which the most
detaild knowledge exists, namely Hercules X-1. In doing so I shall
mainly concentrate on recent developments based on EXOSAT observations
which shed new light onto the 35-day cycle of this source. However, it
may be appropriate to start with a brief summary of the established
properties of Her X-1.

The key parameters of the system are summarized in table II. Let me add a few remarks concerning the history and orientation of the neutron star spin.

There are spin-up/-down phases with scales of weeks discovered by Uhuru, but also long-term changes: Spin-up between 1972 and 79, a spin-down between 79 and 83 and a spin-up since May 84. These spin-up/-down variations strongly support the idea that the source is (on average) in rotational equilibrium with the inner edge of the accretion disk. This

Table II: Key parameters of Her X-1

mass of the neutron star	$1.3\ M_\odot$
mass of its companion	$2.2\ M_\odot$
P_{orbit}	1.7 days
inclination of the orbital plane	$\sim 85^\circ$
P_{rot} (neutron star)	1.24 s
X-ray heating of the companion	A5/F0 $\rightarrow$ B0
polar magnetic field (cyclotron lines)	4×10^{12} G
distance	~ 6 kpc
age	$\sim 10^7$ a
existence of an accretion disk?	yes

requires a radius of the magnetosphere (disk boundary of $\sim 1.8 \times 10^8$ cm or ~ 200 neutron star radii and a magnetic momentum of $\sim 10^{30}$ G cm³ which is in very rough agreement with that derived from the measured polar field strength (4×10^{12} G $\times (10^6)^3 = 4 \times 10^{30}$ G cm³). Turning the argument around one can conclude that the polar field strength of 4×10^{12} G is that of the dipole component and not a "fringe field" as discussed by Ruderman (1985) recently.

As far as the orientation of the neutron star spin is concerned information can be derived from a comparison of the observed spectra and pulse shapes with the prediction of polar cap radiation models: The energy dependence of the 1.24 s pulse profiles suggests that the inclination of the spin axis with the line of sight is ~ 45 to 60° (Nagel 1981a,b). The same range of inclinations is found from an analysis of the variation of the cyclotron line shape as a function of pulsational phase (Voges et al. 1985). Such an inclination implies a substantial tilt ($\gtrsim 45^\circ$) of the neutron star spin with respect to the orbital angular momentum.

3. THE 35-DAY CYCLE OF HER X-1

The 35-day cycle of Hercules X-1 which was discovered by Tananbaum et al. (1972), has not yet been satisfactorily explained. Various mechanisms have been proposed to account for this phenomenon: asymmetric injection of matter and precession of the accretion disk (Katz, 1973); precession of HZ Her and a slaved accretion disk (Roberts, 1974; Petterson, 1975, 1977); precession of the neutron star (Brecher, 1972; Pines,

Pethick, and Lamb, 1973; Lamb et al., 1975); non-synchronous rotation of
HZ Her (Pringle, 1973; Bisnovatyi-Kogan and Komberg, 1975); the non-
synchronous rotation of a star spot on HZ Her (Henriksen, Reinhardt and
Aschenbach, 1973); non-linear oscillations (Wolff and Kondo, 1978,
Kondo, Van Flandern and Wolff, 1983); self-excited oscillations in HZ
Her (Arons, 1973; McCray and Hatchett, 1975); presence of a third star
(Mazeh and Shaham, 1977); occultation by circumstellar gas streams
(Burke, 1976); self-excited non-linear mass flow oscillations in the
accretion disk (Meyer and Meyer-Hofmeister, 1984).

Merely from a phenomenological point of view the model of a precessing accretion disk obscuring the primary X-rays has been most successful in explaining the various details of the X-ray and the optical
behaviour (e.g. Gerend and Boynton, 1976). However, the physical nature
of disk precession has been a matter of controversy and doubts concerning its mechanical feasibility have been raised (Papaloizou and
Pringle, 1982; Kondo, Van Flandern and Wolff, 1983). The slaved disk
models have been criticized for their incompatibility with the observed
eccentricity of the system which is very close to zero, the difficulty
of persistent precession of a fluid body, and the poor quality of the
35-day clock (Chevalier, 1976; Boynton, Crosa and Deeter, 1980).

The results from a 35-day cycle sampling of the X-ray emission with
the EXOSAT observatory have shed new light onto this problem. Our data
show clear evidence for changes of the 1.24 sec pulse profiles with the
35-day phase. Most of these pulse profile changes have already been
noticed (Doxsey et al., 1973; Holt et al., 1974; Boynton and Deeter,
1976; Pravdo et al., 1977; Joss et al., 1978; Staubert et al., 1979;
Gruber et al., 1980; Bai, 1981). However, the quality of our data allows
us to extend these trends in the pulse profile variations to the short-
on state in the middle of the 35-day cycle. We present arguments that
support the interpretation of our results in terms of the precession of
the neutron star, Her X-1.

4. OBSERVATIONS AND RESULTS

The observations were carried out from 1 March to 5 April 1984 with the
low (LE, 0.02-2 keV) and medium energy (ME, 1 - 30 keV) detectors of
EXOSAT at sampling intervals of about four days (for a description of
the payload and detectors see Taylor et al., 1981; de Korte et al.,
1981; Turner, Smith and Zimmermann, 1981). A total of 11 observations
were made and the exposure times varied between 10^4 to 4×10^4 seconds.
Preliminary results have already been reported (Kahabka et al., 1984,
1985; Oegelman et al., 1984, 1985; Voges et al., 1985). These measurements include, for the first time, an extensive and detailed coverage of
the short-on state which appears at phase $\Psi_{35} = 0.67$, where Ψ_{35} is the
phase of the 35-day cycle with respect to the turn-on which is 0.15
phase units earlier than the maximum intensity phase. We denote by
"main-on" and "short-on" the portions of the 35-day cycle between $0.0 \leq
\Psi_{35} \leq 0.4$ and $0.6 \leq \Psi_{35} \leq 0.8$ respectively. The data discussed in
this paper correspond to 5 consecutive observations at $\Psi_{35} = 0.23$,
0.33, 0.67, 0.13, and 0.23.

Using the four main-on observations of the LE pulses, which have a more stable shape in comparison to the ME pulses, we were able to determine a P and $\dot{P}$ value of the 1.24 sec pulsations covering the full 35-day cycle as (Oegelman et al., 1985):

$$P = 1.23779200 \pm (0.00000005) \text{ sec}$$
$$\dot{P} = (-2 \pm 1) \times 10^{-13} \text{ sec/sec}$$
$$\text{at epoch JD 244 5778.56}$$

When this ephemeris was used to examine the phase of the short-on state LE pulses, it was discovered that the minimum was shifted by $180° \pm 10°$ with respect to the main-on state LE pulses. Since we do not have continuous coverage of the pulses during the 35-day cycle, in principle, it is possible that variations of $\dot{P}$ in between the main-on states could cause this shift. However, this is unlikely because the variations of $\dot{P}$ would have to be an order of magnitude larger than the average value, and be just the right amount to produce the $180°$ shift.

Using the same ephemeris, we have investigated the behaviour of the ME pulses throughout the 35-day cycle. The results are shown together with the LE pulse shapes in figure 1. As can be seen from this figure, the soft X-ray component shows the well known roughly sinusoidal pulse shape (e.g. McCray et al., 1982) during the main-on state, with peak positions shifted by ~180° with respect to the ME pulse peaks. As already discussed, during the short-on state the phase of the LE pulse is shifted by 180° with respect to the main-on state LE pulse.

The ME pulse profiles show a double peaked main pulse and an interpulse during the main-on state which confirms the results of other observers (e.g. Joss et al., 1978). A comparison of the pulse profiles observed at $\Psi_{35} = 0.13$ and 0.23 shows a change in the leading edge of the main pulse which has already been noted by Gruber et al. (1980). On the declining flank of the main-on state $\Psi_{35} = 0.33$ we find the well known sinusoidal pulse profile (Joss et al., 1978; Ohashi, 1984). The most striking feature that comes out of our observation is the fact that the pulse profile becomes quite different during the short-on state at $\Psi_{35} = 0.67$ as compared with the main-on state. As can be seen in figure 1, not only the pulse and interpulse show a reversal of their relative importance, but both of the pulses become narrower during the short-on state. This is also reflected in a change of the separations of the double peaks observed both in the main pulse and interpulse. While during the main-on state, the doublet separation is 0.2 for the main pulse during the short-on state the separations decrease to 0.12 and 0.13, for the pulse and interpulse, respectively. Since the pulse profiles change slightly with energy the quoted figures have been derived for a narrow energy interval between 2 and 4 keV. At energies higher than 10 keV the double peak structure becomes less pronounced and the pulse ratios change somewhat. However, the trends mentioned above still persist. A detailed discussion of the variations of the pulse profile with energy will be given elsewhere.

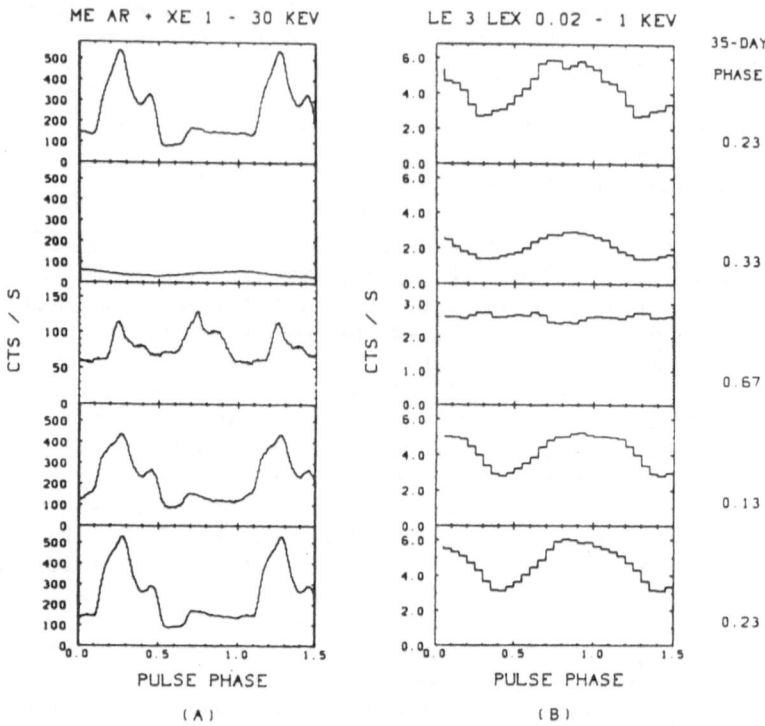

Figure 1. Light curves of the 1.24 sec pulses observed at various phases of the 35-day cycle for: a) hard X-rays (ME) in the range 1 to 30 keV; b) soft X-rays (LE with 3000A Lexan filter) in the range 0.02 to 2 keV. The scale of the ME and LE light curves at Ψ_{35} = 0.67 (short-on state) has been magnified by a factor of four and two, respectively. The typical statistical errors are indicated at the upper right hand corner of each curve. The 35-day phase Ψ_{35} (measured with respect to turn-on), which progresses from top to bottom, is also indicated for each pair of ME, LE pulse shapes.

A comparison of the pulse profiles at phases Ψ_{35} = 0.13 and 0.23 shows a reduction in width of the main pulse. This decrease is also found in a continuous two day observation of Her X-1 (Ψ_{35} = 0.14 to 0.19) which will be reported elsewhere. In summary, the data show continuous changes of pulse width and peak separations during the 35-day cycle.

5. PRECESSION OF THE DIPOLE AXIS

The strong change in 1.24 sec pulse profiles between main-on state and short-on-state is in gross conflict with the widely accepted model which assumes a neutron star with a fixed rotational axis and inclined polar radiation pattern, and which explains the 35 day modulation solely by an absorbing, precessing accretion disk. In this model the pulse and interpulse should be reduced by the same factor during the short-on state. Actually, these relative pulse intensities reverse. Furthermore, the phase separations of the double structures observed both in the pulse and the interpulse change with the 35-day cycle, showing a systematic narrowing with increasing Ψ_{35}. It is difficult to imagine that these variations are produced by obscuration effects during the 35-day cycle as assumed in the precessing accretion disk models. Likewise it is difficult to understand these variations in pulse shape in terms of a modulation of the mass accretion rate during the 35-day cycle. The most straightforward interpretation is that pulse and interpulse represent the two polar beams which we see under changing aspect angles. The data suggest that during the main-on state the "first pole" is seen pole-on while the interpulse is produced by the "second pole" which is barely visible during the main-on state but becomes prominent during the short-on state. This interpretation is supported by the fact that the soft X-ray light curves also show a change of 180° ± 10° between main-on and short-on states indicating that we are seeing the domination of the second pole during the short-on phase.

The observed variation of the ME pulse profiles strongly supports the idea that the primary X-ray beam is of the pencil type. Such a beam pattern is suggested also by the interpretation of the phase-dependent continuum (Pravdo et al., 1977) and cyclotron line feature in hard X-rays (Voges et al., 1982). The fact that both pulse and interpulse show a double peak structure indicates that the pencil beam may have a hollow-cone structure as predicted by slab-type models for the polar radiation (Basko and Sunyaev, 1976; Nagel, 1981a, 1981b). In this case the peak separation would depend on how close the line of sight passes the polar direction when sweeping the beam. At any rate the peak flux observed will depend on this minimum polar angle which we call θ_{ik}. Because the main pulse of the main-on state is very prominent, we estimate that $\theta_{11} \leq 5°$, while for the interpulse we assume $\theta_{12} \sim 75°$. Since fluxes of pulse and interpulse are not too different during the short-on state, the corresponding angles θ_{ik} should not be too different either (θ_{21} $\theta_{22} \sim 20°$). Although these angles are only approximate, they give an idea of the magnitude of the required beam-pattern shifts necessary to account for the data. We note that these angles would have to be modified if, in addition to precession, absorption effects depending on the 35-day phase contribute to the amplitude modulation. Furthermore, the accretion flow geometry may also change with the 35 day precession and influence the pulse shapes (Lamb et al., 1975).

A priori, there are two mechanisms which can change the magnetic dipole axis during the 35-day cycle; forced or free precession. However, to produce the 35-day cycle by forced precession the required torque is 10^6 times larger than that which could be supplied by the accretion disk and this seems quite unlikely.

Free precession of the neutron star as the cause of the 35-day
cycle of Her X-1 has been suggested by several authors (Brecher, 1972;
Pines, Pethick, and Lamb, 1973; Lamb et al., 1975). The precession
ω_p and the rotational ω_r are related by $\omega_p = \omega_r\cos\Phi(I_{\shortparallel} - I_{\perp})/I_{\shortparallel}$, where
$(I_{\shortparallel} - I_{\perp})/I_{\shortparallel} \neq 0$ is the asymmetry in the moments of inertia (oblateness) and Φ is the angle between spin and figure axis. If χ is the an-
gle between the figure axis and the magnetic dipole (beam) axis, the
angle β between spin axis and magnetic dipole axis will vary according
to $\cos\beta = \cos\Phi\cos\chi + \sin\Phi\sin\chi\cos\omega_{ss}t$. The limiting angles are $\beta_{max} =$
$\chi + \Phi$ and $\beta_{min} = \chi - \Phi$.

For a rotating and precessing beam the minimum polar angles Θ_{ik}
will depend on the inclination angle α of the spin axis with respect to
the line of sight and on the angle β. Assuming that the limiting angles
are reached at $\Psi_{35} \simeq 0.15$ and 0.65, we can derive these angles α,
β_{max} and β_{min} from the minimum polar angles Θ_{ik} estimated above. Using
an angle of $i = 85°$ of the disk axis with respect to the line of sight
we find $\alpha \sim +30°$ (tilted towards the observer). Another possible solution is $\alpha \sim -40°$ (tilted away from the observer). We prefer this latter
solution for reasons to be discussed in section 4. In both cases we find
$\beta_{min} \sim 80°$, and $\beta_{max} \sim 130°$ which corresponds to $\Phi = 25°$ and $\chi = 75°$
($= 105°$). We cannot say anything about the sidewards tilt of the spin
axis, at this stage. In figure 2 we sketch the geometry discussed above.

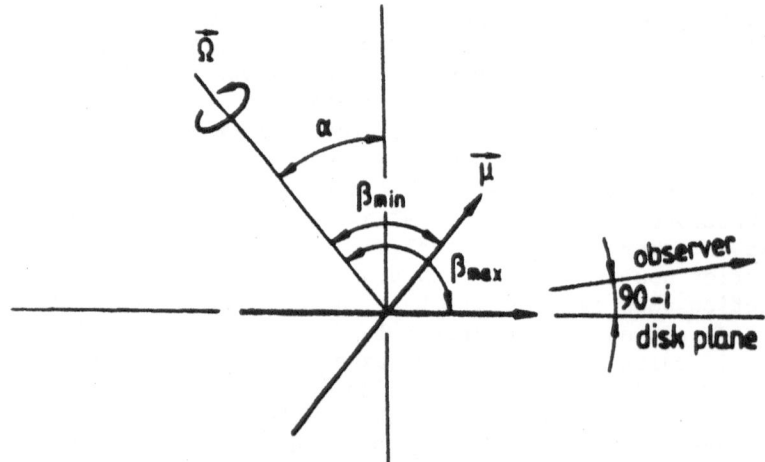

Figure 2. Geometrical parameters of the freely precessing neutron star
as implied by the observations. The spin axis Ω is tilted with respect
to the accretion disk (orbital plane) by $\alpha \sim 40°$. The magnetic dipole
moment μ of the neutron star precesses back and forth between $\beta_{max} \sim$
$130°$ and $\beta_{min} \sim 80°$.

6. THE RESPONSE OF THE DISK AND REPROCESSING REGIONS

Owing to the broad intensity patterns of the poles (FWHM ~ 90°), their precession alone will not be able to provide for the observed, almost complete obscuration of the X-ray source during the off states of the 35-day cycle. There must be an additional mechanism controlled by the neutron star's precession. Prime candidates for such an effect are the Alfven shell and the warp of the inner disk whose geometries should change with precession phase.

Several authors have discussed the soft X-ray flux during the on-state and the iron line emission in terms of reprocessing of the primary beam which is intercepted by an "Alfven shell" located at a distance from the neutron star which is somewhat larger than the corotation radius (McCray, and Lamb, 1976; Basko, and Sunyaev, 1976; Bai, 1980, Basko, 1980). This Alfven shell is thought to represent accreting material floating on top of the magnetosphere from the disk to the poles (McCray, and Lamb, 1976). Bai (1980) has pointed out that the Alfven shell material must be clumpy in order to make it transparent for the primary beam. McCray et al. (1982) favour reprocessing at the inner edge of the accretion disk to explain the soft X-ray flux.

Our present observations yield new constraints on the geometrical and physical conditions of the reprocessing regions in Her X-1 which we shall discuss elsewhere (Kahabka et al., 1986). As far as the history of the reprocesssed component through the 35-day cycle is concerned its intensity is high during the main-on and short-on state (e.g. Kahabka et al., 1985), while during the off-state these components are extinguished along with the primary component. Since the Alfven shell can hardly provide an obscuration of the central source and self-obscuration at the same time, we can argue that the main obscuration mechanism must be something else.

Another observation mechanism is connected with the fact that in the case of a precessing dipole axis, one may expect a change of the shape of the innermost disk as a function of the 35-day cycle. Let us consider a rotating neutron star having the rotation vector Ω tilted by an angle α with respect to the normal of the disk. The magnetic dipole vector μ rotates at an instantaneous angle β around the spin vector. This angle changes slowly throughout the 35-day cycle as a consequence of the precession, between β_{max} and β_{min}. Lipunov and Shakura (1980) and Lipunov, Semenov and Shakura (1981) have shown that the interaction between the accretion disk and the magnetic field of the neutron star, in conjunction with the viscous forces in the disk, will act so as to turn the inner part of the accretion disk along the neutron star's rotational equator. The Lense-Thirring effect (Bardeen and Petterson, 1975) although weaker, will act in the same way. As a result, most of the time the inner disk region is expected to be coplanar with the rotational equator of the neutron star which is fixed in space. In particular, Lipunov, Semenov and Shakura (1981) derive the distance R_{pl} at which the disk will begin to coincide with the equatorial plane of the neutron star. Using Her X-1 parameters we get $R_{pl} \sim 10^{9} \cos\alpha (3\cos^2\beta - 1)^{2/9}$ cm, which is larger than the Alfven radius $R_A \sim 2\times10^{8}$ cm for a range of β angles. It is interesting to note that at "critical angles" $\beta_c = 54.74^{\circ}$

and 125.26° the magnetic torques averaged over a rotation period become
zero. If we consider that, as discussed in the previous sections, the
angle β is varying in the range 80° to 130°, the warped structure
should disappear during certain phases of the 35-day cycle. At other
times, depending on the orientation of the spin vector with respect to
the viewing direction, the radiation from the neutron star may be completely blocked. In order to provide an X-ray obscuration for the observer the tilt angle of the spin-axis has to be negative.

As discussed above, the pulse/interpulse ratio during the main-on
state requires $\alpha \sim -40°$ and $\beta_{max} \sim 130°$. This is close to the critical
angle $\beta_c = 125.26°$ where the warped structure and the obscuration dis-
appears. Therefore, the main-on state can be accounted for by this picture in a consistent way. An obvious explanation for the short-on state
would be that it represents the phase around the closest approach of the
angle β to the other critical angle $\beta_c = 54.74°$. In view of the short
duration of the short-on state we estimate that the required β_{min} is
$\sim 70°$. On the other hand, the ratio of pulse to interpulse observed during the short-on state requires a $\beta_{min} \sim 80°$. The apparent discrepancy
of $\sim 10°$ in the required β_{min} angles may be connected with the fact that
the angular momentum vector has a sideward tilt. Furthermore, the theoretical considerations of Lipunov, Semenov and Shakura (1981) assume a
small tilt of the spin vector ($\alpha < 15°$) while our data require $\alpha \sim 40°$.
An open question is whether, at such large angles, the warped disk remains a continuous surface or "tears" under the magnetic torque effects.

In this scenario the declining phase of the main-on state is produced by a gradual rise of the warped structure when β starts to devi-
ate significantly from β_c. The warped disk would first occult the
neutron star and then gradually obscure the Alfven shell which has a
lateral extent of a few times 10^8 cm. This is in good agreement with
the behaviour observed at the declining flank of the main-on state,
which shows a time lag of ~ 2 days between the disappearence of the
sharp pulse and the gradual decline of the sinusoidal pulsations. The
latter are most likely produced by reprocessing (Thomson scattering) of
the primary beam hitting the Alfven shell (Kahabka et al., 1986).

7. DISCUSSION

We have presented data from a 35-day cycle sampling of the pulse profiles of Her X-1 by EXOSAT. The data show that the beam pattern, as seen
by the observer, changes strongly between the main-on and short-on
state. Such an effect has already been reported by Staubert et al.
(1979) on the basis of hard X-ray (20 - 120 keV) balloon data but with
much lower statistical significance.

This change of pulse profiles is in direct conflict with models
where the 35-day modulation of X-rays is attributed to absorption in the
middle and outer parts of a precessing accretion disk, where the pulse
and interpulse should be modulated by the same ratios. If the modulation
is caused by disk precession the absorption has to take place near the
inner edge of the disk. Bai (1981) has shown that if the inner surface
material of the disk at $r \sim 3 \times 10^8$ cm is at about 10^5 K it will have a

scale height of about 10^6 cm and the emission from the two poles of the neutron star will be attenuated differently as they propagate along our line of sight. However, due to the magnetic torques on the disk and the Lense-Thirring effect (Bardeen and Petterson, 1975) the inner edge of the disk would be coaligned with the rotational equator of the neutron star and a precessing disk would not have a precessing inner edge at $r \sim 3\times10^8$ cm. In addition, it is difficult to explain the observed variation of pulse width and phase separation of pulse peaks in the case of the precessing inner disk model. We would like to point out that a very similar differential absorption for the two beams as discussed by Bai (1981) might also occur in our precessing dipole model since the two poles would have different absorption paths along the line of sight which would also vary with the phase of precession.

A decreasing pulse width could be associated with a variable scattering geometry of the Alfven shell as suggested by Pravdo et al. (1977). But in order to account for the different pulse profiles in the main-on and short-on state a gross change in the Alfven shell geometry would be necessary. Furthermore, it seems difficult to account for variation of the phase separation of the pulse peaks.

In principle, it is conceivable that 35-day clock mechanisms which employ a phase dependent mass accretion rate (e.g. Crosa, and Boynton, 1980; Meyer, and Meyer-Hofmeister, 1984) may also modulate the beam pattern by changing the geometry of the accretion column. However, as indicated by the optical data, there are no large variations of the heating of HZ Her and hence the $\dot{M}$ that produces the X-ray luminosity should be fairly constant (Deeter et al., 1976). In addition, the stability of the pulse arrival phases during the 35 day cycle indicate that there are no gross changes of $\dot{M}$ on the time scale of 35 days.

A direct interpretation of the reduction of the pulse widths, peak separations and the increase of the second pole amplitude with respect to the main pole during the short-on state is that we are observing the beam pattern at angles that differ by about 50° between the main- and short-on states of the 35-day cycle. Consequently, we conclude that the 35-day clock must be intrinsically associated with the free precesssion of the neutron star rather than the structural periodicities of the accretion disk.

There have been two major objections to a precessing neutron star. One objection, due to Shaham (1977), is that if vortex pinning exists in the crust at a level required to explain pulsar glitches, the precession frequency should be much shorter (10^2 to 10^3 sec) and there should be phase coherence and damping time-scale problems. The second objection, by Boynton, Crosa, and Deeter (1980), is that the poor stability of the 35-day clock, which is of the order of a few per cent, argues against such a precise mechanism as the neutron star precession. One possible solution to both of these problems may be that the angular momentum of the pinned component is small enough to allow the domination of the oblateness term in the precession frequency, but large enough to cause a variation in the 35-day period amounting to few per cent, via vortex creep. We estimate that a ratio of pinned angular momentum to total angular momentum around 10^{-9} would account for the observations. It has been pointed out by Shaham (1985) that while early calculation of the

pinning energy of vortices to the neutron stellar crust (Alpar, 1977)
indicated very high pinning barriers ($\delta\Omega < 10$ sec^{-1}), recent calculations of the superfluid gap energy as well as recent work on the Vela and
Crab pulsar glitches (Alpar et al., 1984) indicate much lower barriers,
of order $\delta\Omega < 0.1$ sec^{-1}. With such low barriers, any $> 1°$ wobble will
cause massive unpinning and the superfluid will behave as essentially
unpinned; this will make a 35 day wobble period, stemming from crustal
oblateness, quite feasible, even if occasional pinning-unpinning events
may superpose frequency noise on it.

In the model discussed, the precessing beams and the warped inner
disk obscuration will produce illumination changes on HZ Her, the companion star, as they vary with the 35-day cycle phase. Additional sha-
dowing effects are expected from the Alfven shell whose geometry will
depend on the direction of the magnetic dipole (Kahabka et al., 1986).
The general optical effects expected from this modulated heating will be
similar to that produced by a precessing, tilted accretion disk which
has been employed so far in this context (Gerend and Boynton, 1976). We
therefore believe that the large body of optical observations can be explained in principle by our model although this remains to be proven in
detail. One of the distinct requirements of the above interpretation is
that the rotation axis of the neutron star must be tilted with respect
to the binary plane axis by an angle of the order of 30 to 40 degrees. A
more detailed analysis of the optical data may be used to test this requirement and obtain more precisely the direction of the tilt, in three
dimensional space.

Apart from the optical effects there are several effects in X-rays
which may require a feedback loop including HZ Her and processes taking
place at the outer rim of the disk: marching pre-eclipse dips, preferred
turn-ons of the 35-day on-state at orbital phases 0.2 and 0.7, and the
sharp turn on of the on-state. A comprehensive explanation for these
effects has been proposed by Crosa and Boynton (1980). They assume a periodic mass transfer from HZ Her to the disk which is triggered by the
passage of the X-ray shadow over the most strongly X-ray heated face of
the HZ Her surface. We note that this or similar feedback mechanisms are
not only compatible with our scenario, but that the inclination of the
inner disk naturally provides the disk asymmetry needed in their scheme
(Boynton, Crosa and Deeter, 1980).

Besides Her X-1 there are two other pulsating X-ray sources showing
long periods, which may be due to neutron star precession as well:
LMC X-4 with 30.5 days (Skinner et al., 1982; Lang et al., 1981, Pietsch
et al., 1985) and possibly SMC X-1 with 60 days (Gruber and Rothschild,
1984). In the case of LMC X-4 it is known that the inclination of the
accretion disk with respect to the line of sight is rather large, 90-i ~
24° (Kelley et al., 1983). The very low pulsed fraction of only a few
percent that only increases during flares and the sinusoidal pulse shape
(Pietsch et al., 1985) may indicate that in LMC X-4 only reprocessed radiation is visible. Possibly, some of the long period transients (Priedhorsky, Terrell and Holt, 1983; Priedhorsky and Terrell, 1983a, 1983b,
1984), and SS433 (Abell and Margon, 1979), may also belong to this class
of objects.

REFERENCES

Abell, G. O., and Margon, B. 1979, Nature 279, 701
Alpar, M. A. 1977, Ap. J. 213, 527
Alpar, M. A., Anderson, P. W., Pines, D., and Shaham, J. 1984, Ap. J.
    278, 791
Arons, J. 1973, Ap. J. 184, 539
Bai, T. 1980, Ap. J. 239, 999
Bai, T. 1981, Ap. J. 243, 244
Bardeen, J. M., and Petterson, J. A. 1975, Ap. J. Letters 195, L65
Basko, M. M., and Sunyaev, R. A. 1976, MNRAS 175, 395
Basko, M. M., and Sunyaev, R. A. 1976, Astron. Zh. 53, 950
Basko, M. M. 1980, Astr. Ap. 87, 330
Bisnovatyi-Kogan, G. S., and Komberg, B. V. 1975, Astron. Zh. 52,457
Boynton, P. E., and Deeter, J. E. 1976, Ap. J. Letters 210, L133
Boynton, P. E., Crosa, L. M., and Deeter, J. E. 1980, Ap. J. 237, 169
Brecher, K. 1972, Nature 239, 325
Burke, J. A. 1976, Ap. J. 209, 556
Chevalier, R. A. 1976, Ap. J. Letters 18, 35
Cordova, F., Mason, K. O. 1983, in Accretion-Driven Stellar X-ray
    Sources, eds. W. H. G. Lewin, E. van den Heuvel, Cambridge
    University Press
Crosa, L., and Boynton, P. E. 1980, Ap J. 235, 999
Deeter, J., Crosa, L., Gerend, D., and Boynton, P. E. 1976, Ap. J. 206,
    861
Doxsey, R., Bradt, H. V., Levine, A., Murthy, G. T., Rappaport, S., and
    Spada, G. 1973, Ap. J. Letters 182, L25
Gerend, D., and Boynton, P. E. 1976, Ap. J. 209, 562
Gruber, D. E., et al. 1980, Ap. J. Letters 240, L127
Gruber, D. E., and Rothschild, R. E. 1984, Ap. J. 283, 546
Henriksen, R. N., Reinhardt, M., and Aschenbach, B. 1973, Astr. Ap. 28,
    47
Holt, S. S., Boldt, E. A., Rothschild, R. E., Saba, J. L. R., and
    Serlemitsos, P. J. 1974, Ap. J. Letters 209, L131
Joss, P. C., Fechner, W. B., Forman, W., and Jones, C. 1978, Ap. J. 225,
    994
Kahabka, P., Pietsch, W., Truemper, J., Voges, W., Kendziorra, E., and
    Staubert, R. 1984, in X-Ray Astronomy, '84, Int. Symp. on X-Ray
    Astronomy, Bologna, Italy, eds. M. Oda, and R Giacconi, p. 193,
Kahabka, P., Oegelman, H., Pietsch, W., Truemper, J., and Voges, W.
    1985, Space Sci. Rev. 40, 355
Kahabka, P. et al. 1986, in preparation
Katz, J. I. 1973, Nature Phys. Sci. 246, 87
Kelley, R. L., Jernigan, J. G., Levine, A., Petro, L. D., and Rappaport,
    S. 1983, Ap. J. 264, 568
Kondo, Y., Van Flandern, T. C., and Wolff, C. L. 1983, Ap. J. 273, 716
Korte, P. A. J. de et al. 1981, Space Sci. Rev. 30, 495
Lamb, D. Q., Lamb, F. K., Pines, D., and Shaham, J. 1975, Ap. J. Letters
    198, L21
Lang, F. L. et al. 1981, Ap. J. Letters 246, L21
Lewin, W. H. G., Joss, P. C. 1983, in Accretion-Driven Stellar X-ray

Sources, eds. W. H. G. Lewin, E. van den Heuvel, Cambridge University Press

Lipinov, V. M., and Shakura, N. I. 1980, Sov. Astr. Letters 6, 14

Lipinov, V. M., Semenov, E. S., and Shakura, E. S. 1981, Sov. Astr. 25, 459

Mason, K. O. 1985, Space Sci. Rev. 40, 99

Mazeh, T., and Shaham, J. 1977, Ap. J. Letters 213, L17

McClintock, J.E. 1986, Proceedings of the ESA Workshop on "The Physics of Accretion onto Compact Objects", Tenerife, April 1986

McCray, R., and Hatchett, S. 1975, Ap. J. 199, 196

McCray, R. A., and Lamb, F. K. 1976, Ap. J. 204, L115

McCray, R. A., Shull, J. M., Boynton, P. E., Deeter, J. E., Holt, S. S., and White, N. E. 1982, Ap. J. 262, 301

Meyer, F., and Meyer-Hofmeister, E. 1984, Astr. Ap. 140, L35

Nagel, E. 1981a, Ap. J. 251, 277

Nagel, E. 1981b, Ap. J. 251, 288

Oegelman, H., Kahabka, P., Pietsch, W., Truemper, J., Voges, W., Kendziorra, E., and Staubert, R. 1984, in X-Ray Astronomy '84, Int. Symp. on X-Ray Astronomy, Bologna, Italy, eds. M. Oda, and R. Giacconi, p. 197

Oegelman, H., Kahabka, P., Pietsch, W., Truemper, J., and Voges, W. 1985, Space Sci. Rev. 40, 347

Ohashi, T. 1984,in X-Ray Astronomy '84, Int. Symp. on X-Ray Astronomy, Bologna, Italy, eds. M. Oda, and R Giacconi, p. 187

Papaloizou, J., and Pringle, J. E. 1982, MNRAS 200, 49

van Paradijs, J. 1983, Accretion-Driven Stellar X-ray Sources, eds. W. H. G. Lewin, E. van den Heuvel, Cambridge University Press

Petterson, J. A. 1975, Ap. J. Letters 201, L61

Petterson, J. A. 1977, Ap. J. 218, 783

Pietsch, W., Pakull, M., Voges, W., and Staubert, R. 1985, Space Sci. Rev. 40

Pines, D., Pethick, C. J., and Lamb, F. K. 1973, Ap. J. 184, 271

Pravdo, S. H., Boldt, E. A., Holt, S. S., and Serlemitsos, P. J. 1977, Ap. J. Letters 216, L23

Priedhorsky, W. C., Terrell, J., and Holt, S. S. 1983, Ap. J. 270, 233

Priedhorsky, W. C., Terrell, J. 1983a, Ap. J. 273, 709

Priedhorsky, W. C., Terrell, J. 1983b, Ap. J. 303, 681

Priedhorsky, W. C., Terrell, J. 1984, Ap. J. 280

Pringle, J. E. 1973, Nature Phys. Sci. 243, 90

Rappaport, S. S., Joss, P.C., 1981, in Accretion Driven Stellar X-ray Sources, eds. W. H. G. Lewin, E. van den Heuvel, Cambridge University Press

Roberts, J. W. 1974, Ap. J. 187, 575

Ruderman, M., 1985, Proceedings of the NATO-ASI "High Energy Phenomena and Collapsed Stars", Cargèse, September 1985

Shaham, J. 1977, Ap. J. 214, 251

Shaham, J. 1985, private communication

Shulman, S., Friedman, H., Fritz, G., Henry, R. C., and Yentis, D. J. 1975, Ap. J. Letters 199, L101

Skinner, G. K., Bedford, D. K., Elsner, R. F., Leahy, D., Weisskopf, M. C. and Grindley, J. 1982, Nature 297, 568

Staubert, R., Kendziorra, E., Pietsch, W., Reppin, C., Truemper, J., and
    Voges, W. 1979 in X-Ray Astronomy, eds. W. Baity, and L. Peterson
    (Pergaman; Oxford), p. 489
Tananbaum, H., Gursky, H., Kellog, E. M., Levinson, R., Schreier, E.,
    and Giacconi, R. 1972, Ap. J. Letters 174, L143
Taylor, B. G., Andresen, R. D., Peacock, A., and Zobl, R. 1981, Space
    Sci. Rev. 30, 479
Tennant, A., Fabian, A. C., Shafer, R. 1986, M. N. R. a. S. 221, 27 P
Turner, M. J. L., Smith, A., and Zimmermann, H. U. 1981, Space Sci. Rev.
    30, 513
van der Klis, M. 1986, Proceedings of the ESA Workshop on "The Physics
    of Accretion onto Compact Objects", Tenerife, April 1986
Voges, W., Pietsch, W., Reppin, C., Truemper, J., Kendziorra, E.,
    Staubert, R. 1982, Ap. J. 263, 803-813
Voges, W., Kahabka, P.,Oegelman, H., Pietsch, W., and Truemper, J. 1985,
    Space Sci. Rev. 40, 339
Wolff, C. L., and Kondo, Y. 1978, Ap. J. 219, 605

COSMIC-RAY ACCELERATION IN UHE AND VHE GAMMA-RAY SOURCES

K. Brecher
Department of Astronomy
Boston University
Boston, MA 02215
U.S.A.

ABSTRACT.   VHE (10^{12} eV) and UHE (10^{15} eV) gamma-ray emission has
been reported (1) from the x-ray source Cygnus X-3, as well as from
several known x-ray binaries, including Her X-1, LMC X-4, Cen X-3 and
4U 0115 + 63.   These observations, if subsequently found to be
statistically significant, imply the acceleration of high energy
cosmic rays in these sources. In this paper, we discuss briefly some
of the models so far proposed for cosmic ray acceleration in these
sources.

1. INTRODUCTION

Any model of the acceleration of cosmic rays in the VHE and UHE
gamma-ray sources detected to date would have to satisfy the following
criteria:
    (a) Gamma-rays with energies of up to 10^{16} eV have been reported,
implying initial particle energies (hadrons) of 10^{17} eV, assuming that
the gamma-rays arise from pion decay.
    (b) The total cosmic-ray luminosity of these sources must be at
least 10^{38} erg/sec, perhaps as high as 10^{39} erg/sec in the case of Cyg
X-3.
    (c) UHE particle production may be the major energy loss
mechanism for these sources (though non-relativistic bulk gas ejection
may also be important).
    (d) The spectrum of accelerated particles could be monoenergetic
(2) even though the gamma-ray spectrum has an E^{-2} power-law photon
distribution.
    (e) Particle acceleration must be fast (seconds or less), in
order for the particles to escape the acceleration region without
major energy loss before hitting the presumed gas target giving rise
to the observed VHE and UHE gamma-rays.
    Three kinds of models have been proposed to date to account for
the purported cosmic ray fluxes. In the following sections, we review
very briefly some features of these models. All have been discussed

23

K. E. Turver (ed.), Very High Energy Gamma Ray Astronomy, 23–25.

at length in a variety of papers, so only the general features of
these models are considered here.

2. Pulsar Acceleration

Pulsars are kr.own to accelerate particles to high energies, in the
case of the Crab nebula, electrons with energies of at least 10^{11} eV.
It has been suggested (3) that a similar mechanism may apply to Cygnus
X-3, a source with a 4.8 hour gamma-ray periodicity, but with a
possible shorter underlying (pulsar?) periodicity in the 1 - 100 ms
(perhaps even 12.59 ms). While such a model could be made to fit the
properties of Cygnus X-3 (owing to the unknown magnetic field as
well), it cannot fit the observed luminosity of the four other
reported VHE and UHE gamma-ray sources because of their longer
observed pulse periods. At least for these sources an alternative
energy source, derived from accretion rather than rotational energy,
is required.

3. Shock Acceleration.

There are a number of natural reasons to consider this possibility.
First, we are dealing with accreting binary systems (at least for four
of the candidate sources, and probably for Cyg X-3 as well). Second,
the ultimate energy for the accelerated particles comes from
accretion, rather than from rotation, allowing for a long lived
source. The most developed model (4) assumed spherical accretion onto
a magnetized neutron star. A standing shock near the polar cap can
accelerate protons to high energies. However, these particles suffer
severe energy losses by synchrotron radiation and other processes.
The maximum accelerated particle energy is achieved by equating the
acceleration to loss times and, for reasonable parameters, gives a
maximum energy of less than 10^{16} eV. If the accelerated protons hit
other protons in the acceleration region, neutrons can be formed which
then escape. The model is clever, and may even apply to these
systems.

4. Unipolar Induction

This model in a sense combines some of the features of pulsar
acceleration models with the shock acceleration model, in that the
particles are accelerated by a parallel electric field, but the
ultimate energy source is from accretion. Basically, we (5) assume
that the accreting magnetized neutron star magnetic field exists out
to the Alfven surface, at which point the accretion disc contains some
r and z components of poloidal magnetic field. The Keplerian motion
of particles in the accretion disc moving though the magnetic field
gives rise to an induced electric field. The potential drop across
the disc $V \propto B^{-3/7}L^{5/7}$, where L is the accretion luminosity. For weak
enough magnetic fields ($B \approx 10^9$ gauss) and strong enough accretion
rates ($L \approx 10^{39}$ erg/sec), a potential drop of 10^{17} volts can develop.
If the Alfven surface lies just above the neutron star surface, the
accelerated particle luminosity can equal the accretion luminosity.

Since particles, not photons, carry off the energy, the total accretion luminosity can exceed the normal (photon) Eddington luminosity by a factor of 10 - 100, thus allowing for the high observed non-thermal luminosity of Cygnus X-3.

5. Conclusions

In conclusion, though none of these models is known to operate with certainty in any of the accreting binary systems, the lack of models cannot now be used as a reason to disbelieve the observations. If these systems are indeed gamma-ray sources, we may at last have found the origin of high energy cosmic rays within our own Galaxy.

6. References

(1)   These Proceedings provide the best summary to date of the observational situation. However, see also the Proceedings of the 19th ICRC held in La Jolla, August, 1985.
(2) Hillas, A. M., Nature, **312**, 50, 1984.
(3) Eichler, D. & Vestrand, W.T., Nature, **306**, 613, 1984.
(4) Kazanas, D. and Ellison, D., Nature, **319**, 380, 1986.
(5) Chanmugam, G. and Brecher, K., Nature, **313**, 767, 1985.

IN DAYS OF YORE

JOHN V JELLEY
Nuclear Physics Division
AERE
Harwell
Oxon.   OX11-ORA
U.K.

ABSTRACT.   An account is given of the early work on the detection of optical Cherenkov radiation from extensive cosmic ray air showers (EAS). A brief discussion is likewise presented of the discovery of radio pulses from such showers.

INTRODUCTION

It is now over thirty years since our small group in the Nuclear Physics Division at AERE, Harwell, first detected the Cherenkov light flashes associated with extensive air showers (EAS).

As I have not myself been in the field for at least twenty years, I have little to contribute on the research aspects.   In view of this it seemed that it might be of interest to some of you present, to hear a personal, and therefore somewhat biased, account of the very early days in the field, during what was clearly a most exciting time.

Likewise, and again from an historical and personal view, I intend to describe the early work, twelve years later, in which three groups cooperated successfully in the first detection of radio pulses associated with the EAS.   Although I realise this theme lies off the mainstream of this Meeting, it may be that some of you could be stimulated to see if there are indeed any radio techniques which could be used to supplement those at present adopted for point-source γ-ray searches, particularly at the higher energies.

HOW IT ALL BEGAN; THE VERY EARLY DAYS

It was in 1953[1] that my colleague Galbraith and I discovered the light pulses from the night sky, which we soon demonstrated were correlated with cosmic-ray EAS.   Although we did not exactly stumble on the discovery, it was nevertheless the result of a combination of accidental circumstances.   By 1953 I had already been working on Cherenkov radiation in other areas of CR research, and indeed I had heard of this phenomenon as far back as 1948.

27

K. E. Turver (ed.), Very High Energy Gamma Ray Astronomy, 27–37.
© 1987 by D. Reidel Publishing Company.

It was during that year, as a research student at Cambridge, that
Dr T E Cranshaw casually mentioned over tea one day, in the Cavendish
laboratory, that he had been reading some Russian papers by
Cherenkov[2], Frank and Tamm[3], and others, in which they had
shown both experimentally and theoretically that fast charged particles
could produce faint light in solid and liquid media; this, moreover,
occurred when the particles were assumed to be travelling at essentially
constant velocity. Ingrained into me from my training as an
undergraduate in the Physics Department at Birmingham University, was
the impression that electromagnetic radiation could result only from the
acceleration of electric charges, as in Bremsstrahlung and Synchrotron
Radiation. I was therefore puzzled by this news and went off to read
the Russian papers, coming away with at least a partial understanding of
the theory of Frank and Tamm. In the immediate post-war years most
Russian physics appeared in the "Journal of Physics" of the USSR which,
at that time, was published in English, this being long before journal
translations became available. Cranshaw and I were working in nuclear
physics at that time, and so we thought no more about the topic.

Joining AERE Harwell, in late 1949, I began work in
Dr B Pontecorvo's group, which was engaged in various problems in CR
physics, notably in designing a large EAS array. It so happened that
the liquid scintillator had just been developed in the USA, by the group
at MIT. Pontecorvo, realising the significance of this technical
achievement, had asked me to look into the question of what
concentrations of the active element of these scintillators, para-
terphenyl, were really required, because this compound was at the time
only available in the US, and was very expensive.

In these investigations, I was measuring the light yield, in
pulses, from individual sea-level CR particles (mainly μ-mesons) passing
through solutions of para-terphenyl in benzene, solutions of gradually
reduced concentration. I found, to everyone's surprise, that some
light pulses were recorded even in pure benzene. It was then that I
recalled that casual conversation with Cranshaw in Cambridge. If
indeed Chernkov radiation was responsible, then light pulses should be
observable, and at much the same rate and intensity, from almost any
pure liquid. This was soon shown to be the case, and from this, the
first photo-electronic detection of Cherenkov radiation generated by
single charged particles traversing pure liquids[4], stemmed the
development of Cherenkov detectors of a variety of forms[5] which
have, ever since, been widely used in high energy particle physics and
CR research, both in the laboratory, and in space.

It was, however, an even more casual and accidental encounter
which led to our discovery of the light pulses from the night sky. One
day in 1953 Professor Blackett was visiting Harwell, and, during his
short stay of a few hours, came to talk with us about the then current
programme in the CR group. Blackett, hearing of our work on Cherenkov
radiation in water, quite casually mentioned that as far back as 1948 he
had shown[6] that there should be a contribution to the light of the
night sky, amounting to about 10^{-4} of the total, due to Cherenkov
radiation produced in the upper atmosphere, from the *general* flux of
incoming CR particles. Since his paper was presented at a Royal

Society Meeting devoted mainly to optical emissions from aurorae, it is little known among CR physicists.

An interesting point, and one that is seldom appreciated, is that as far as we know, this was the first time that anyone had suggested that Cherenkov radiation would in fact occur in gases, all the previous work concentrating on the effects in solids and liquids.

Blackett was only with us for a few hours, and neither he nor any of us ever mentioned the possibility of *pulses* of Cherenkov light, from EAS.   It was a few days later that it occurred to Galbraith and myself that there was a real possibility that such pulses might exist and be detectable.   Neither of us, in 1986, can recall exactly which of us first thought of it, or even how we started to discuss it.

The very simple experiment which was a success on the very first night, was carried out in total ignorance of the intensities or pulse rates we might expect, since we had carried out no theoretical calculations, and had very little background on which to base our frivolity! We knew that showers of energies about 10^{15} - 10^{16} eV were detectable with the simple EAS array built by Cranshaw and Galbraith[7], this being sited in a field about 1 km from the rather brightly lit area around Harwell.   We knew only that such showers produced about 10^5 - 10^6 electrons and positrons on the ground, that they fell dominantly in an area of a few hundred m^2, that this burst of particles lasted about 10 ns., and that the optimum signal/noise would be expected if the bandwidth of the amplifier was chosen to match as closely as possible the reciprocal of this, namely about 100 MHz.

In great excitement we therefore set up an experiment with equipment of extreme simplicity.   In those days, not long after the war, there were countless shops in Lisle Street and High Holborn in London crammed with ex World War II optical and electronic equipment. We soon purchased some 25 cm f/0.5 parabolic signalling mirrors.   We mounted one of these at the bottom of a dustbin, standard Harwell stores issue, and clamped an EMI photomultiplier, itself a relatively new device, to a retort stand inside the dustbin, with its photocathode placed at the focus of the mirror.   Coupling this to the fastest amplifier available at the time (0.03 µs rise-time) and pointing the contraption to the zenith sky, at a dark site near Harwell, we were rewarded with success on the very first night.

Large single light pulses were observed at once, at a rate of between one and two per minute;   these were bandwidth-limited, distributed randomly in time, possessed a fairly broad zenith-angle dependence, and had a pulse-height spectrum similar to the numbers-vs.-energy spectrum of CR primaries.

After simple control experiments, with the lid on the dustbin and a small filament lamp inside, we soon showed that these were indeed light pulses from the sky.   The claim that these pulses were associated with EAS was soon substantiated, when a few nights later we observed coincidences between the light flashes and shower particles detected on the EAS array.   This original experiment was subsequently described in a letter to Nature.[1]   A photograph of the original apparatus is shown in Fig.1.

Figure 1: The original Cherenkov light receiver built by Galbraith and Jelley, comprising a 25 cm f/0.5 rear-silvered ex World war II signalling mirror, and a 50 mm EMI photomultiplier , mounted in a dustbin.

Looking back, it was remarkable that we were successful, first, because the pulses were seen, not on a triggered time-base, but on a continuously running one, and secondly, because the counting-rate was fortuitously appropriate for such visual detection on the CRT. Had the rate been, say, 1 hr^{-1} or less, it is doubtful whether we would have pursued the experiment, and had it been several per min. or faster, the pulses might well have been picked up by some meticulous astronomer with a wide-field telescope and suitable counting electronics.

We soon demanded greater sensitivity and larger fields of view. In that carefree past we were rapidly able to procure 60 and 90 cm diameter mirrors. Quite soon, at our instigation, numerous 1.5 m ex-Army searchlights were rolling down the roads from surplus equipment depots in Coventry, and shortly afterwards some even crossed the Irish Sea. Some early recordings of pulses with such a receiver are shown in Fig.2.

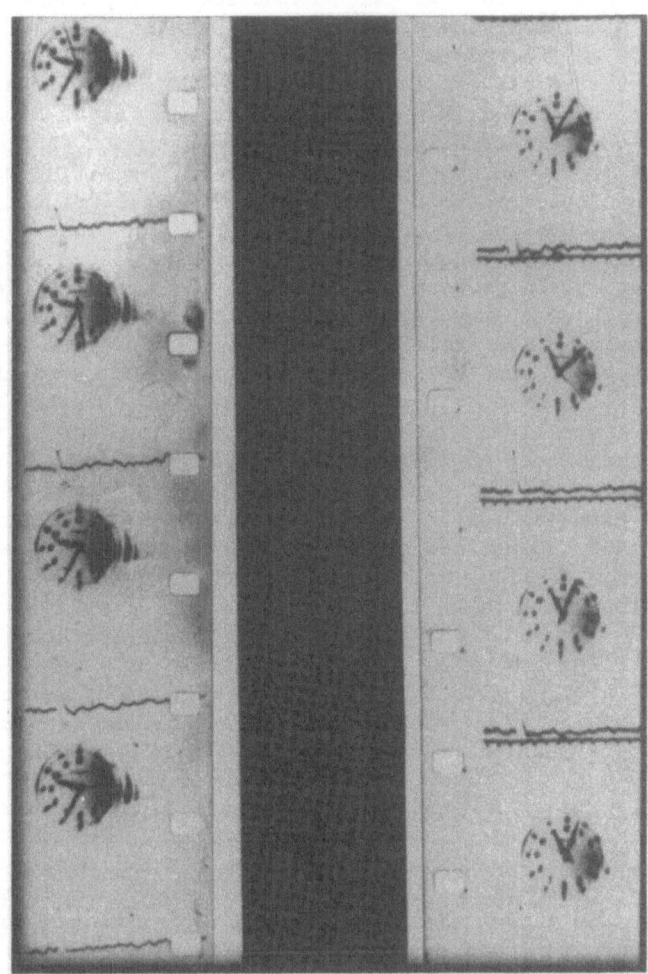

Figure 2: CRT recordings of light pulses obtained with one of the early receivers, showing the Cherenkov signals superimposed on the night sky noise. Overall bandwidth 25 MHz, limited by the fastest amplifier then available.

I think it is fair to say that at that time we were more
interested in the topic as a facet of radiation physics than we were in
the potential of the technique which might some day perhaps be of
importance to studies of the primary CR, or as a tool in γ-ray
astromony.   It may therefore also be of some historical interest that
in 1954 the author took a small 'light receiver' to the Royal Greenwich
Observatory, at Herstmonceux.   This was strapped to the barrel of their
6-inch (15.24 cm) refractor and was used to search for CR from the Crab
nebula.   With Dr.T.Gold, the then Assistant to the Astronomer Royal,
several nights of observations were expended on this project, of which
we were tolerably hopeful of success at the time.   Since by then it was
appreciated that the continuum radiation from the nebula was probably
due to the synchrotron process, it seemed likely that the Crab was a
source of ultra high energy CR, and hence a source of γ-rays.   This
attempt to detect γ-rays from this source was, to say the least, naive,
and little was known then of how small was the fraction of the primary
CR in the form of γ-rays.   It is also important to remember that this
pre-dated, by several years, the discovery of the pulsar in the Crab.
       During the winter of 1953, and into 1954, we pursued studies of
these light flashes[8][9], which revealed that the "pool" of light
on the ground extended out to $\geqslant$ 100 m, covering an area considerably
greater than that encompassed by the electron-photon cascade.   We also
carried out rough absolute calibrations of the light-intensities in the
flashes, finding these to be about 3 photons.cm^{-2}.$flash^{-1}$, and,
in addition, confirmed that the index in the power-law spectrum was
similar to that for the primary charged-particle CR.
       After struggling with the atmospheric conditions associated with a
typical English winter, we were able in 1954, through encouragement from
Sir John Cockcroft, to take a substantial load of equipment to the Pic
du Midi Observatory in France (altitude 2860 m) where we were blessed
with superb observing conditions, both as regards atmospheric
transparency and also very low background light.   Even these
experiments, however, could only be described as exploratory, since the
portable EAS array (with Geiger trays) was incapable of defining the
core locations and shower energies with the precision we would have
liked.   However, this was the first time that C-light-receivers were
operated at a really first class site, and it was soon appreciated that
with suitable equipment, and care, it was possible to obtain remarkably
consistent figures for counting rates, stable from hour to hour and from
night to night.
       While at the Pic, we attempted crude spectrophotometry with
gelatin filters, and a search for polarization of the light.   Both
these experiments were unsuccessful, for reasons which are all now too
obvious.

At this point I would like to mention a peripheral interest which arose immediately after our discovery in 1953[1].   Just at that time Professor Hanbury Brown was developing the Intensity Interterometer, at Jodrell Bank, using two 150 cm searchlights, in the first measurement of the angular diameter of a (non red-giant) star: I recall that this was carried out on Sirius.   Hanbury Brown was quite alarmed by the thought of the presence of *any* source of light correlated over distances corresponding to the separation of his searchlights. I was therefore peremptorily summoned to Jodrell to discuss this problem! Although the EAS light flashes are in fact quite large on the nanosecond time-scale, they would not show up on his long integrations, and the potential headache went away.

Shortly after this, interest was awakened in other countries, and the first experiments reported from abroad were carried out in the USSR[10], where the main findings of our work were confirmed. These experiments were then followed up by a number of groups in the USA, and also in Ireland.

I paricularly draw attention to the work in Ireland, because a close association grew up between UCD, Dubin, and our own group at AERE, and much of the somewhat later work was carried out under a collaborative programme.   I have always been grateful to Professor Neil Porter and his group for their efforts in this.

During the latter period of this exploratory work, in the late fifties, considerable effort was applied to improving the sensitivity of the installations, to increase the counting rates, and to lower the energy thresholds of the EAS detected. I will mention, but briefly, some of the developments taking place at that time.

Since the yield of Cherenkov light varies as λ^{-2} (in the form of photon-numbers-flux per unit wavelength interval), while the continuum spectrum of the night sky decreases with decreasing λ, the value of attempting to work in the UV became obvious, and a number of experiments were carried out in attempts to exploit this situation; I refer to the work which involved colour filters and UV converters[11], and some observations using solar-blind photomultipliers[12].

There were also tentative estimates of the feasibility of working in the infra-red and μ-wave regions of the spectrum, with the aim of being able to work in daytime, or at least during twilight or on bright moonlit nights.   While up to 1966 (see the general review (13), pps 143-146), the possibilities in these spectral regions seemed marginal, it may well be that with the latest detectors now available, a re-think along these lines may be justifiable.

All the light receivers in use up to 1961 relied on optical systems using either a single PM tube or at most just a few channels. A remarkable collaborative experiment was, however, carried out between UDC (Dublin) and MIT(USA) in 1961[14], in which direct two-dimentional images of the Č-flashes on the celestial sphere were obtained, using image intensifiers, phosphor storage, and a TV camera read-out system.   This isolated but unique experiment was in effect the forerunner of the multi-element detector on the 10 m flux collector on Mount Hopkins, about which we will have heard during this Meeting.

Returning to the memories of those times, we used to ponder over whether the observed light pulses were indeed due to Č-radiation.   It is interesting to note that even to this day there has never been observational proof that they are so due!   The broad lateral distribution of the particles in an EAS, together with the fact that the multiple Coulomb scattering angles of the shower particles exceed the Č-angle, by at least an order of magnitude, excluded the possibility of detecting the characteristic polarization of the radiation. Furthermore, it proved to be exceedingly difficult to study either the spectral distribution or to measure the absolute light fluxes, with pulses of such short duration.   Nevertheless, we are now all confident of the assumption that Č-radiation is responsible, if only because the alternative radiation mechanisms suggested do not appear to fit the observations[15]; these alternatives included ionization and recombination radiation, synchrotron radiation in the Earth's field, and the low energy tail of the Bremsstrahlung spectrum.

I have mentioned this, to illustrate that we were at the time sufficiently concerned about it to design a laboratory experiment to detect and measure the light from *single* relativistic particles in air, in the absence of the night-sky background, and under controlled conditions.   This experiment[16], and two others carried out at about the same time[17,18], revealed that Č-radiation in gases from single particles was indeed observable, and that the light yield was in agreement with the classical theory of Frank and Tamm.

In the mid-fifties, concurrent with the experiments outlined above, there were also of course developments on the technical side, one of which was the introduction of coincidence techniques, with means of continuously monitoring the chance-rates.   Equally important were developments of ways to stabilise the sensitivities of the detectors against variations of sky-brightness during long runs[19], especially important in our poor climate, yet equally important even at good sites, during drift scans close to the Galactic Plane, or any regions of sky containing bright stars.

So much for the "early days", an era of purely exploratory experiments.   The potential of the technique as the basis for detecting ultra high energy (UHE) γ-rays became obvious at a relatively early stage, when on the one hand the features of large effective collecting areas and high angular resolutions of quite simple light receivers were fully appreciated, and, on the other, it was realised that unlike the sources of the charged-particle CR primaries, UHE γ-rays would be expected to arise from point sources.   The application of the technique for the detection of γ-ray induced EAS was, I believe, first suggested in 1961[20] and an early review on the topic appeared in 1963[21].

UHE γ-ray astronomy was in its infancy in the sixties, and the early equatorially-mounted searchlight mirror systems, such as the one described in (21), were very primitive compared to the systems, current and planned, about which we have heard at this Meeting.

An important consideration on the theoretical side was whether UHE γ-rays produced in celestial sources would suffer any significant absorption while traversing the vast distances involved.   Pair

production by photon-photon collisions between the γ-rays and various
background radiations[22] appeared to be a serious possibility, but
it was shown[23,24] that, in the energy region 10^9-10^{11} eV
covered by the existing and planned programmes, this process would be
quite negligible, even for extra-Galactic sources, and also for the
local regions close to such objects as pulsars[25].

THE EARLY WORK ON RADIO EMISSION

After the early Cherenkov work many suggestions were made on how one
might detect EAS by some radio technique, either through a direct
emission mechanism, or by a radar reflection process. Such ideas were
stimulated by the great strides that had been made in radar during
World War II, by the simplicity and low cost of even quite large antenna
arrays, and above all, by the fact that observations would not be
restricted to daylight hours.

The conception of detecting showers by radar goes back to the
early work of Blackett and Lovell in 1940[26]. While they failed to
detect cosmic-rays by the echo technique, the scene was nevertheless set
for the discovery of the daylight meteor showers. The radar technique
was revived by Suga[27] and further studies followed somewhat
later[28].

Coming now to the possibilities of direct emission, it seemed at
the outset that the yield of incoherent Č-radiation from an EAS in the
microwave region of the spectrum was below detectability simply because
the beamwidth of an antenna of adequate aperture was so small compared
with the angular size of the Č-image on the sky.

Looking at the problem from a purely energetic point of view we
were however still optimistic. For example, a 10^{16} eV shower
expends most of its energy, in ionization, before it reaches the ground,
taking about 50 μs to traverse the atmosphere, thus dissipating at a
power level of about 30 watts. Any time-compression of the radiation by
relativistic effects would represent an enhancement of the pulse power.
Considering that even a modest receiver in the metre-wave band has a
sensitivity of about 10^{-14} watt, there seemed a reasonable chance
that by some mechanism a fraction $\geqslant 10^{-15}$ of the EAS energy might
appear in the radio band. With these thoughts in mind, and realising
that a sizeable fraction of the shower energy is at any one time in the
form of Bremsstrahlung, I wondered what fraction of this might appear in
the radio band.   Looking at the classical treatment of Bremsstrahlung
for screened nuclei, due to Weizsäcker[29], it appeared that the
spectrum was flat, down to zero frequency.

It was on the basis of this argument that Sir Francis (then Dr)
Graham Smith and I carried out a simple experiment at the Mullard Radio
Astronomy Observatory in Cambridge: this I think was around 1956. We
built a small EAS Geiger array at Harwell, took it to Cambridge and ran
it in conjunction with a zenith-viewing dipole array at 151 MHz with a
bandwidth of a mere 100 kHz. The experiment was unsuccessful and was
soon abandoned. It was soon found that our basic assumption about the
spectrum was unjustified since effects of both multiple

scattering[30] and refractive index[31] were shown to reduce the
Bremsstrahlung yield to zero, as $\omega \to o$.   These points have been
subsequently discussed elsewhere[32].

The whole question of radio detection lay dormant as far as we
were concerned, until Askaryan[33] proposed that charge excess in an
EAS could produce coherent Cherenkov radiation, perhaps in the metre
band, at radio frequencies.

This rekindled interest and by the cooperation of three groups, at
Harwell, Jodrell Bank and UCD Dublin, we successfully detected the first
radio pulses from EAS[34].   UCD provided the data and design for the
EAS array, Harwell built the portable battery-operated EAS system and
recording gear, and Jodrell custom-built a large broadside array for the
receiver at 45 MHz.

The radiation mechanism was subsequently put on a sound footing by
Kahn and Lerche[35] at the University of Manchester, while an
alternative treatment was later presented by Allan[36].

Subsequent work in this field is too extensive to discuss in this
talk of reminiscences but I would like to list the main groups which
became involved, taken mainly from papers in ref(37):- AERE Harwell,
Jodrell Bank (Nuffield Radio Astronomy Laboratory), Haverah Park
(Imperial College and the University of Leeds), Moscow State and Kharkov
State Universities, Chacaltaya (Bolivia: University of Michigan),
Penticton (the University of Calgary and the Dominion Radio
Astrophysical Observatory) and at Glencullen (Ireland, University
College of Dublin).

REFERENCES

1. Galbraith, W. and Jelley, J.V. (1953) Nature 171, 349
2. Cherenkov, P.A. (1934) Dokl.Akad.Nauk. (SSSR) 2, 451
3. Frank, I.M. & Tamm, Ig. (1937) Dokl.Akad.Nauk. (SSSR) 14, 109
4. Jelley, J.V. (1951) Proc.Phys.Soc. A64, 82
5. Hutchinson, G.W. (1960) Prog.Nuc.Phys. 8, 195
6. Blackett, P.M.S. (1948) Rep.Gassiot Comm. of the Royal Society on 'Emission spectra of the night sky and aurora', p.34
7. Cranshaw, T.E. & Galbraith, W. (1954) Phil.Mag. 45, 1109
8. Jelley, J.V. & Galbraith, W. (1953) Phil.Mag. 44, 619
9. Galbraith, W. & Jelley, J.V. (1955) J.Atmos. and Terrestial Phys. 6, 250 and 304.
10. Nesterova, N.M. & Chudakov, A.E. (1955) Zh.Eksp.Teor.Fiz. 28, 384
11. White, J., Porter, N.A. and Long C.D. (1961) J.Atmos. and Terrestial Phys. 20, 40
12. Charman, W.N. and Jelley, J.V. (1965), unpublished.
13. Jelley, J.V. (1967) Prog. Elem.Particle & C.R.Phys.IX, 143-146
14. Hill, D.A. & Porter, N.A. (1961) Nature 191. 690
15. Jelley, J.V. Ref. 13 above, pps 60 and 61.
16. Barclay, F.R. & Jelley, J.V. (1955) Nuovo Cimento 2, 27
17. Ascoli Balzanelli,A and Ascoli,R (1953) Nuovo Cimento 10, 1345
18. Ascoli Balzanelli, A and Ascoli, R. (1954) Nuovo Cimento 11, 562
19. Fruin, J.H. & jelley, J.V. (1968) Canad.J.Phys. 46, S1118
20. Zatsepin, G.T. & Chudakov, A.E. (1961) Zh.Eksp.Teor.Fiz. 41, 655
21. Jelley, J.V. & Porter, N.A. (1963) Quarterly J.of the R.A.S. 4, 275
22. Nikishov, A.I. (1961) Zh.Eksp.Teor.Fiz. 41, 549
23. Gould, R.J. & Schréder, G. (1966) Phs.Rev.Lett. 16, 253
24. Jelley, J.V. (1966) Phys.Rev.Lett. 16, 479
25. McBreen, B. (1969) Nature 224, 893
26. Blackett, P.M .S. and Lovell, A.C.B.(1940) Proc.Roy.Soc.London Series A, 177, 183
27. Suga, K. (1962) Proc.5th Interam.Seminar Cosmic Rays, 2, XLIX
28. Matano, T et al (1968), Canad.J.Phys. 46, S 255
29. Weizsäcker, C.F. von (1934) Z.f.Physik. 88, 612
30. Migdal, A.B. (1954) Dokl.Akad.Nauk. SSSR 96, 49
31. Feinberg, E.L. and Pomeranchuk (1956). Nuovo Cimento Sppl.yo 3, 652
32. Jelley, J.V. (1967) Physics Letters 25A, (No.5), 346
33. Askaryan, G.A. (1962) Soviet Phys. JETP(English Transl.) 14, 441
34. Jelley, J.V. Charman, W.N., Fruin, J.H., Smith, F.G., Porter, R.A., Porter, N.A., Weekes, T.C., and McBreen, B. (1966) Nuovo Cimento 46,649
35. Kahn, F.D. and Lerche, I. (1966) Proc.Roy.Soc.(London) Series A. 289, 206
36. Allan, H.R. (1968). Canad.J.Phys. 46, S 234
37. Contributors to the Proceedings of the 10th International Conference on Cosmic Ray, Calgary, Alberta, Canada (1968). Canad.J.Phys. 46, Papers EAS 59 to EAS 69.

TeV AND PeV SOURCE STATUS

P. V. Ramana Murthy
Tata Institute of Fundamental Research
Colaba
Bombay 400 005
India

ABSTRACT. In this Rapporteur paper, the status of the various
celestial sources of TeV and PeV gamma rays is presented. This is not
a review paper; rather it is a report based on mostly the presentations
made at the workshop by the various groups based on their respective
observations.

1. INTRODUCTION

During the 4 years' time that elapsed since the previous International
workshop held at Ootacamund, India [1] the subject of gamma ray
astronomy at TeV and PeV energies has witnessed increased activity and
attracted much wider attention than during the earlier years. Now
there are many more claims of detection of gamma ray signals from the
various celestial objects. Also there are many more groups active in
the field. See Ramana Murthy and Wolfendale [2] and Protheroe [3]
for recent reviews.
    The techniques employed to detect gamma rays at these high energies
are well-known [4] . For the TeV energy region, which is also called
the very high energy gamma ray region, one employs the Atmospheric
Cerenkov Technique covering the energy range of approximately 0.1 TeV
to 10 TeV. For the PeV energy (ultra high energy) region, one uses
the air shower technique covering the energy range of approximately
0.1 to 10 PeV (1 PeV = 10^{15} eV). All the observations are made by ground-
based detector systems. For both the energy regions, the charged
primary cosmic ray induced showers constitute the background that is
difficult to eliminate. Indeed the signal-to-noise ratio can be as low
as 10^{-3}. In this enormous background of cosmic ray induced showers one
identifies a gamma ray signal as a spike in an otherwise uniform distribution of events with respect to either time or spatial direction or
both.
    The claimed sources are not monolithic; they range from isolated
pulsars through X-ray binaries to extra-galactic objects spanning a
distance scale from 500 Kpc to 4.5 Mpc. Among all these, the Cyg X-3
system stands out as a very special object for a number of reasons (i)

39

K. E. Turver (ed.), Very High Energy Gamma Ray Astronomy, 39–51.
© *1987 by D. Reidel Publishing Company.*

it exhibits enormously energetic and intense activity as evidenced
by the occurrence of giant radio bursts (ii) emission of PeV gamma
rays attention to which was first drawn by the Kiel group [5] (iii)
the claimed emission of neutral radiation capable of producing muons
detected in deep underground detectors [6] and (iv) exhibition of a
12.59 ms periodicity in its emission of TeV gamma rays first claimed
by the Durham group [7] . Indeed the importance of and the interest in
this object is reflected in the fact that the organizers of the workshop
arranged a session solely devoted to it ; see the Rapporteur paper by
Watson [8] . Nevertheless, for the sake of completeness brief comments
will be made on this object in this report as well
       Most of the groups active in the field with the notable exception
of those from the USSR were represented at the workshop. There were
reports on the status of the various objects observed by each group. In
this report each group has also   presented values for the chance
probability i.e. the probability that chance fluctuations in the background could simulate the result, as computed by them. It is not clear
if the stated probabilities were always multiplied by the number of
degrees of freedom inherent in the data analysis that led to the claim.
Some of the results were also presented in the contributed papers at
the workshop. The present paper is based on the totality of information
presented at the workshop.
       In this paper, we did not mention in most cases the observed values
of fluxes of gamma rays for two reasons: (i) Most observed fluxes are
around 10^{-11} and 10^{-14} cm^{-2} s^{-1} at 1 TeV and at 1 PeV respectively
within a factor of 3 and (ii) in most cases the sources were claimed
to be variable. It is not clear what significance one can attach to a
time-averaged flux when the source is variable over varying time scales.

2. GALACTIC SOURCES

2.1 Crab pulsar/nebula (2CG 184-05)

2.1.1. <u>TeV gamma rays</u>. The Durham group observed [9] two episodes
lasting for 15 m each during 150 hrs of observation, during which the
Crab pulsar emitted TeV gamma rays with the same periodicity as at radio
frequencies (chance probability, P_{ch}= 2.10^{-8}). The Tata group too
observed [10] a similar episode $(P_{ch}$ = 4.10^{-4}) during 170 hrs,
though with some differences. The Haleakala collaboration has presented
at the workshop evidence for 3 episodes lasting for 2000 to 4000 s each
during 30 hrs of observation; $P_{ch} \sim 10^{-5}$.
       The question arises whether, besides the bursts of pulsations, the
pulsar steadily pulsates albeit at a lower flux level. The Durham
group claims that there is a persistent weak emission of pulsed TeV
gamma rays at a flux level of 7.10^{-12} cm-2 s^{-1} $(P_{ch} \sim 2.10^{-4})$. The Tata
group, however, places a 3 sigma upper limit for such an emission at a
flux level of 10^{-12}cm-2 s-1.
       Vishwanath [11] has analysed the data base of the Tata group
and concluded that the Crab pulsar emits pulsed TeV gamma rays over time
spans $\sim$ 1 min. This kind of activity is reported to be confined to

∿ 2% of the total observation time. Graham Smith thought that variability in the emission at high energies (incoherent processes), in
contrast to radio emission (coherent processes), is difficult to understand . However, since the emission is only at detection threshold,
it is conceivable that whenever the background fluctuates downwards,
one sees the emission though the source itself may not be variable
(Vishwanath). Also in the model of Cheng et al. [12] , approximately
0.01 of the generated power in the outer magnetosphere gaps may appear
as TeV gamma rays resulting from Compton-synchrotron processes and there
may be fluctuations in the synchrotron light levels which are strongly
dependent on the local magnetic fields (Orford).

      Cawley et al [13] claimed an year ago that there was an unpulsed
emission of TeV gamma rays from the Crab nebula/pulsar at a high
(5.6 sigma) significance level. Now at this workshop the authors had
withdrawn this result as some D.C. background effects were not properly
treated in their earlier analysis.

2.1.2 PeV gamma rays. There are no new results presented at this
workshop. The Lodz group [14] claimed PeV gamma ray emission from
the Crab pulsar/nebula on the basis of a spike in the R.A. scan at the
Crab position. (see Fig.1). The Tien-Shan group supported this claim.
Also the Fly's eye group [15] claimed seeing the signals on one
occasion but not during a second time when they looked for it. In none
of these claims, a periodicity test (P≃ 33 ms) was possible. Also the
angular resolution was very poor, lowering the sensitivity of detection.
On the other hand there were several reports (Haverah Park, Akeno,
Dugway) claiming not to have been able to detect any signal. The upper
limits placed by some of these later observations are a factor of 50
lower than the claims made by the earlier group; see Fig.2, taken from
Watson [16] .

2.2 Vela pulsar (2 CG 263-02)

2.2.1 TeV gamma rays. Being a southern sky object Vela pulsar was
observed only by a few groups. Grindlay et al [17] and Bhat et al [18]
claimed in the past that it emits pulsed (P ∿ 89 ms) TeV gamma rays
at a 4 to 4.5 sigma level. At this workshop, the Tata group (P.N. Bhat
et al., see their contributed paper in this volume) presented evidence
for pulsed emission at 4 sigma level, provided a cut is made to select
only the lower energy showers; see Fig.3. As seen in the figure, the
TeV peak coincides in phase with the optical main pulse rather than
with the sub-GeV gamma ray peak as expected in the model of Cheng et al
  [12] . Taking all the degrees of freedom into account, the authors
claimed that P_{ch} ∿ 10^{-4} .

2.2.2 PeV gamma rays.   The Potchefstroom group did not see any
unpulsed emission from this object. Their time-keeping was not good
enough over long time spans of observation, preventing them to carry
out any periodicity analysis.

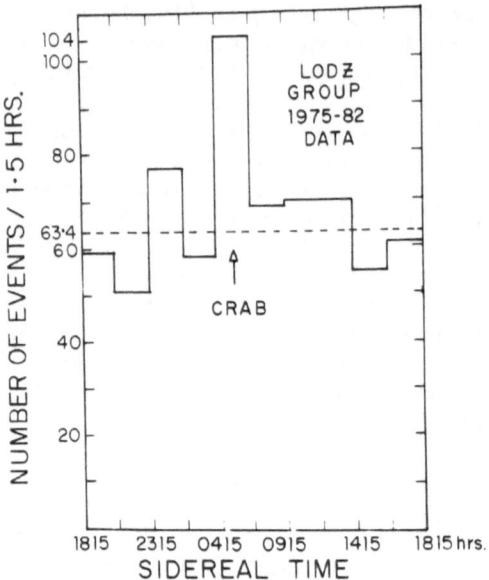

Fig.1: Lodz group[14] sees 104 showers at PeV energies in the bin
       containing Crab, compared to an average of 63.4 showers per
       bin

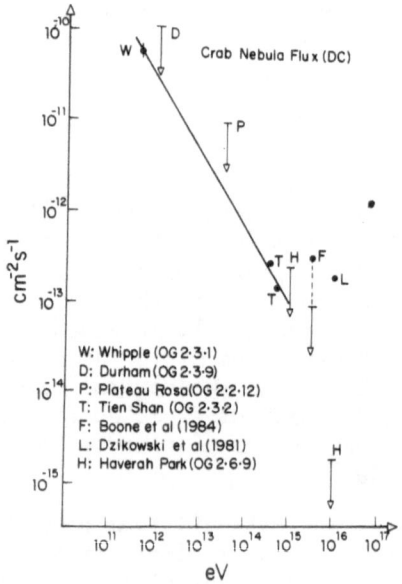

Fig. 2: Watson's [16] compilation of unpulsed TeV/PeV gamma ray
        emission by Crab. Note that the Whipple Collaboration (W)
        have withdrawn their result at 0.4 TeV at this workshop.

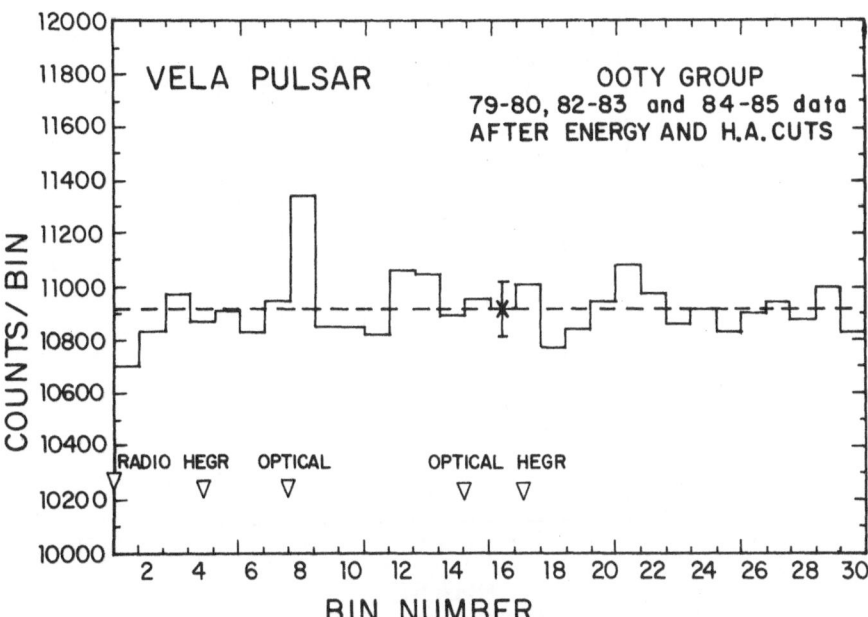

Fig.3: Phasogram of TeV showers in the direction of the Vela pulsar as obtained by P. N. Bhat et al. (see contributed paper, this volume). The 4 sigma peak in the 8th bin coincides with the optical main pulse and not with that of the GeV gamma rays.

2.3. PSR 1937 + 21 (1.5 ms pulsar)

The Durham group have observed this pulsar during July, August and September 1984 and claimed that it does emit pulsed TeV gamma rays during each month; $P_{ch} \sim 10^{-4}$. Timing uncertainties do not allow the authors to determine the relative phase of TeV gamma ray signal with respect to the radio signal. There were no reports at PeV energies.

2.4 PSR 1953 + 29 (6 ms pulsar)

The Durham group had already published [19] their claim of having detected pulsed emission of TeV gamma rays by this object at $P_{ch} = 3.10^{-5}$. No new reports were presented at the workshop; nor were there any reports at PeV energies.

2.5 PSR 1802-23 (2CG006-00)

This is a suspected 112.5 ms pulsar - one of the three - in the rather wide (2° diam.) error circle of the COS-B source, 2CG 006-00. The Potchefstroom group [20] have claimed detecting 6 bursts of pulsed TeV emission from this object at $P_{ch} = 5.10^{-5}$. The light curve is shown in Fig.4. There were no reports at PeV energies by any group on this object.

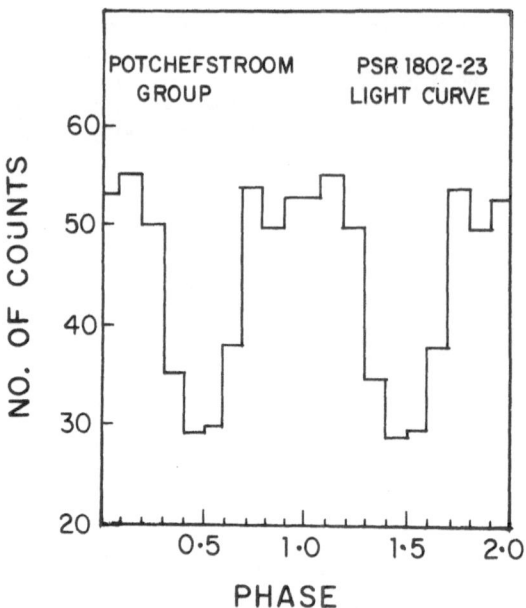

Fig. 4: The light curve in TeV gamma rays from the 112.5 ms pulsar
PSR 1802-23 as seen by Raubenheimer et al. [20] .

2.6 CYG X-3

2.6.1. TeV gamma rays. Cyg X-3, an X-ray binary, was first shown by
the Crimean Astrophysical Observatory group (Stepanian et al.) to be
pulsating in TeV gamma rays with a 4.8 hr period. No new results were
presented at the workshop on the 4.8 hr periodicity of TeV gamma rays.
For a review of the earlier claims, see Weekes [21] . The 12.59 ms
periodicity presumably due to a pulsar, of TeV gamma rays first claimed
by the Durham group [7] drew much attention at the meeting. The
Whipple Observatory collaboration subjected their 83-85 data to a
Rayleigh test whenever the observations satisfied $0.55 < \phi_{4.8\ hr} < 0.70$
for periods in the range $12.586 < P < 12.596$ ms. They did find
3 episodes just outside the expectation on chance basis, one each, on
25 May, 12 Oct. and 15 Nov.1985. Taking into account all the degrees of
freedom, the authors did not think that any of the episodes could be
claimed as a signal. In particular they did not observe any statistically
significant excess in the trigger rate on time scales of 1,2,3...8 min.
The Haleakala collaboration reported that during a 100 s interval on
12 Oct. 85, the average shower rate of 42/100 s went up to 67/100 s -
a signature to search for 12.59 ms periodicity as recommended in Ref.7.
When they subjected the 60 s data (embedded in the 100 s) to Rayleigh
test in the range $12.57 < P < 12.60$ ms, they found a peak (i.e. low chance
probability ; $P_{ch} \backsim 5.10^{-4}$) at P = 12.5925 ± 0.0005 ms, a value close

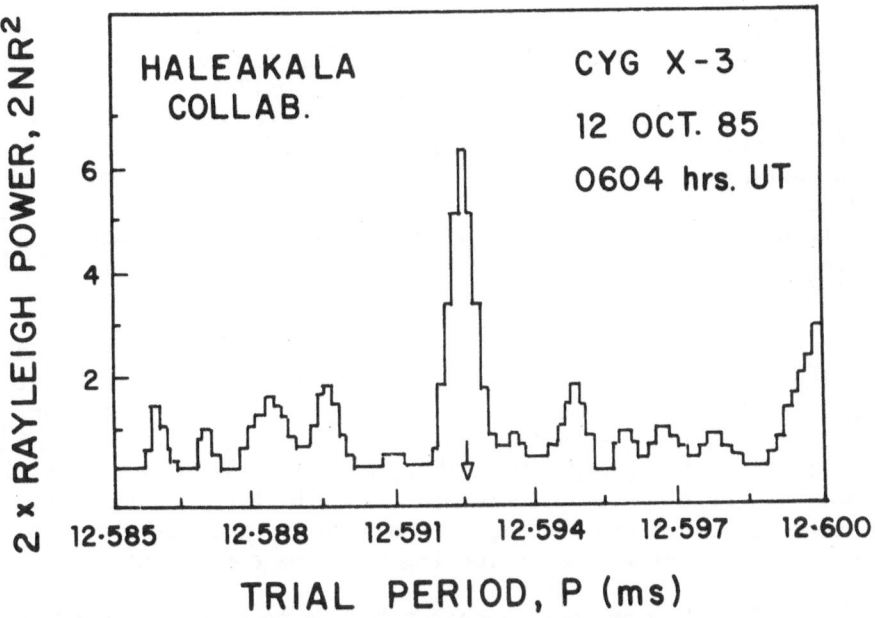

Fig.5: Rayleigh test by the Haleakala collaboration on TeV gamma rays from Cyg X-3 detected at 0604 hrs UT on 12 Oct.85 shows a peak at 12.5925± 0.0005 ms . The phase in the 4.8 cycle of this episode is 0.74.

to the one reported in Ref.7; see Fig.5. Soon afterwards the same group observed a 12 min. long episode of TeV gamma ray emission with nearly the same period. The Durham group too has reported seeing a similar episode on the same date. All the observations on the 12.59 ms periodicity are summarised in the Table below.

TABLE : A summary of 12.59 ms periodicity in TeV observations

GROUP	Date	UTC hrs.	Φ 4.8 hr.	P_{chance}	Period(ms)
Durham	12 Sept.83	0519	0.625	3.10^{-7}	12.5908±0.0003
	12 Oct.85	1930 ± 0030	0.55 ± 0.10	3.10^{-6} *	12.5928
Haleakala	12 Oct.85	0604	0.74	5.10^{-4} *	12.5925±0.0005
Whipple	12 Oct.85	0612 to 0619	0.76	3.10^{-5} *	12.5876

* Chance probability was not multiplied by all degrees of freedom.

A few comments are in order. The episodes observed by the Haleakala and Whipple collaborations, though close to each other, did not coincide in time. It is to be noted that the Haleakala group did not see any increase in the shower rate during the Whipple episode (0612 to 0619 hrs UT) nor did the Whipple group see any excess count rate during the Haleakala episode (0604 hrs. UT). The Durham group did not furnish a more precise time of occurrence of their 12 Oct.85 episode; such an information would help one establish the 4.8 hr phase dependence of the millisecond period which in turn would be valuable to any model builder. Also none of the groups presented the light curves. It would be bizarre if they looked vastly different and reassuring if identical.

2.6.2 PeV gamma rays. Referring the reader to a more detailed Rapporteur paper on the subject by Watson [8] in this volume, we merely mention here that todate the strongest evidence for the pulsed (P = 4.8 hr) emission of PeV gamma rays by this object is from the Kiel group [5]. Results from the other groups (Haverah Park, Akeno, Tata etc.) are either consistent with or at least not contradicting the Kiel results. Muon content in the showers and the signal phase in the various claims may pose some difficulties if one is looking for some consistency.

2.6.3. Muons. Though muons per se do not belong to gamma ray astronomy, the muons associated with Cyg X-3 are thought to be produced by some kind of neutral radiation which may be gamma-ray-like in character - hence

their relevance here. In the past, Soudan and NUSEX [6] groups, operating detectors designed to discover proton decays in deep underground locations claimed having detected a 4.8 hr. periodicity in the muons deep underground pointing approximately (but not exactly) in the direction of Cyg X-3. These claims, sensational as they were, led to a plethora of suggestions regarding new types or new interactions of particles; see De Rujula [22] for a review. Subsequently two other groups, Frejus and Kamioka, published [23] not seeing any such radiation. The KGF and IMB groups too did not see any muon signal. At this workshop a Swiss group (Bern) reported having operated a muon telescope deep underground and not seeing any muon signal. Their 90% C.L. upper limit to the muon flux is just below the line connecting the Soudan and NUSEX fluxes as a function of depth.

The Soudan group has reported at the workshop seeing an excess of muons during the period 24 Sept.-7 Oct.85 (when Cyg X-3 was known to be very active at Radio wavelengths) in the direction of Cyg X-3, compared to other times.

NOTE: The reports at the Workshop by the Haleakala group of a 12 ms periodicity in their Cygnus X-3 data are not confirmed in later and fuller reanalyses described in their paper - see p. 105. EDITOR.

2.7. HER X-1

2.7.1 <u>TeV gamma rays</u>. The Durham group was the first to have claimed
[24] seeing a burst (3 min) of pulsed TeV gamma rays from this X-ray
binary with a period of 1.24 s, a value that was in the ball-park of
what was expected from X-ray data. Chance probability was 7.10^{-5}. Later
the Whipple group had seen 7 episodes of pulsation during 1984 and 85
   [25]   . In these later observations, the pulsations lasted for longer
times, typically of 40 m duration. The Her X-1 also exhibits 1.7 d and
35 d cycles in the X-ray data. The Whipple group has noted that the
various episodes observed by them occur only during either the high or
the low 'on' states in the 35 d cycle while there is no preference for
any particular phase in the 1.7 d cycle. Indeed the episode on 16 June
85 started before the commencement of the X-ray eclipse and extended
for 70 min. well into the X-ray eclipse; P_{ch} = 4.10^{-5} for this particular
episode. This indicates that the gamma ray production site is not coincident with that of X-rays. According to the Whipple group, 7% of all
data exhibits the 1.24 s periodicity. The Haleakala collaboration too
has claimed seeing TeV pulsations (P = 1.24 s) from this object at
$P_{ch} \backsim 2.10^{-3}$.

2.7.2. <u>PeV gamma rays</u>. There was one claim published [26] by the
Fly's eye group for pulsed (P = 1.24 s) emission of 500 GeV gamma rays
by Her X-1 over a 40 min. span. It is interesting that the Durham group
   [27]   who were observing the object at the same time, did not find
any evidence for pulsations in the TeV gamma rays. There have been no
reports on any further PeV energy observations on this object.

2.8 4U0115+63

This object too is an X-ray binary. It was first reported by the Durham
group [28] to be pulsating (P = 3.61 s) in TeV gamma rays at a
P_{ch}=5.10^{-6} level. Subsequently the Whipple collaboration confirmed the
result at P_{ch}= 10^{-2}. At this workshop the Haleakala collaboration
claimed to have observed 3 episodes of pulsed emission during 40 hrs.
of observation at P_{ch} $\backsim$ 10^{-4} . There were no reports of any observations at PeV energies.

2.9 Vela X-1

2.9.1 <u>TeV gamma rays</u>. Vela X-1 is known to be an X-ray binary exhibiting
periodicities of 283 s and 8.96 d. The Potchefstroom group claimed to
have seen steady pulsations in TeV gamma rays with a period of 283 s
from this object at P_{ch}= 2.10^{-5}. They also reported to have seen
unpulsed emission at the same level of significance.

2.9.2. <u>PeV Gamma rays</u> . The Adelaide group published [29] their claim
of having detected pulsed (P = 8.96 d) PeV gamma rays from this object
at P_{ch} $\backsim$ 10^{-4}. At this workshop, the Potchefstroom group claimed
seeing it in pulsed (P = 8.9 d) PeV gamma rays at $P_{ch} \backsim 10^{-5}$.

2.10 Geminga (2CG195+04)

This object, the second brightest gamma ray source in the sky (after the Vela pulsar) at $E \gtrsim 100$ MeV has had a chequered history. There have been claims and counter-claims that the object pulsates (P ≉ 60s) in X-rays and 100 MeV-1 GeV in gamma rays; see Bignami et al. [30] and Buccheri et al [31] for details. In the TeV energy range Zyskin and Mukanov [32] and Kaul et al [33] claimed seeing pulsations at P ⌣ 60s, while Helmeken and Weekes [34] , Cawley et al [35] and Bhat et al [36] did not see any. For a critique see Reference 36 and for a summary of results, Fig.6. It appears that there is no independently confirmed evidence that this object pulsates in TeV gamma rays with a period around 60 s. There have been no reports on PeV gamma ray observations of this object.

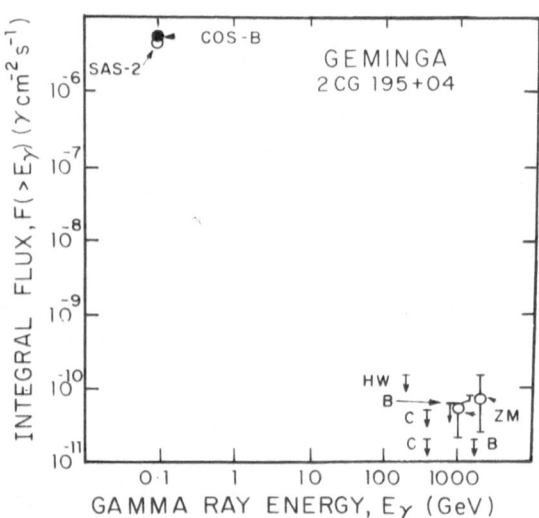

Fig.6: Fluxes of gamma rays from Geminga as a function of energy, as published by the various authors; see the text. The fluxes at 100 MeV and the higher upper limit by Cawley et al.(∈) refer to unpulsed emission and all other points to pulsed emission.

3. EXTRA GALACTIC SOURCES

3.1. LMC X-4

The Adelaide group published [37] seeing pulsed emission (P= 1.41 d) of PeV gamma rays from this source at $P_{ch} \sim 0.01$. No further observations were reported; nor were there any observations at TeV energies.

3.2. M31 (Andromeda nebula, NGC224)

The Durham group published [38] their claim of having detected TeV gamma rays from this object by drift scan at $P_{ch}= 0.01$. No further observations were reported; nor were there any observations at TeV energies.

3.3 Cen A (NGC5128)

Grindlay et al [39] claimed several years ago detection of TeV gamma rays from this object. There have been no further reports. No reports have been made on PeV observations.

4. DISCUSSION AND SUMMARY

As detailed in the previous two sections, there are claims for a dozen celestial objects emitting gamma rays at TeV and/or PeV energies. At their face value most, if not all, claims appear to be valid as the probability of chance fluctuations to mimic the signal is low even when one considers all the degrees of freedom inherent in the analysis. Nevertheless there are some nagging questions that are often raised by well-meaning and considerate colleagues. These are:(i) Why isn't there even a single source seen at a 10(or higher) sigma level ? As shown in Fig.7, all the claims are at less than 6.8 sigma level. (ii) Why is the luminosity vs distance plot of all the claims lie so close to $L \propto D^2$ line (i.e. why are the observed fluxes confined to such a narrow range?)? How does one source 'know' what the others are doing? (iii) Why doesn't one see a claimed source every time one tries to ? If one postulates source variability, then, why doesn't the source fluctuate a little higher to let one see it as a 10 sigma source? On the other hand, it must be said in favour of the claimants that one could not produce a 5σ peak if one analysed the data with a wrong period even in tens of thousands of trials.

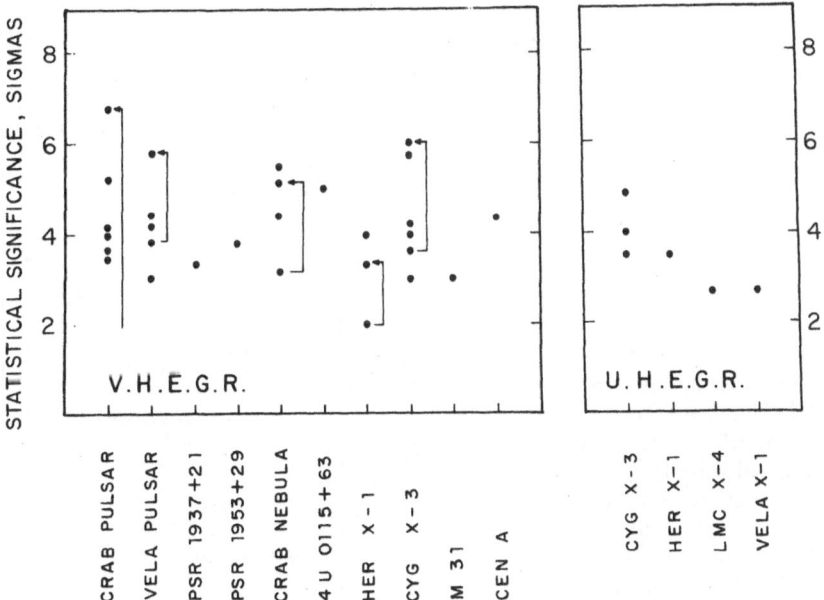

Fig.7: Statistical significance of the various TeV/PeV source claims (Ramana Murthy unpublished). The upper points connected by arrows refer to specially selected data segments and the lower points to the same data when no selection is made.

Nor did one see any peak in a blank region of the sky.At this time a fair appraisal of the field(in the author's view) would be that the case for the reality of the sources is not gold-plated to convince even the worst sceptic nor the evidence is so flimsy as to be ignored altogether.Most detector systems are operating probably at threshold levels. It comes as no surprise, therefore, that the participants at the workshop were unanimous in stressing the need to reduce the proton shower background at hardware and/or software levels. Several groups (Crimean,Durham,Haleakala, Leeds, Whipple etc.) are actively engaged in finding a reliable algorithm to achieve this.

A source can be claimed to be variable if it was established to be one beyond doubt a few times, at least once. Another way to approach the issue would perhaps be for two independent groups operating in a common longitudinal belt range to observe a given source simultaneously. Even a single group can operate two arrays far apart such that they see the same source but not the same showers,as was demonstrated by Bhat et al.[10].Importance of such observations cannot be over-emphasized in the context of lack of blindingly intense sources and the preponderence of variable sources.

With many more groups having embarked on TeV/PeV gamma ray observations and the older groups putting in more effort,it is reasonable to expect that by the time of the next workshop (1989-90?)one would have established some sources at significance levels of 10 sigma or greater and staked claims for several tens of sources at 5 sigma level.

Fig.8 : Luminosity-distance diagram of the various TeV source claims (Ramana Murthy, unpublished). The dotted line, $L \propto D^2$, is drawn to aid the eye. A source seen at $x\sigma$ can be seen at $2\,x\sigma$ if its luminosity increases by the vertical separation between two horizontal dashes shown as $x\sigma$ and $2x\sigma$.

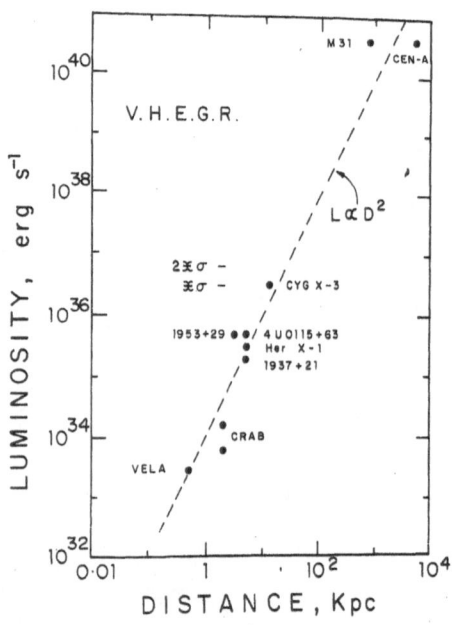

REFERENCES

1. Proc. International w/s on very high energy gamma ray astronomy
   Ed: P.V. Ramana Murthy and T. C. Weekes. Published by Tata Institute
   of Fundamental Research,Bombay 5, India (1982)
2. P. V. Ramana Murthy and A.W. Wolfendale, GAMMA RAY ASTRONOMY,
   Cambridge University Press, U.K. (1986).
3. R. J. Protheroe, Proc. Astron. Soc. Australia (1986).
4. N. A. Porter and T. C. Weekes, Smiths.Astrophy.Obs.Rep.381 (1978)
   P.V.Ramana Murthy, NON SOLAR GAMMA RAYS,p.71, Pergamon Press(1980).
5. M. Samorski and W. Stamm, Ap.J.Letters, 268, L17 (1983).
6. M. L. Marshak et al. Phys. Rev. Lett. 54, 2079 (1985).
   G. Battistoni et al. Phys. Lett. 155B, 465 (1985).
7. P. M. Chadwick et al., Nature, 318, 642 (1985).
8. A. A. Watson, Rapp. paper on Cyg X-3, this volume.
9. Gibson et al., Nature, 296, 833 (1982).
10. P. N. Bhat et al., Nature, 319, 127 (1986).
11. P. R. Vishwanath, Contributed paper, this volume.
12. K. S. Cheng et al, Ap.J., 300, 522 (1986).
13. M.F.Cawley et al,19th Int. Cosmic Ray Conf.(La Jolla), 1,131(1985).
14. T. Dzikowski et al., Jour. Phys. G. 9, 459 (1983).
15. J. Boone et al., Ap. J., 285, 264 (1984).
16. A. A. Watson, Rapp. paper 19th Int.Cosmic Ray Conf.(La Jolla) (1985).
17. J. E. Grindlay et al., Ap. J. 201, 82 (1975).
18. P. N. Bhat et al., Astron. Ap., 81, L3 (1980).
19. P. M. Chadwick et al., Nature 317, 236 (1985).
20. B. C. Ranbenheimer et al., to appear in Ap.J. Letters (1986).
21. T. C. Weekes, Proc. 6th Astrophys. meeting, XXI st Recontre de Moriond
    Savoie, France (1986).
22. A. De Rujula, CERN preprint TH 4267/85 (1985).
23. Y. Oyama et al., Phys. Rev. Lett., 56, 991 (1986).
    Ch. Berger et al., Phys. Lett., 174 B, 118 (1986).
24. J. C. Dowthwaite et al., Nature, 309, 691 (1984).
25. P. W. Gorham et al., to appear in (a)Ap.J. and (b)Ap. J. Lett.(1986).
26. R. M. Baltrusaitis et al., Ap.J. Lett. 293, L69 (1985).
27. P. M. Chadwick et al.,19th Int.Cosmic Ray Conf.,(La Jolla),1,254(1985)
28. P. M. Chadwick et al., Astron. Ap., 151, L1 (1985).
29. R. J. Protheroe et al., Ap. J. Lett., 280, L47 (1984).
30. G. F. Bignami et al., Nature, 310, 464 (1984).
31. R. Buccheri et al., Nature,316, 131 (1985).
32. Y.L.Zyskin and D. B.Mukanov,18th Int.Cosmic Ray Conf. (Bangalore)
    1 ,122(1983); 19th Int.Cosmic Ray Conf.(La Jolla),1,177 (1985).
33. R. K. Kaul et al., 19th Int. Cosmic Ray Conf.(La Jolla),1,165(1985).
34. H.F. Helmken and T. C. Weekes, Ap.J., 228, 531 (1979).
35. M. F. Cawley et al., 19th Int. Cosmic Ray Conf.(La Jolla)1,173(1985).
36. P.N. Bhat et al, to appear in Astronomy and Ap.,(1986).
37. R. J. Protheroe and R. W. Clay, Nature, 315, 205 (1985).
38. J. C. Dowthwaite et al, Astron. Ap., 136,LT4 (1984).
39. J. E. Grindlay et al., Ap.J.Lett., 197,L9 (1975).

Cygnus X-3

A A Watson

Department of Physics, University of Leeds, LEEDS LS2 9JT

Abstract

The contributions to the Cygnus X-3 session of the Durham NATO Workshop on Ultra High Energy γ-ray astronomy are described. At TeV energies emission at 4.8^h and 12.6 ms is now well established. Much further work on PeV emission is needed and the question of underground muons remains enigmatic.

1 Introduction

The Cygnus X-3 workshop, which was chaired by Ramana Murthy (Tata) and had Chardin (Saclay) and Ruddick (Minnesota) as discussion leaders, proved to be one of the liveliest of the meeting. The main topics discussed were (i) the TeV signals modulated at 4.8^h and 12.6 ms, (ii) the evidence for PeV emission with 4.8^h modulation and (iii) particles detected underground from the direction of Cygnus X-3 which also show the 4.8^h modulation. There was some discussion of models of the Cygnus X-3 system but, as these are reviewed in the papers by Brecher and Hillas elsewhere in this volume, I will make little mention of them here. Instead two additional points relating to PeV detections, which were raised but not fully discussed, will be mentioned.

2 4.8^h modulation at TeV energies

A number of speakers (notably Ramana Murthy and Chardin) expressed a lack of conviction about the evidence for 4.8^h modulation at TeV energies because, as yet, no 6 or 7 σ detections have been reported and there is a lack of simultaneous observations by independent groups. The second criticism can only be met if two groups are operating at similar latitudes and similar longitudes (within about 20° in each case) so that usefully long periods of overlap observation can be attempted. While weighing the first criticism it

K. E. Turver (ed.), Very High Energy Gamma Ray Astronomy, 53–61.
© 1987 by D. Reidel Publishing Company.

must be remembered that the total annual observing time of a TeV
telescope is less than 1000 hours and that a particular source can be
observed during only a few moonless periods each year. With present
telescope sensitivities it is, in my view, unrealistic to expect a group
to devote the necessary time to get 6 or 7 σ results on a single
object when there is a sky full of objects to be explored.
Independent detections of Cygnus X-3 at about the 4 σ level are, to
me, quite convincing because of the coincidence in phase of the
maximum of emission as found by different observers. A summary[1]
of the phases of maximum and their significance is shown in figure 1,
which is based on relatively recent observations (1980 - 84) all
analysed with the van der Klis-Bonnet-Bidaud ephemeris.

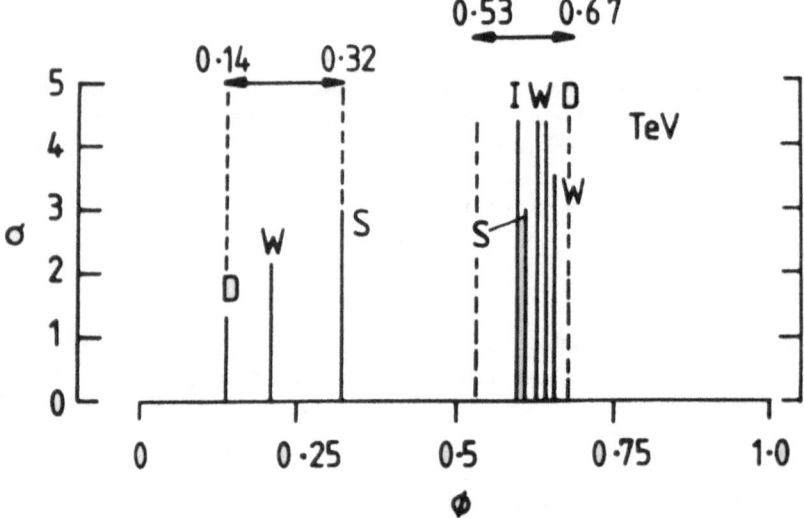

Figure 1   Summary of phase, and significance, of emission
maximum as recorded at Dugway (D), Whipple Observatory (W), Crimean
Astrophysical Observatory (S) and at JPL (I). All observations were
made between 1980 and 1984. The vertical dotted lines indicate the
extremes of phase (near 0.25 and 0.65) observed in all experiments
from ~ 1 TeV to ~ 10^{14}eV. All data have been analysed with the van
der Klis-Bonnet-Bidaud ephemeris. References and other details can
be found in[1].

The only telescopes presently operating, which are able to
observe Cygnus X-3, are those at the Whipple Observatory (approx
latitude, longitude 32°N, 111°W), Crimean Astrophysical Observatory
(45°N, 34°E) and Mt Haleakala (20°N, 155°W). The Durham instrument,
which was sited at Dugway (41°N, 112°W), was geographically close to
the Whipple Observatory but the possibility of simultaneous
observations at the two sites was restricted because of climatic
variability.

There is a wealth of evidence that Cygnus X-3 exhibits time variations on a variety of time scales and at a range of energies. Two of the best documented occurrences at TeV energies are by the Whipple observatory group[2,3] who have reported occasions on which strong modulation has been observed in one moonless period but not in the one immediately following. Simultaneous, or nearly simultaneous, observations by two groups are much needed here. Variations on 19 day and 34 day cycles have been reported by the Durham group[4] whose flux estimate at the cycle peak of 2×10^{-9} cm^{-2} s^{-1} at 1 TeV is in good agreement with the reports from other groups when allowance is made for uncertainty in the estimates of telescope energy thresholds.

3   The 12.6 ms Pulsar

Although a high sensitivity telescope, such as that in Haleakala, may produce a 7 σ signal on the 4.8^h period without excessive use of the available observing time, it is not surprising that little progress towards that end has been made in the last year. At the La Jolla Conference the Durham group[5] announced the detection of a 12.5908 ms periodicity in the TeV γ-ray emission from the Cygnus X-3 system. Naturally the Whipple and Haleakala groups have attempted to confirm this detection and highlights of this meeting were reports of preliminary analysis of data obtained by these groups: these analyses appear to provide confirmation of the original Durham detection. The initial claim was further bolstered by analyses of Durham data taken in 1981, 1982 (at Dugway) and 1985 (in Durham).

During the Cygnus X-3 workshop, Turver (Durham) reviewed the evidence for 12.6 ms periodic emission presented by the three groups. All groups observed Cygnus X-3 on 12 October 1985, soon after the on-set of the very large radio burst: all groups observed unusual enhancements of emission, lasting for several minutes, from the direction of Cygnus X-3 and the following results were reported:

(i)   The Durham group, observing close to $\phi = 0.65$ (the time of maximum X-ray emission in the 4.8^h cycle), found sufficient excess counts to make a period search meaningful. A period of 12.5927 ms was detected with high confidence (chance probability 10^{-4}). During the outburst, the period was found to shorten ($\dot{p} \sim -8 \times 10^{-10}$ s/s). Similar behaviour observed during the October 1983 burst was used for the initial pulsar analysis[5].

(ii) The Haleakala group observed two bursts around $\phi_1 = 0.74$ and $\phi_2 = 0.83$. The periods found independently from the events of the two bursts are consistent with the Durham period when allowance is made for the negative p which was also observed during the second Haleakala burst.

NOTE: The reports at the Workshop by the Haleakala group of a 12 ms periodicity in their Cygnus X-3 data are not confirmed in later and fuller reanalyses described in their paper - see p. 105. EDITOR.

(iii)   The Whipple Observatory group observed a burst at
ϕ = 0.72. The period found from the events of the burst is
not consistent with that of the other groups but Fegan
(Dublin/Whipple) commented that if shorter time windows
were used the period might come into line with those of the
other groups.

This remarkable sequence of observations on 12 October was made
shortly after the on-set of the second radio burst from Cygnus X-3
which started about 7 October and reached 17 Jy at 11 cm on
11 October 1985.

Although the Durham and Haleakala groups find p_{local} < 0 during
the burst, the Durham group (using observations from 1981 to 1985)
report a secular p = + 10^{-14} s/s which, on conventional pulsar theory,
leads to a lifetime of about 7000 years for the object. The negative
local p is hard to understand within the framework of binary models
of Cygnus X-3 which suppose that the neutron star/pulsar is between
the observer and the companion star when ϕ ~ 0.65.

While further confirmations of the 12.6 ms pulsar are clearly
desirable a very convincing case for its existence has now been made
as a result of the 12 October 1985 observations.

4   PeV Observations of Cygnus X-3

While the future of TeV astronomy, through observations of
Cygnus X-3 (and of many other sources) is now clearly established,
the same is not true for PeV astronomy. More than three years after
the initial Kiel claim[6], in which a 4.4 σ DC signal, and a very
significant peak in a phase interval, $\Delta\phi$, of 0.1 at ϕ = 0.35, was
reported from only 31 events, that result remains the most convincing
single piece of evidence for emission at PeV energies. At this meeting
the Haverah Park group[7] described a re-analysis of their data
(1979-82), which had been used[8] to claim confirmation of the Kiel
result, using a new X-ray derived ephemeris due to Mason (MSSL,
private communication). The sharp peak initially seen in the Haverah
Park phaseogram is weakened and in later years, although the weak
DC signal (~ 3 x 10^{-14} cm^{-2} 5^{-1} above 10^{15} eV) appears to persist
through to 1985, the phase of emission moves between a region near
ϕ = 0.25 and a region near ϕ = 0.65. The Akeno group (Kifune et al,
these proceedings) presented new results from a low energy threshold
run from December 1985 - May 1986. No muon-poor selection was
possible with these data but peaks of 2.1 and 2.7 σ are seen above
the background in adjacent bins ($\Delta\phi$ = 0.05) centred on ϕ = 0.65. The
background has not been finally calculated for this experimental
period and may peak near ϕ = 0.65. However fluctuations in it are
thought to be insufficient to explain the full excess.

A preliminary result from the Ooty group was reported at the
Taos meeting in July by Tonwar (Tata). For data collected over the 2
years, June 1984 - June 1986, this group observed a DC excess of 3 σ
in a 4° x 4° box centred on Cygnus X-3 for events with electron
number N > 5 x 10^5. For slightly smaller events a significant DC

enhancement (2.5 σ) is found only when rather old showers (s > 1.4) are selected. A phase analysis shows a rather broad light curve with a peak near ∅ = 0.6. These results have not yet been published.

Chardin, throughout this meeting, has persistently raised the question of the validity of the background calculations used by the various groups, in particular Kiel, Haverah Park and Akeno. Although there is undoubtedly reason for concern in some of the reported backgrounds used for the phaseogram interpretations, there is no obvious reason why spurious enhancements would tend to occur at similar phases when measurements are made at different experimental sites and over different time scales. Because of the long exposure time required to accum. ulate any events with an air shower array, the background is unlikely to be very non-uniform throughout the epoch-folded histogram. Although caution is certainly needed, the phase compilation shown in figure 2, for all reported data above 10^{15} eV, strongly hints at PeV emission near 0.25 and 0.65. In this figure the dotted lines mark bounds set by observations at lower energies (10^{12} - 10^{14} eV : see[1] for a review) after allowance has been made for the Mason ephemeris which moves the bounds (to greater ∅) by ~ 0.04. Each independent report near 1 PeV is represented by a unit vector plot at the phase of maximum intensity in the phaseogram reported by each group.

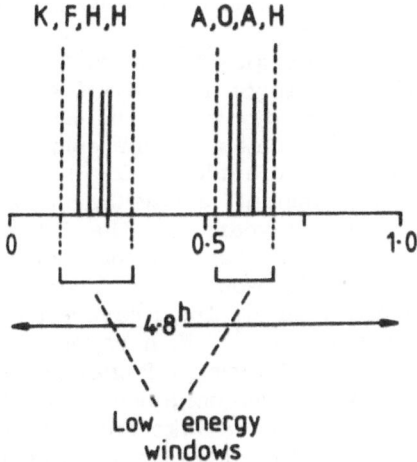

Figure 2   Summary of position of maximum phase for experiments in which PeV emission has been reported from Cygnus X-3. To be non-controversial, unit vectors are plotted at each ∅. Low energy windows, found from the bounds of the observations between ~ 1 TeV and ~ 10^{14}eV are shown. Observations are from Kiel (K), Fly's Eye (F), Haverah Park (H: for independent data sets), Ooty (O) and Akeno (A: independent data sets). References can be found in the text or in[1].

5 What is the evidence on time variability of Cygnus X-3?

A number of speakers (Lamb (Iowa), Hermsen (Leiden), Cawley (Whipple) and Weekes (Whipple)) raised the question of the time variability of Cygnus X-3 on time scales longer than 4.8^h. Certainly at X-ray energies there is considerable evidence for variations in the amplitude of the 4.8^h modulation of the order of a factor 3 over 40 days observation (9,10) and other variations have been discussed by Priedhorsky and Terrill[11] based on Vela satellite observations. While a factor of 3 is readily detectable with recent X-ray satellites, for most experiments at TeV and PeV energies such a change can put the source below the detection sensitivity of the telescopes. As mentioned earlier, there have been two particularly well documented cases in which the Whipple Observatory group[2,3] observed strong signals in one month but not in the following month. In both cases the signals between $\phi = 0.6 - 0.7$ were strong in the first month (20% enhancement at 3.6 σ[2], and 8.5% enhancement at 4.4 σ[3]). In both cases the following observation period revealed no excess above background and constancy can be rejected at the $\sim 2 \sigma$ and 3σ levels respectively: changes of the order of 3 or 4 in the intensity have thus been detected. Similar changes have been reported by the Crimean group[12,13].

The Durham group[4] have reported evidence of a 19 day amplitude variation of about 20% but there is also evidence from the TeV experiments for the 34 day variation sometimes claimed at X-ray energies[14].

There is evidence at PeV energies[7] and at TeV energies[3] that the favoured phase of emission can be near $\phi \sim 0.25$ and/or $\phi \sim 0.65$.

A most extreme form of intensity variation has been deduced by the Kashmiri (Gulmarg) group[15] using a very simple dual-photomultiplier system. They claim evidence for a long-term reduction in the luminosity of the PeV emission: assuming an exponential decay they derive a time constant of 2.3 years. If the PeV and TeV emissions are linked[16] then the corresponding decrease in TeV emission over the years since 1972 would have been ~ 400 : such a decrease can certainly be excluded by the measurements from the Crimean Astrophysical Observatory[17]. With regard to the experimental arrangement used at Gulmarg, Fegan pointed out that any wide angle optical receiver (even a coincidence system) which operates without padding lamp servo controls to take out sky brightness variations (as in this case) must be judged with "absolute caution".

6 Are there methods which can be used to enhance sensitivity to PeV γ-rays?

The PeV γ-ray signals are so weak, and the collecting areas so far operating so small ($\sim$ that of a TeV telescope), that it is clearly advantageous to search for methods which might help reject the charged cosmic ray background: the muon content and the shower age are two such parameters.

6.1   The muon content of showers

Because the (γ,p) cross-section is about 300 times smaller than the (p,p) cross-section, it was pointed out long ago that showers initiated by γ-rays would be poor in muons by comparison with nucleon/nucleus initated showers. However, when this supposition was tested on the Kiel Cygnus X-3 events[18] it was found that the muon content was very similar to that of normal showers of the same size. It must be recalled that the Kiel detector, although under 800 g cm^{-2} of absorber, is not a tracking detector and that all events had cores falling with 30 m of it: thus 'punch through' of very low energy photons cannot be completely ruled out. By contrast the earliest Akeno observations of Cygnus X-3[19] showed a signal only when muon-poor events were selected: here the muon detectors were always more than 50 m from the core. At Haverah Park the only attempt to study the muon content of Cygnus X-3 events so far was made when the signal itself was rather weak[20] and the search was thus inconclusive.

In this context, although strictly not within the scope of the Cygnus X-3 workshop, it is of interest that the Potchefstroom group have reported a detection of Vela X-1 at PeV energies when events deficient in muons are selected. Few details were available at this meeting although the 4 m^2 muon detector (threshold 1 GeV) was probably far from the core of the events so selected. It is also worth recalling that an enhancement of TeV γ-ray selection was achieved in observations of CEN A when events containing muons were rejected[21].

6.2   The Shower Age

Viswanath (Tata) raised the question of the use of large s, the shower age, as a γ-ray signature. The Kiel group[6] selected showers with s > 1.1 in an attempt to enhance the γ-ray content of the events observed. A similar selection was adopted by the Adelaide group for their study of Vela X-1[22] but not for LMC X-4[23]. For this latter source they argue that the broad s-distribution found for inclined events precludes the use of the s-parameter. The Ooty group (Tonwar, Taos meeting, unpublished) used the age parameter to enhance their DC signal, finding it to be more effective for small showers (N > 10^5) than for showers with N > 5 x 10^5. This has prompted Reid (Leeds) to suggest that what is actually happening is that near the threshold of an array, flat s events are of higher primary energy than steep s events and thus, if the γ-ray spectrum, is flatter than the charged particle spectrum, the γ-ray component will be enhanced.

It is very important to investigate the effectiveness of s as a selection parameter as it may have potential for use in large arrays where muon detectors would be uneconomic. Surprisingly little attention has been given to the calculation of this parameter, nor to the precision with which it can be measured: this is a topic needing further work.

7   Underground Muons from Cygnus X-3

The observations of underground muons from the direction of
Cygnus X-3, and which exhibit the 4.8^h periodicity, continue to attract
attention. Observations by the Soudan group at the time of the
October 1985 radio flare were described by Ruddick (Marshak et al, to
be published). About a week before the flare (24 September –
1 October) the group observed an enhancement of the underground
muon signal in the phase bin $0.725 < \phi < 0.750$ which lies in the same
region of the 4.8^h period as previous enhancements reported by
Soudan[24] and NUSEX[25]. They presented their observations in the
form of a plot of the events/week falling within this phase interval.
For 5 weeks from mid-September 1985 the weekly counts are 2, 3, 7, 4
and 2 (with the radio flare occurring in the 4th week) against
estimated background counts of 1.5, 1.3, 1.7 1.7 and 1.9. Chardin
pointed out that the efficiency of the Soudan detector (which was
turned off for some months in mid-1985) was higher in late 1985 than
earlier on. From a plot given to me by Ruddick it appears that the
estimated background was ~ 0.6/week in early 1985, while for late 1985
it had risen to 1.0/week. In the early part of the year 5 events were
detected in 22 weeks while about 13 were expected: in the later part
of the year (and into 1986) 30 events were observed in 24 weeks and
about 24 were expected. The background question clearly requires
still further examination.

A phenomenological model was described by Ruddick[26] which
explains the underground muon observations in terms of the
production of a long-lived neutral particle at Cygnus X-3 which
interacts within the earth to produce a massive secondary particle
which decays to at least one muon.

Acknowledgements

I would like to thank Ted Turver for inviting me to be a
rapporteur at this NATO Workshop, NATO for financial support and all
who spoke at the Cygnus X-3 session for provision of written
comments and other material. Some of the material referred to has
not yet been written up, and I apologize to authors for any
misrepresentations of their views.

References

[1] Watson A A, Proceedings 19th Int Cosmic Ray Conference La Jolla
     9, 111, 1985

[2] Weekes T C et al, Astron Astrophys 104, L4, 1981

[3] Cawley M F et al, Ap J 296, 185, 1985

(4) Dowthwaite J C et al, Astron and Astrophysics 126, 1, 1983

(5) Chadwick P M et al, Nature 318, 642, 1985

(6) Samorski M and W Stamm, Ap J 268, L17, 1983

(7) Eames P V J et al, 1987, these proceedings

(8) Lloyd-Evans J et al, Nature 305, 784, 1983

(9) van der Klis M and Bonnet-Bidaud J M, Astron Astrophys 95, L5, 1981

(10) Bonnet-Bidaud J M and van der Klis M, Astron Astrophys 101, 299, 1981

(11) Priedhorsky W and J Terrell, Ap J 301 886, 1986

(12) Fomin V P et al, Proc 17th Int Cosmic Ray Conf, Paris 1, 28, 1981

(13) Vladimirsky B M et al Proc 13th Int Cosmic Ray Conf, Denver 1, 456, 1973

(14) Molteni D et al, Astron Astrophys 87, 88, 1980

(15) Bhat C L et al, Ap J 306, 587, 1986

(16) Hillas A M, Nature 312, 50, 1984

(17) Stepanyan A A et al, Proc Int Wshp on VHE Gamma Rays Ootacamund (Bombay: Tata Inst) p43, 1982

(18) Samorski M and Stamm W, Proc 18th Int Cosmic Ray Conf (Bangalore) 11, 244, 1983

(19) Kifune T et al, Ap J 302

(20) Blake P R et al, Proc 19th Int Cosmic Ray Conf, La Jolla 1, 66, 1985

(21) Grindlay J et al, Ap J 197, L9, 1975

(22) Protheroe R J and R W Clay, Nature 315, 205, 1985

(23) Protheroe R J et al, Ap J 208, L47, 1984

(24) Marshak M et al, Phys Rev Lett, 54, 2079, 1985 and 55, 1965, 1985

(25) Battistoni et al, Phys Lett, 155B, 465, 1985

(26) Ruddick K, Phys Rev Lett, 57, 531, 1986

SOURCE MECHANISMS (PULSARS)

K J Orford,
Department of Physics,
University of Durham,
Durham DH1 3LE, UK.

ABSTRACT.
The workshop on Source Mechanisms (Pulsars) was of great benefit
to experimental gamma-ray workers in understanding the latest
attempts to produce coherent models of pulsar emission. These new
models are able to explain the presence of high energy photons in
a natural way, and provide predictions on pulse position, width
and energy spectrum which may guide new observations.

The Workshop was chaired by R C Lamb. F Graham Smith and J
Truemper led the discussion.

The Workshop began with the Chairman asking the participants
to contribute questions which they felt should be discussed. The
questions which the participants provided may be listed under
three headings:

(a) What can energetic gamma-rays tell us about the physics
of pulsars ?
(b) Could there be any radio quiet, gamma-ray loud pulsars ?
(c) Are transient phenomena expected ?

The overall theme of the workshop was whether or not pulsar
theories could usefully help to direct the observing programme of
VHE observers.

The first question, from P Goret, was the cause of a
considerable discussion of current pulsar theories:
There are two proven gamma-ray pulsars :- the Crab and
Vela. COS-B has measured these very well. If the source
mechanisms are the same at 1GeV as at 1TeV, then of what
use are TeV gamma-ray measurements of pulsars ?
This provocative question was initially answered by F G Smith.

It is important to know whether or not the emission region
is the same for all energies emitted by the pulsar. Given the

K. E. Turver (ed.), Very High Energy Gamma Ray Astronomy, 63–70.
© 1987 by D. Reidel Publishing Company.

geometry is known, the more interesting question, touching on the
ultimate source of all high energy emission, is what is the
energy spectrum of gamma-rays and where does it turn over ?

In the light of this answer, the participants were given a
very brief resume of current pulsar theory as it concerns
gamma-ray emission. The limits on the magnetic field through
which a TeV gamma-ray may travel to escape from the production
region make it very unlikely that such gamma-rays that are seen
at the Earth originate from near the neutron star surface. Given
that assumption, the only current models which are capable of
providing significant VHE gamma-ray emission are based upon
magnetospheric gaps a considerable fraction of the distance from
the pulsar to the radius of light cylinder. The model proposed by
Cheng, Ho and Ruderman (1,2) was chosen as an example for
discussion. In this model, the flow of charge from some regions
of the magnetosphere to outside the light cylinder will be
supplied by the creation of electron pairs via the
materialisation of gamma rays. The possible locations of stable
outer magnetospheric gaps are shown in Figure 1 (lettered A to C).

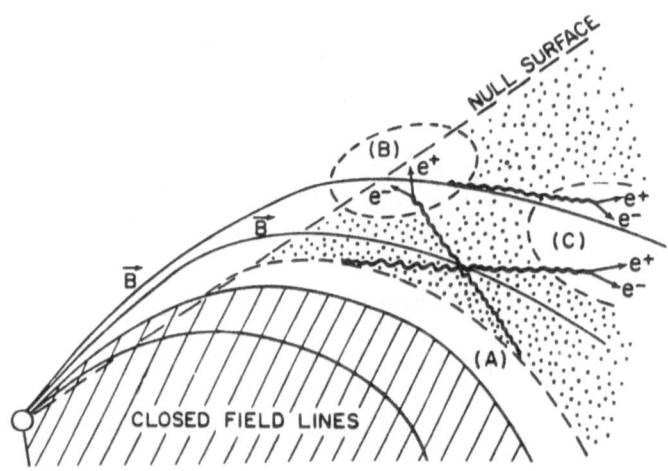

Figure 1. Possible sites for outer magnetospheric gaps. (Cheng,
Ho and Ruderman).

The gap lettered A was used to illustrate the way VHE
gamma-rays could be produced by (a) the acceleration of electrons
by the large electric field along the gap, (b) the radiation of
synchrotron photons by these electrons, and (c) the conversion of
these lower energy photons to VHE gamma-rays by inverse Compton
collisions with oppositely charged electrons streaming along the
gap in the opposite direction. If this picture as proposed by
Cheng, Ho and Ruderman is correct, then a number of consequences
follow for VHE gamma-rays. First, as pointed out by the authors,

the spectrum at TeV energies should reflect the synchrotron spectrum about 1000 times lower in energy. This synchrotron spectrum may be limited at an energy of several GeV. The VHE spectrum may then have a cut-off at several TeV, depending on the exact conditions in the gap. Secondly, there will be TeV gamma-ray emission both outwards from the gap and inwards, due to the opposite charges flowing in the gap. If the magnetic field experienced by the gamma-rays in traversing the centre of the magnetosphere is sufficiently low then TeV gamma-rays may be seen twice per neutron star revolution, both arising from emission from the same polar region. Thirdly, the emission may be in a fan-shaped beam, in a large range of latitude angles as shown in Figure 2 (due to Cheng Ho and Ruderman).

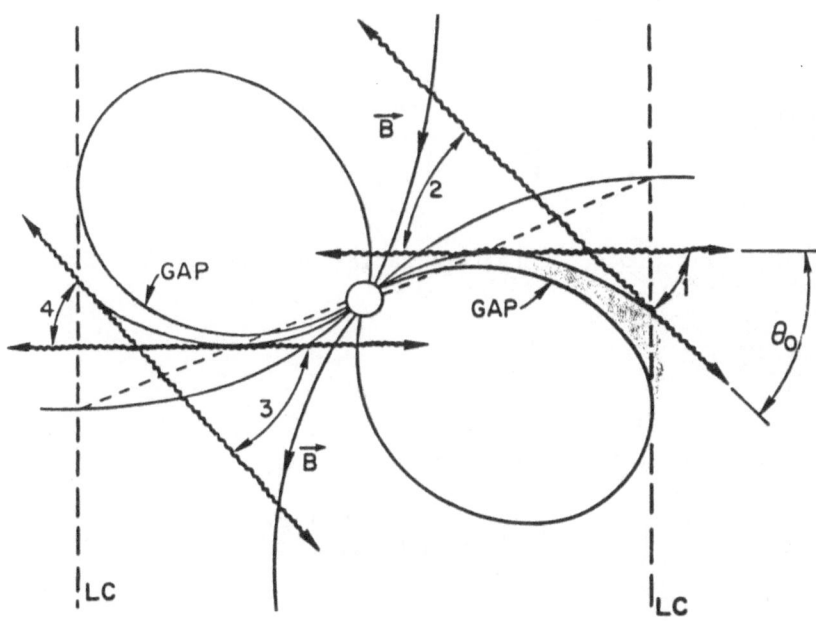

Figure 2. Model outer magnetosphere showing probable gap and beaming geometry. (Cheng, Ho and Ruderman).

A specific prediction of interest to VHE experiments is that a key parameter for outer gap formation is the voltage drop which may be produced. This is proportional to the product ΩB_n^2 and may be sufficient in PSR1509-58, PSR0540-693 and PSR1937+214 for the production of high energy gamma-rays. It is interesting to note that there has been a possible detection of this last, the millisecond pulsar, reported at this workshop (3).

K.Brecher outlined the mechanism for a possible alternative to the outer gap emission model. In polar gap models there is an electric field

$$E_{11} = \underline{\frac{v \times B}{c}} \sim \underline{\frac{\omega \lambda}{c}} \ B_p$$

The expected scale height of the polar cap is:

$$1 \sim 5 \times 10^3 \left(\frac{R_c}{10^6 cm}\right)^{2/7} \left(\frac{P}{1s}\right)^{3/7} \left(\frac{B_p}{10^{12}}\right)^{-4/7} \ cm$$

where $R_c \sim c/\omega R^{1/2}$. R is the radius of curvature of the magnetic field lines in the polar gap. Particles passing through the gap are accelerated to an energy

$$\epsilon \sim e \ E.dx \sim e\lambda E_{11}$$

giving a particle Lorentz factor

$$\gamma_e = \underline{\frac{e\lambda}{m} \frac{E_{11}}{c^2}} = \underline{\frac{e\omega}{m} \frac{\lambda^2}{c^3} B_p}$$

These particles will produce curvature radiation with characteristic energy

$$\epsilon_r = \underline{\frac{3}{2} \frac{\hbar c}{R_c}} \ \gamma_R^3$$

However photons with energy ϵ_γ satisfying

$$\epsilon_\gamma > \epsilon_{abs} = 10^{10} \left(\frac{B}{10^{12}G}\right)^{-1} \left(\frac{z}{10^6 cm}\right)^2 \left(\frac{P}{1s}\right) \ eV$$

will be absorbed before leaving the pulsar magnetosphere (z = distance from the centre of the star). The photons will form pairs, the pitch angle ψ of which will be

$$\psi = 10^{14} \ eV.G/(\ \epsilon_r \ B_p \ )$$

The maximum synchrotron radiation energy is

$$\epsilon_s = 0.5 \ \hbar \ \omega_B \ (\epsilon_c \ / \ 2mc^2 \ )^2 \ \psi$$

where $\omega_B = eB/mc$.

If $\epsilon_* > \epsilon_{abs}$, the synchrotron radiation will also be absorbed and an e+/e- cascade will develop in the magnetic polar gap. If the Lorentz factor γ_p of the created pairs has $\gamma_p > 1/\psi$ the pairs will emit most of their energy, eventually getting down to $\gamma \sim 1/\psi$.

The cascade will result in photon emission peaking at an energy $\sim \epsilon_{abs}$. For the millisecond pulsar, this will be about 10^{11} eV.

F G Smith was asked about the validity of particular models for high energy photon production and stated that any model proposed so far could have some correct elements but that no model is yet internally consistent. K Brecher agreed and pointed out that no model yet can balance the current flow in and out of the magnetosphere satisfactorily, but that individual features of models may nevertheless be correct in broad outline. He further suggested that, independently of the details of how and where radiation is produced, some general statements may be made about pulsar emission which will be true. In particular, he made a plea for observations of the 5 ms pulsar. This resulted from a scaling argument based on a very important early paper by Pacini (4) on incoherent synchrotron emission (ISR). The basic assumption is that radiation is emitted via ISR from near to the radius of light cylinder. The synchrotron luminosity is

$$L_{sync} = \frac{e^2}{c} \left(\frac{eB}{mc}\right)^2 \gamma^2 n V$$

where γ is the Lorentz factor of the emitting particles, n is the particle density and V is the emitting volume. If the assumption is made that $V \propto r_{CR}^2 \Delta$ then

$$L_{sync} \propto (\gamma \Delta)(n \, r_{CR}^2 \, \gamma) \, B_{II}^2$$

Assuming that the magnetic field is approximately dipolar then

$$B_\perp \propto B \propto B_o \, r_{CR}^{-3}$$

and

$(n \, r_{CR}^3 \, \gamma) \propto B_o^2 \, \omega^4$, the total energy flux of particles, then the ISR luminosity scales as

$$L_{ISR} \propto (\gamma \Delta) \, B_o^4 \, \omega^{10}$$

This high dependence of luminosity on ω has successfully predicted the ratio of the optical emissions of the Crab and Vela

pulsars, and the rate of decrease of the Crab Pulsar's optical
flux.

The scaling of pulsar parameters using the above ideas leads
to two propositions which should be correct in general terms.
1. For a fixed rate of total energy emission, the ratio of
the radiated lumiosity to particle and low frequency dipole
radiation increases with increasing angular frequency,
2. For fixed age, the proportion of incoherent synchrotron
radiation increases with increasing angular frequency.

On these very general grounds, which are independent of the
detailed mechanisms for X and gamma-ray production, it may be
possible to rank pulsars in terms of their angular frequency and
its derivative as shown in figure 3.

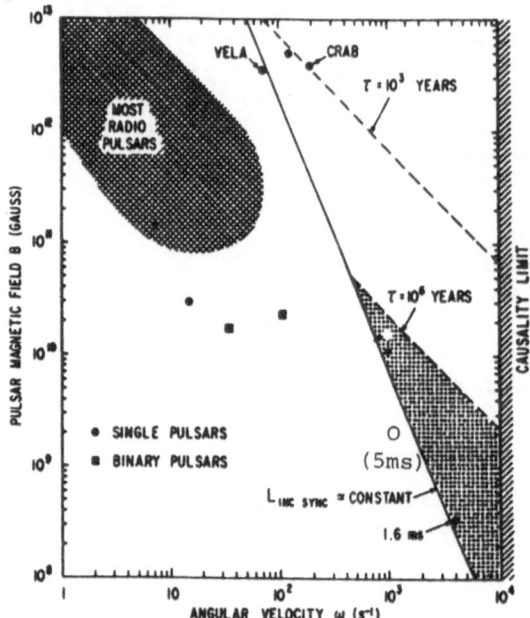

Figure 3. Possible location on B/ω plot of gamma-ray emitters.
The dotted region possesses sufficient age to have the likelihood
to be seen, and sufficient $B^2\omega$ to possibly emit VHE gamma-rays.

The lower right shaded region contains those high spin rate
pulsars which are more luminous than the VELA pulsar. The 5 ms
pulsar is seen to be outside this region, but since it is nearer
by an order of magnitude than the norm it may be expected to be
visible via ISR.

Turning to the question of the possible existence of radio quiet, gamma-ray loud pulsars, F G Smith suggested that these could exist. On purely geometrical grounds, it could be possible for a pulsar to emit gamma-rays into a fan beam spanning up to ~ 90 degrees of latitude, with almost 100% chance of the Earth being in the beam. If the radio emission came from the polar cap then the radio may be emitted in a cone of half angle 10-15 degrees, with a chance of being missed of ~ 25%. On these simple arguments, the proportion of gamma-ray pulsars which may not produce detecable radio pulses may be ~ 25%. It is too early to state whether this view is approximately true because of the low number of gamma-ray pulsars so far discovered, but is is interesting to speculate that the strong gamma-ray source Geminga may be a gamma-ray pulsar although it is not seen as a radio pulsar. W hermsen was asked if any of the COS-B gamma ray sources had been scanned for pulsar emission. He replied that indeed they had, but the low counting rate together with the long accumulation times rendered the COS B experiment relatively insensitive to pulses with an unknown period. In contrast to that experiment, he outlined the COMPTEL payload which is expected to have much greater sensitivity. It could produce up to 10 pulsars at the 5 SD level and up to 50 pulsars at the 3 SD level.

The possibility of transient emission from pulsars was the subject of some discussion. F G Smith started this part of the workshop with the opinion that there were no obvious mechanisms which could easily cause the Crab and Vela pulsars to have strong transient bursts of emission. However, it was suggested that VHE transient emission was possible although the mechanism must not allow lower energy gamma-ray and X-ray transients.

A possible mechanism could be a small instability in the e+/e- production in the outer gap. The production of VHE gamma-rays is very sensitive to the local gap magnetic field and the absorption to small density fluctuations in the flow of low energy secondary photons. The energy spectrum above 1 TeV may be very steep, approaching an inverse Compton cut-off of ~ 6 TeV. Any small variation in this cutoff could therefore produce a large variation in flux above a fixed threshold near 1 TeV without materially affecting lower energy gamma-rays or X-rays. J Truemper pointed out that the infrared cut-off in the Crab pulsar's spectrum may be explained if the local cyclotron frequency in the emission region is located in the short infrared range. The Doppler boosted incoherent beamed radiation would then start at infrared frequencies. The result of this may be that the low energy cut-off may not be affected in a similar way to the TeV cut-off and TeV transients may be possible without any other energy range being noticably affected.

Summary

   The main points to emerge from the workshop of interest to
experimenters were:
      1. Pulsar theories, both detailed and simple, have little
      difficulty in explaining how TeV gamma-rays could be
      produced, at least for pulsars with a sufficiently high
      voltage drop, which is proportional to $\Omega^2 B$. The production
      of PeV gamma-rays is more difficult.
      2. Of all the pulsar parameters, angular frequency is likely to
      give the best indication of possible TeV emission. On this
      criterion the 5 ms pulsar may be a good target for observations.
      3. Some fast pulsars may be luminous gamma-ray sources but give
      no detectable radio pulses.
      4. Pulsed TeV transients may be possible.

   The general feeling of the workshop participants was that
although new theoretical insights were neither generated nor
expected, the experimental workers in the field of gamma-ray
astronomy were encouraged to make observations on pulsars and
noted that some features of these observations could be both
unique to the wavelength range and valuable indicators of pulsar
emission mechanisms.
   The chairman thanked all participants, especially the
discussion leaders, for their contributions to the workshop.

References
1 Cheng K.S., Ho C., Ruderman M., Ap.J. 300, 500-521 (1986).
2 Cheng K.S., Ho C., Ruderman M., Ap.J. 300, 522-539 (1986).
3. Chadwick et. al., this Workshop, p 159.
4. Pacini F., Ap.J.Lett. 163, L17-L19 (1971).

SOURCE MECHANISMS (ACCRETION OBJECTS)

A. M. Hillas
Physics Department
University of Leeds
Leeds LS2 9JT
England

ABSTRACT.   Substantial problems remain in understanding the acceleration
of particles to energies needed to explain the gamma-ray observations
(particularly above 10^{14} eV). Dynamo action by an accretion disc
(Brecher) seems confined to large distances from the neutron star in
many cases, where little of the gravitational energy is tapped. Accreting matter probably has too low a density to sustain shock acceleration
in a 10^{12} gauss field. Some neutron stars spin too slowly for a pulsar
like action to work, though Cyg X-3 could well be simply a fast pulsar
seen through gas. Protheroe has shown that magnetic focusing of quasi-
monoenergetic protons may greatly influence the orbital phase pattern
of the gamma ray signal.

The workshop was chaired by J. Quenby. He and K. Brecher and J. Trümper
led the discussion. Brecher gave a discussion of acceleration involving
fields generated by a rotating accretion disc, and problems of super-
Eddingtonian accretion rates, but this will not be covered in detail in
this report as much of it was later included in his invited lecture.

1. SPECIAL PROBLEMS IN ACCOUNTING FOR VHE/UHE GAMMA RAYS

i)   The particles generating the gamma-rays take a huge total power:

$$L_{uhe} > L_{X-rays} \qquad \text{at least some of the time.}$$

(How does gravitational infall not put most of the released energy
into thermal radiation?)

ii) Individual particles gain extraordinary energies:

$> 10^{16}$ eV   in Cyg X-3, Vela X-1, LMC X-4*
$> 10^{15}$ eV   in Her X-1*

(* these have yet to be confirmed by independent observations)

iii) The phases of emission of the gamma-rays have to be explained

K. E. Turver (ed.), Very High Energy Gamma Ray Astronomy, 71–80.
© 1987 by D. Reidel Publishing Company.

iv) Gamma-ray pulses may be seen when the neutron star is hidden
      behind the companion (Her X-1, Whipple; Vela X-1, Potchefstroom).

      To set the scene, a brief sketch will be given of the provisional
picture of these sources which has formed the background to current
discussions.

2. INITIAL PREJUDICES

2.1. The gas target picture

In the accreting sources, or sources emitting above the TeV range, it
is currently assumed (following Vestrand and Eichler, 1979, 1982) that
somewhere near a neutron star particles are accelerated to high energy
and emitted in a wide range of directions, and that the gamma-rays are
produced only when the particles encounter clouds of gas. The gamma
rays must escape pair production in any local magnetic field, and the
m.f.p. for this process is less than 1 km if the transverse component
of magnetic field strength $B > 10^4$ gauss/E_{15} (E_{15} being the photon
energy in PeV). As one expects that $B \gg 10^4$ gauss in the acceleration
region, PeV gamma rays presumably emerge from a separate region. If the
gamma rays appear mainly at particular orbital phases (as in Cyg X-3,
Vela X-1, LMC X-4, probably 4U0115+63) one may suppose gas targets to be
available at particular places in the orbiting system - for instance as

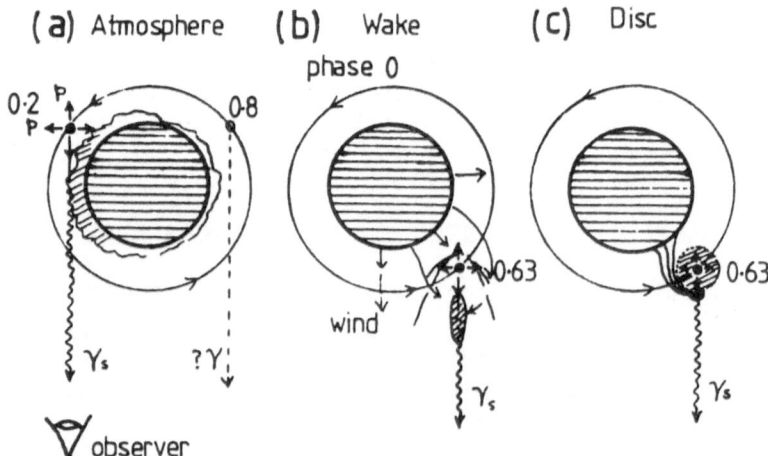

Figure 1. Possible gas targets which may appear in the path of protons from the n-star at special orbital phases (e.g. 0.2, 0.63). The
neutron star is the source of possibly isotropic protons. (a) Atmosphere of companion serves as target, (b) accretion wake (accretion
from wind), (c) structure attached to accretion disc. In case (a) the
target may appear near phase 0 (rather than 0.2 and 0.8) if the n
star is not eclipsed, as seems to be the case in Cyg X-3.

in figure 1. At these high energies the accelerated particle is assumed
to be a proton, which generates π^0s, and hence gamma-rays, rather than
an ·electron, as the latter suffers severe energy losses.

2.2. Consequent power of sources

The power emitted as widely-spread protons is very high if the gamma-ray
pulses are indeed due to the occasional availability of a target.

Cygnus X-3: Consider photons above 10^{15} eV: averaged over years an
energy flux $\sim 3\times10^{-10}$ erg cm^{-2} s^{-1} has been seen above 1 PeV (after correcting for losses due to interactions with the primeval radiation in
12 kpc - about a factor 4). The emission usually occurs in a small phase
interval, about 0.02 to 0.05, so it is presumed that we should have seen
20-50 times this flux overall (say >20 times) if a suitable gas target
had been available all the way round the orbit. And at best 10% of the
proton energy will be converted into gamma rays. So if the proton emission is indeed widespread, and if Cyg X-3 is 12 kpc away, the power emitted in the form of protons above about 5×10^{15} eV is

$$L_p > 4\pi(3.9\times10^{22})^2 \times 10 \times 20 \times 3\times10^{-10} \text{ erg s}^{-1}$$
$$= 1.1\times10^{39} \text{ erg s}^{-1}(\text{if isotropic})$$
$$\text{or} \quad > 6\times10^{38} \text{ erg s}^{-1} \quad \text{if emitted into half a sphere.}$$

This is to be compared with the X-ray luminosity: $L_X \simeq 10^{38}$ erg s^{-1}.

Hercules X-1: Fluxes reported during active periods by the Durham,
Whipple Observatory and Fly's Eye groups differ: a long-term average is
less well-known here. But during the active periods one finds roughly,

in the TeV range $L_p \sim 10^{37} - 10^{38}$ erg s^{-1} (during $\frac{1}{2}$-hour bursts)

PeV $L_p > 10^{38}$ erg s^{-1} sometimes.

cf. $L_X = 0.7\times10^{37}$ erg s^{-1}.

4U0115+63: $L_p > L_X$ (TeV).

It seems that we have to account for at least a large part of the energy
generation going into protons of extremely high energy.

3. POSSIBLE ACCELERATION PROCESSES

Three acceleration processes featured in the workshop discussions. No
new processes were proposed.

3.1. Statistical acceleration at an accretion shock

If the matter falling towards the pole of a neutron star forms a collisionless shock, one may have very rapid statistical acceleration of
protons that move back and forth across the shock front. As several
authors have proposed that it is possible for collisionless shocks to
convert a very large fraction of the bulk kinetic energy into relativistic particle energy, this is an attractive possibility.

Most previous authors discussing shock acceleration have concluded
that one can regard the process as a form of Fermi acceleration, in
which protons gain energy as they bounce between magnetic scattering ag-

ents carried in by the relatively converging gas streams on the two sides of the shock front. Upstream, these irregularities can be generated by streaming of particles away from the shock. The e-folding time for energy gain is, for a strong shock,

$$t_{acc} = (\lambda_{upstream} + 4\lambda_{downstream}) \, c \, /v_{infall}^2$$
$$\sim 5\lambda_{down} \, c \, /v_{infall}^2$$

where λ is a scattering mean free path. Eichler and Vestrand (1985) and Kazanas and Ellison (1986) have considered the application of this acceleration process to accreting neutron stars (having Her X-1 and Cyg X-3 in mind), and find that the maximum energy reached is limited by the competition with energy loss due to proton synchrotron radiation, which has an energy loss time constant of

$$t_{synch} = 4.4\times10^{-12} \, E_{15}^{-1} \, B_{12}^{-2} \text{ sec,}$$

where B_{12} is the magnetic field strength in 10^{12} gauss units. This yields a maximum proton energy of

$$E < 0.06 \, (r/R_*)(r_L/\lambda) \, B_{*12}^{-1/2} \text{ PeV,}$$

if the shock is at a distance r from the centre of the neutron star (radius R_*, polar surface field $B_{*12}\times10^{12}$gauss), and r_L is the gyroradius of these protons (of energy E). Even making an apparently very optimistic assumption that the scattering m.f.p. is as small as the gyroradius, one falls well short of PeV protons unless the field strength is far below 10^{12} gauss. (In Her X-1 and 4U0115+63 at least, an apparent X-ray cyclotron line shows that $B_* > 10^{12}$ gauss.)

In a 10^{12} gauss field, the gyroradius is 3cm at 1 PeV, so that in escaping several km through the magnetosphere after acceleration, the protons face even more severe synchrotron losses. Kazanas and Ellison propose that the escaping particle beam is essentially neutrons, formed by nuclear collisions in the plasma in the accelerating region. The energy spectrum of nucleons will be approximately $E^{-2}dE$, and a photon spectrum of very similar form would be generated - very reasonable.

In the workshop, Quenby suggested that useful insight might be gained from work on the acceleration of non-relativistic particles in oblique shocks in turbulent gas (with application to solar flares) by Decker and Vlahos (1985). In an oblique shock, with magnetic field

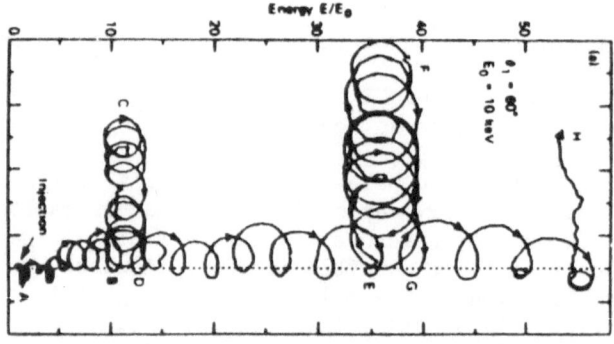

Figure 2. Trajectory of a (non-relativistic) proton drifting along an oblique shock front, being scattered back to the front, gaining energy.

lines oblique to the shock front and sharply deflected at the front (B changing there), particles drift for a while along the front, gaining energy in a $\underline{v} \times \underline{B}$ electric field, perhaps more quickly than by the Fermi scattering mechanism where $\underline{B}$ is perpendicular to the front. Numerical trajectory calculations (e.g. figure 2) showed that an appreciable number of protons were quickly scattered back to the front when they started to wander away, so that Quenby proposed that some particles are kept continually at the shock by strong scattering, so that they move along the front at the full ("∇B") drift speed and experience the $\underline{v} \times \underline{B}$ electric field , gaining energy rapidly until further gains are prevented by synchrotron losses. This suggested that protons might reach 10^{14} eV, roughly, in 10^{12} gauss fields.

This seems to the rapporteur to be equivalent to the "optimistic" assumption by earlier authors that scattering is as strong as possible ($\lambda = r_L$), the resulting energy being much the same as the estimate quoted above for this case. It is apparently another example of apparently different shock acceleration mechanisms giving much the same result. The maximum energy still falls short of 10^{16} eV, and I understand that already in the simulations, the energy in the turbulent component of the field was a large fraction of the magnetic energy.

This seems to me to raise a very serious difficulty in the application of shock acceleration in the accretion flow near a neutron star. In a 10^{12} gauss field, the infalling gas has an energy density only 10^{-5} that of the magnetic field, near the star. (The ratio varies as $r^{-2.5}$.) It is thus very hard to see how the infalling plasma can distort the field appreciably (a) to make the field topology highly twisted so as to produce very strong scattering, and (b) to form a magnetic shock at all (as v_{Alfven} is large) - unless the matter falls in as small very dense bullets, but the nuclear collision m.f.p. then becomes too short.

3.2. Accretion disc acting as dynamo

Chanmugam and Brecher (1985) proposed this as the accelerator, and Brecher has outlined the ideas in an invited lecture, so much of the detail will not be repeated here. It is supposed that a plasma disc orbiting the neutron star is threaded by a perpendicularly directed magnetic field $\underline{B}$, probably carried in by the plasma from the surroundings of the companion star. If the orbital velocity is $\underline{v}$, one may expect an electric field $\underline{E} = \underline{v} \times \underline{B}$ to appear, directed radially within the disc. If B is large, and falls off only slowly with r, a large potential may be available. To optimise this, the authors supposed B to be equal to the strength of the neutron star's dipole field at the inner edge of the disc (expected to be near the Alfven radius r_A, where the magnetic energy density equals the kinetic energy density - at about 200 R_* for Her X-1), but to fall off as slowly as $r^{-1/2}$ (even though the magnetic energy then considerably exceeds the kinetic energy in the disc). This yields a potential difference of

$$V = 1.4 \times 10^{13} \, (B(r_A)/10^5 \text{ gauss})(r_A/10 \text{ km})^{1/2} \ln(r_{outer}/r_A) \text{ volts}$$

$$\simeq 2.8 \times 10^{14} \, B_{*12}^{-0.43} L_{38}^{0.71} \text{ volts},$$

where accretion provides $L_{38} \times 10^{38}$ erg s^{-1}. The full space charge distribution and the form of the electric field have not been calculated, so it is still not certain that this potential will be realised or what is the direction in which the particles are ejected.

Brecher emphasised that for the mechanism to be important B_* must not be large - say $<10^9$ gauss - as it is necessary for the disc to extend in close to the neutron star, and a teragauss stellar field pushes the Alfven radius out to ~ 100 stellar radii. This means that (a) only 1% of the available gravitational energy has been released within the disc, so that $L_D \ll L_X$, and (b) B is small in the disc so that one falls far short of 10^{16} volts. This is not promising for Her X-1 or 4U0115+63 but might be relevant to Cyg X-3 if that has $B \sim 10^8$ gauss at the neutron star surface, combined with a high accretion rate. Brecher envisages a roughly radial acceleration of protons in the region above the disc, and a gas stream feeding the disc might provide a target for gamma production.

One difficulty at present is to tap a large part of the accretion power (mostly released very close to the neutron star), at least in the cases mentioned above. Another is with the direction of emission. If the orbital velocity is almost Keplerian, and there is a B_z (normal) field component present in the disc, there must indeed be a radial electric field vB_z produced by a radial charge separation: outside the disc $\underline{E}$ would form arches joining the charges, but with a B_z magnetic field, effective acceleration would only be found at angles very far from the plane of the disc. As in the case of a pulsar magnetosphere, a fully self-consistent field calculation is not yet available.

Brecher pointed to the Jupiter system for examples of related acceleration mechanisms at work, on a much smaller scale: Io provides a conducting blob that results in a 300 kV potential; and closer in, an accreting particle disc (charged by uv) may be responsible for otherwise unexplained 10 MeV electrons.

The source LMC X-4 was briefly discussed. Brecher suggested that a low B_* combined with a very high accretion rate ($\sim 10^{41}$ erg s^{-1}) could push the accretion disc close to the surface and explain the Adelaide observations of 10^{16} eV gamma rays carrying an exceptionally high luminosity. (The Eddington limit might be far exceeded if very little of the accretion power were emitted as X-rays, but as u.h.e. protons.) However, Trümper pointed out that LMC X-4 is a very well-studied X-ray binary that appears to be in rotational equilibrium with its accretion disc, requiring $B \sim 5 \times 10^{13}$ gauss. Such an extreme accretion luminosity would be unacceptable. (The X-ray astronomers may therefore be casting doubt on the reality of this gamma ray source.)

Trümper also referred to Her X-1. Modelling (at Garching) of the very detailed X-ray observations of the 35-day cycle indicates that the cycle is not caused by changes in the outer disc (as in the postulated precession), but that the neutron star is precessing and this will twist the innermost part of the disc on this time scale. If this disc motion is responsible for the absence of gamma-rays as well as X-rays during the "X-ray low", this would indicate that the proton accelerator is not in the disc region, but closer to the neutron star.

3.3. Spinning neutron star causing acceleration, as in a pulsar.

Again, $\underline{v} \times \underline{B}$ can generate a large potential difference: 10^{18} volts could easily be obtained in fast-spinning objects if the star were surrounded by a vacuum; but again self-consistent $\underline{E}$ and $\underline{B}$ fields in the presence of gas are not known. A plasma disc around the star could modify the fields of an isolated star, as has been suggested by Michel (1982). This process was not discussed in any detail in this workshop, as another workshop had been devoted to pulsar acceleration processes. Nevertheless, it may be important even for gamma-ray emitters other than those at present classified as pulsars - notably Cygnus X-3.

The 12.6 ms spin period of Cyg X-3 is relevant here, for if Cyg X-3 does not have a weak magnetic field, the inner edge of a surrounding disc (Alfven radius) will occur where the orbital angular velocity is far less than the angular velocity of the magnetosphere, so matter will be ejected rather than accreted, probably giving rise to the jets inferred from infra-red observations. Regarded simply as a very fast pulsar, with a field of a few teragauss, the proton luminosity of 10^{39} erg s^{-1} and the particle enrgies of 10^{17} eV would not be particularly surprising. Thus Cyg X-3 may well not be an accreting source.

3.4. Other acceleration processes

Though not specifically discussed, some X-ray observations raise the possibility of another mode of acceleration. Trümper noted that rotational locking appears to occur in several X-ray binaries (e.g. Her X-1 and LMC X-4): the neutron star has episodes of spin-up and spin-down and is apparently co-rotating with the inner part of the accretion disc (the Alfven radius is near the co-rotation radius). (Sco X-1, though not yet detected as a v.h.e. gamma source, also shows interesting bimodal behaviour, and emits jets.) There must be an important magnetic coupling between the inner disc and the neutron star, with the disc sometimes getting ahead, and sometimes behind. It is thus possible that lines of force become twisted by shear, and then reconnect, generating large e.m.f.s. One might guess, though, that this would lead to more erratic and briefer episodes of emission than observed.

4. NEW IDEAS ON THE FORMATION OF THE GAMMA-RAY PULSES

The bursts of gamma-rays seen at particular orbital phases - though not always rigidly fixed - in Cyg X-3, Vela X-1, and perhaps 4U0115+63, may indeed map out concentrations of gas of suitable target thickness at particular places in the orbit. Protheroe suggested that the directions in which radiation is detected may, however, be considerably affected by the presence of a magnetic field. There is some support for this from the observations of gamma rays during eclipse in Her X-1 and Vela X-1.

Protheroe considered the case of Cygnus X-3, and illustrated the possibilities with a greatly idealised magnetic field configuration.

Firstly, Protheroe examined the type of companion star that might just fill its Roche lobe, alongside a $1.4 M_{\odot}$ neutron star in Cyg X-3. Either a main sequence star of about $0.5 M_{\odot}$ or a helium star evolved by

stripping the outer hydrogen layer by mass transfer, and having a mass
of $3-4\,M_{\odot}$ would be possible. (The former seems more likely if one is to
agree with the Durham observation of $\dot{P}/P$) He noted that a K dwarf would
typically have much of its surface covered by variable 500-3000 gauss
fields, and though the magnetic field of a helium star was unknown,
other hot stars (e.g. Ap) had dipole fields of 1-30 kgauss. There might
thus be a magnetic field $\sim 10^3$ gauss. For illustration, he calculated
trajectories of protons emitted isotropically from a neutron star and

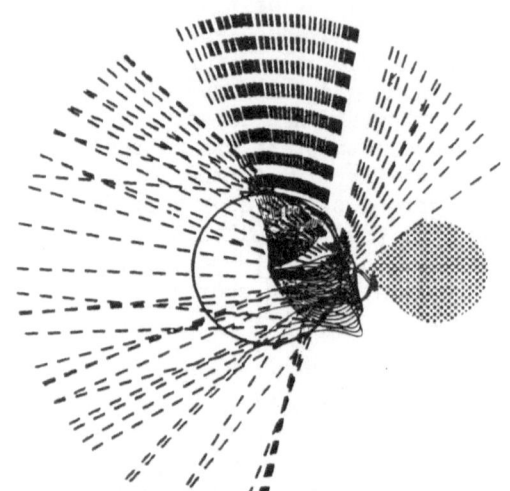

Figure 3. Examples of proton
trajectories in magnetic dipole
field of companion star.

magnetic moment/mag.rigidity

= 200 $R_{\odot}^3$ gauss / PV

binary separation = 1.8 $R_{\odot}$.

then travelling through a dipole field originating in the companion,
until they escaped or else struck a fat rim of varying height running
all round the edge of the accretion disc, such as White and Holt have
proposed to explain the X-ray variations in the binary 4U1822-37, whose
X-ray variation is similar to that of Cyg X-3. Thus he has ignored any
complications due to the accretion disc modifying the extended magnetic
field of the companion, and any fringing field of the neutron star.
Figure 3 illustrates some of the trajectories obtained, though it is
difficult to show clearly the bunching that a distant observer would see
as this involves detection of particles emerging only at a particular
angle of elevation. Figure 4 illustrates some of the simulated gamma
ray signals as a function of orbital phase. The proton emission is
taken to be isotropic. Evidently the rim of the disc, in this model,
has irregularities that cause pulsing even without a field, but the
field can cause major changes in the pattern of intensity observed, and
can itself lead to bunching in the directions.
    If magnetic deflections are indeed considerable, and yet a narrow
beam is seen, the proton beam has to be approximately mono-energetic,
requiring a dynamo rather than a statistical acceleration process. The
gamma-ray energy spectrum would the have to be formed by cascading in
the target (e.g. by a rapid synchrotron dump of the energy of an electron-positron pair generated by a u.h.e. photon: Hillas (1984)).

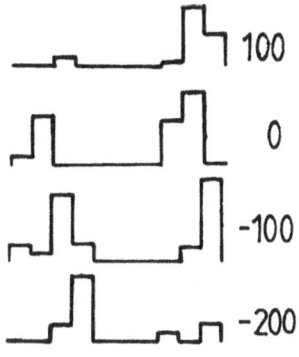

Figure 4. Gamma-ray intensity as function of orbital phase, as strength of magnetic dipole field from companion is varied. Figures (100 to -200) give the ratio of dipole moment (units gauss $\times R_{\odot}^3$) to magnetic rigidity of protons (in PV units). The protons are emitted isotropically from the neutron star and may strike a White-Holt type of variable thick rim to the accretion disc, forming gamma rays. (Protheroe)

5. CAN THE PROPOSED MECHANISMS EXPLAIN THE OBSERVED SOURCES?

The following table comments on the gross features of the problem for each of the three acceleration processes discussed above, for several sources.

Object	Shock acceleration	Disc dynamo	Pulsar dynamo
Cyg X-3	E_{max} problem unless $B_* < 3 \times 10^9$	OK if $B_* < 3 \times 10^8$	OK if $B_* \sim 10^{12}$ (non-accreting)
LMC X-4	E_{max} problem	disc too far out (B high)	spin too slow
Vela X-1	E_{max} problem	OK if B weak (but accretes from wind)	Much too slow spin
Her X-1	OK if Fly's Eye obs. not confirmed	disc too far out (& accel.close to star?)	E_{max} OK? but not total power
4U0115+63	OK at TeV	B too high (but wind accreter?)	E_{max} OK but not total power

Further general problems for these mechanisms may be noted here:
Statistical acceleration: Is B too rigid to scatter particles (by turbulence or kinks) close to the neutron star, as remarked above?
Disc dynamo: The neutron star spin modulates the signal in most cases. This suggests that acceleration occurs on field lines attached to the neutron star poles.
Pulsar dynamo: We require acceleration of protons: it may not be easy to inject them into the right place in the field (comment by de Jager). However, where else does one find Galactic accelerators capable of

accelerating protons to above 10^{19} eV, as required to explain Galactic hadronic cosmic rays?

It is clear that we still have far to go to explain the u.h.e. gamma-ray sources.

REFERENCES

Chanmugam, G and Brecher, K, 1985. Nature, 313: 767.

Decker, R B and Vlahos, L, 1985. 19th Int. Conf. on Cosmic Rays, La Jolla, 4: 10.

Eichler, D and Vestrand, WT, 1985. Nature, 318: 345.

Hillas, A M, 1984. Nature, 312: 50.

Kazanas, D and Ellison, D C, 1986. Nature, 319: 380.

Michel, F C, 1982. Revs. Mod. Phys., 54; 1.

Vestrand, W T and Eichler, D, 1979. Particle acceleration mechanisms in astrophysics, 281 (AIP conference proceedings 56. Ed. J. Arons et al.)

TECHNIQUES OF TeV ASTRONOMY

W. F. FRY
*Department of Physics, The University of Wisconsin
Madison, WI 53706, USA*

The contribution to astrophysics from high energy γ ray studies utilizing air Cherenkov techniques has been very significant. During the first few years only a few sources were reported and even the majority of these signals were a few standard diviations above the background of cosmic rays. In those times the question was often asked as to why there were no sources which gave signals well above the background and it was even suggested by skeptics that indeed no real signals had been observed. At this conference it became clear that this is no longer the case with data being presented on a much longer number of sources and at a greatly increased statistically significant level. It has become increasingly clear that the sporadic nature of many of the sources has restricted the statistical significance of many sets of these data.

One of the most important questions asked at this workshop was, "How can we improve the air Cherenkov technique so that we will achieve at least one order of magnitude in the statistical significance of the result"? There seemed to be at least three quite different approaches for improvements over the existing facilities. These are (1) a proposal to utilize the spatial and angular parameters of the air showers as obtained from imaging techniques, (2) to increase the mirror area and (3) increase the sensitivity by going to the single photon level with fast timing.

In addition to these general ideas for improvement, another substantative issue was discussed. Is it desirable to concentrate on reducing the threshold energy with a corresponding substantial increase in statistics or would it be better to increase the threshold energy by perhaps two orders of magnitude?

Certainly this choice depends upon many unknown properties among which is the energy spectrum of the astrophysical source. Some new technique is needed to compensate for the decreased signal rate at the higher energy. The first paper in this session was a scheme presented by the Wisconsin Group for a very high energy telescope facility.

K. E. Turver (ed.), Very High Energy Gamma Ray Astronomy, 81–90.
© *1987 by D. Reidel Publishing Company.*

I. A 5 x 10^{13} Gamma Ray Telescope

The underlying idea of the proposal is to utilize the increase in photon density at ground level with the increase in the incident gamma ray energy. Montecarlo results suggests that indeed the photon desity in the pancake is roughly proportional to the incident gamma ray energy. This increased visible photon density makes it possible to greatly reduce the area of the collecting mirror with a corresponding reduction of the singles rate from star light. This lower singles rates (it will be assumed that all phototubes will operate at the single photon level as in the Haleakala telescope) would permit a large reduction in the number of tubes that must be in coincidence in order to eliminate the night sky background. As a result, the entire unit can consist of only one small mirror with a few tubes in fast coincidence. This small unit could be on a single small mount.

The known properties of the Haleakala telescope will be used to calculate quantative parameters of each of these units. Some operational constants of the Haleakala telescope are summarized in Table I.

Table I
Summary of the Properties of the Haleakala Telescope

Number of mirrors	6
Area of each mirror	1.82 m^2
Mirror area/phototube	0.6 m^2
Singles counting rate at single photon level	$\sim 3x10^6$ Hertz
Width of Timing gate	$\sim$ 6 n.s.
Rate of 7 fold coincidence	$\sim$ 0.4 Hertz
Acceptance angle/full width	0.5°
Cross-over multiplicity between night sky and C.R. showers	7
Estimated gamma ray threshold	300 GeV

Assuming that the pancake photon density is directly proportional to the gamma ray energy, the data in Table I would suggest that at 5 x 10^{13} e.v. the mirror area per tube for an opening angle of 1.25° would need be only $10^{-2}m^2$ to give a three fold coincidence on a shower of energy 5 x 10^{13} e.v. The cosmic ray proton induced shower rate would be about 0.1 Hertz for 20 independent units.

Each one of these units consisting of three phototubes could utilize one mirror of area $\sim 3 \times 10^{-2}m^2 (R = 10cm)$ placed on a simple pedestal with a alt - azimuth mount. Each of the 20 units should be separated by about 50 to 60 meters from its neighbors so that each unit would utilize independent target areas in the sky.

Such a facility would accumulate statistics at a rate of about a fifth of the present Haleakala telescope at an energy two orders of magnitude above the Haleakala Telescope.

II. Results on Imaging Studies by the Mt. Hopkins Group.

Whipple Observatory Collaboration Report

The atmospheric Cherenkov imaging technique is being pursued at the Whipple Observatory in the U.S.A. and at the Crimean Astrophysical Observatory in the U.S.S.R. Only results from the Whipple Observatory were reported at this meeting.

In principle the recording of the image of each Cherenkov light flash makes possible the identification of gamma rays amongst the more numerous proton background. It is potentially the way in which the atmospheric Cherenkov technique can achieve the flux sensitivity that is limited only by the electron background and by photon counting statistics. Part of the improvement that imaging offers is the improved angular resolution: this is independent of any difference in the character of gamma ray and proton showers. Preliminary results from imaging experiments have been reported (Stepanian et al. 1985, (La Jolla), Cawley et al. 1985, (La Jolla)); in both cases single cameras were used with pixel sizes 0.5°.

Elsewhere in this workshop Hillas describes the simulated image from both kinds of shower. Examples shown in figure A and B illustrate the fundamental differences. The proton shower images are more diffuse and show secondary maxima due to local muons. It can be seen that there is structure in the images down to scales of 0.25°. In figure C the distribution of gamma ray images from a point source is shown where four parallel gamma rays fall, 100m away, along the X and Y axes.

The Whipple Observatory group have compared the measured image parameters (defined in fig. D) from the background (presumably protons) and those predicted in the simulations of Hillas. There is good agreement between the two (figure E) indicating that this method (using arrays of phototubes to form a crude camera) is practical. However the experimenters noted that care must be taken with calibration and noise treatment and that the method is much more complex than the traditional single element Cherenkov detector. The Monte Carlo simulations also predict that there should be a clear separation of gamma rays and protons based on the measurements of the same image parameters (Hillas 1985 (La Jolla)). This is illustrated in figure F. This clear distinction between the gamma ray and proton showers has not yet been verified experimentally. Cuts based loosely on these simulated differences do enhance the percentage gamma–ray content of signals from the reported sources. But there is no consistent cut that works for all zenith angles, observing conditions, source energy spectra, etc. Also no source has yet been detected with such a high excess in its raw data signal that it can be used as a standard on which to base the data selection algorithm. Because the use of data selection criteria, a postiori, introduces many extra degrees o freedeom, particularly when there are many parameters to choose from, the Whip-

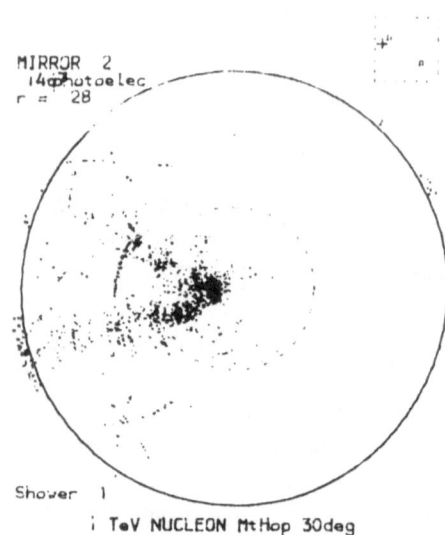

Shower 1

i TeV NUCLEON MtHop 30deg
(azim 47W) MH31DN2 24Jul86
observed at (x,y) = (21, 19)
muon radiation in red

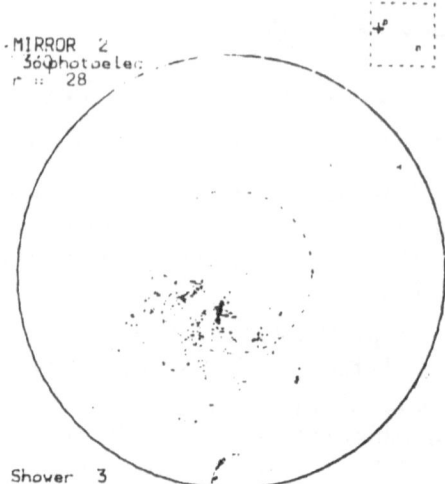

Shower 3

i TeV NUCLEON MtHop 30deg
(azim 47W) MH31DN2 24Jul86
observed at (x,y) = (21, 19)
muon radiation in red

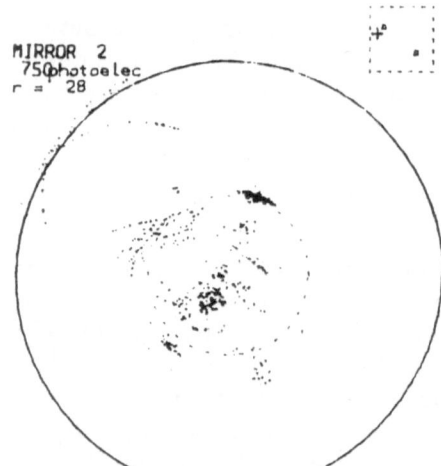

Shower 2

i TeV NUCLEON MtHop 30deg
(azim 47W) MH31DN2 24Jul86
observed at (x,y) = (21, 19)
muon radiation in red

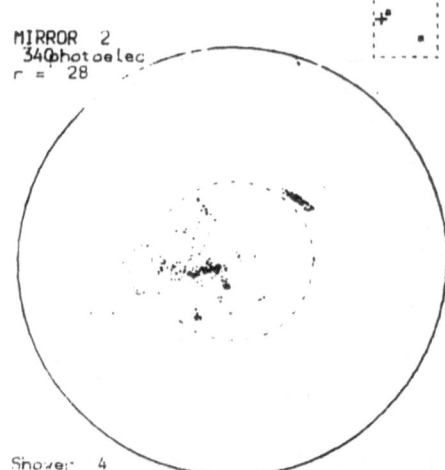

Shower 4

i TeV NUCLEON MtHop 30deg
(azim 47W) MH31DN2 24Jul86
observed at (x,y) = (21, 19)
muon radiation in red

FIGURE A

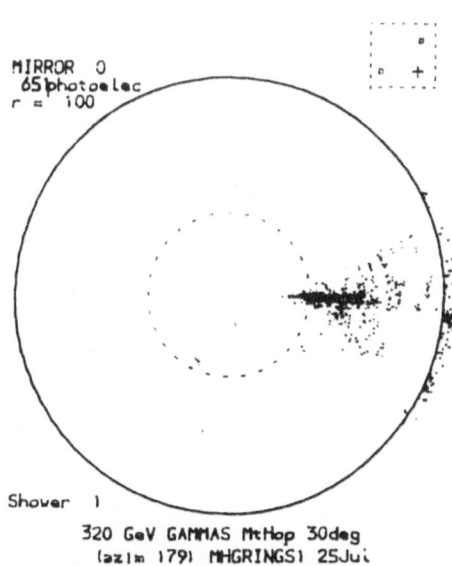

MIRROR 0
  65 photoelec
r = 100

Shover 1
320 GeV GAMMAS MtHop 30deg
(azim 179) MHGRINGS) 25Jul
observed at (x,y) = (-100, 2)
muon radiation in red

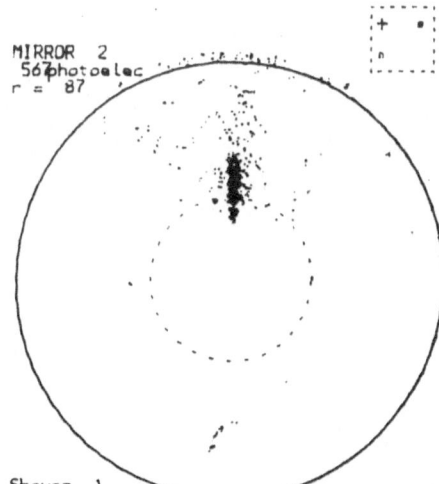

MIRROR 2
  56 photoelec
r = 87

Shover 1
320 GeV GAMMAS MtHop 30deg
(azim 179) MHGRINGS) 25Jul
observed at (x,y) = (-2, ,-87)
muon radiation in red

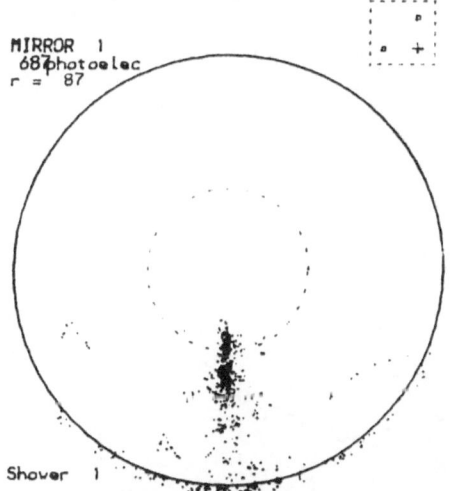

MIRROR 1
  68 photoelec
r = 87

Shover 1
320 GeV GAMMAS MtHop 30deg
(azim 179) MHGRINGS) 25Jul
observed at (x,y) = (2, 87)
muon radiation in red

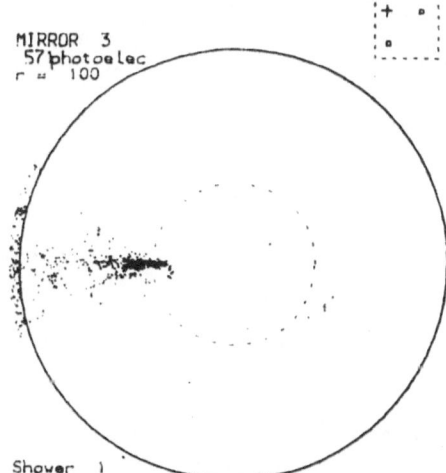

MIRROR 3
  57 photoelec
r = 100

Shover 1
320 GeV GAMMAS MtHop 30deg
(azim 179) MHGRINGS) 25Jul
observed at (x,y) = (100, -1)
muon radiation in red

FIGURE B

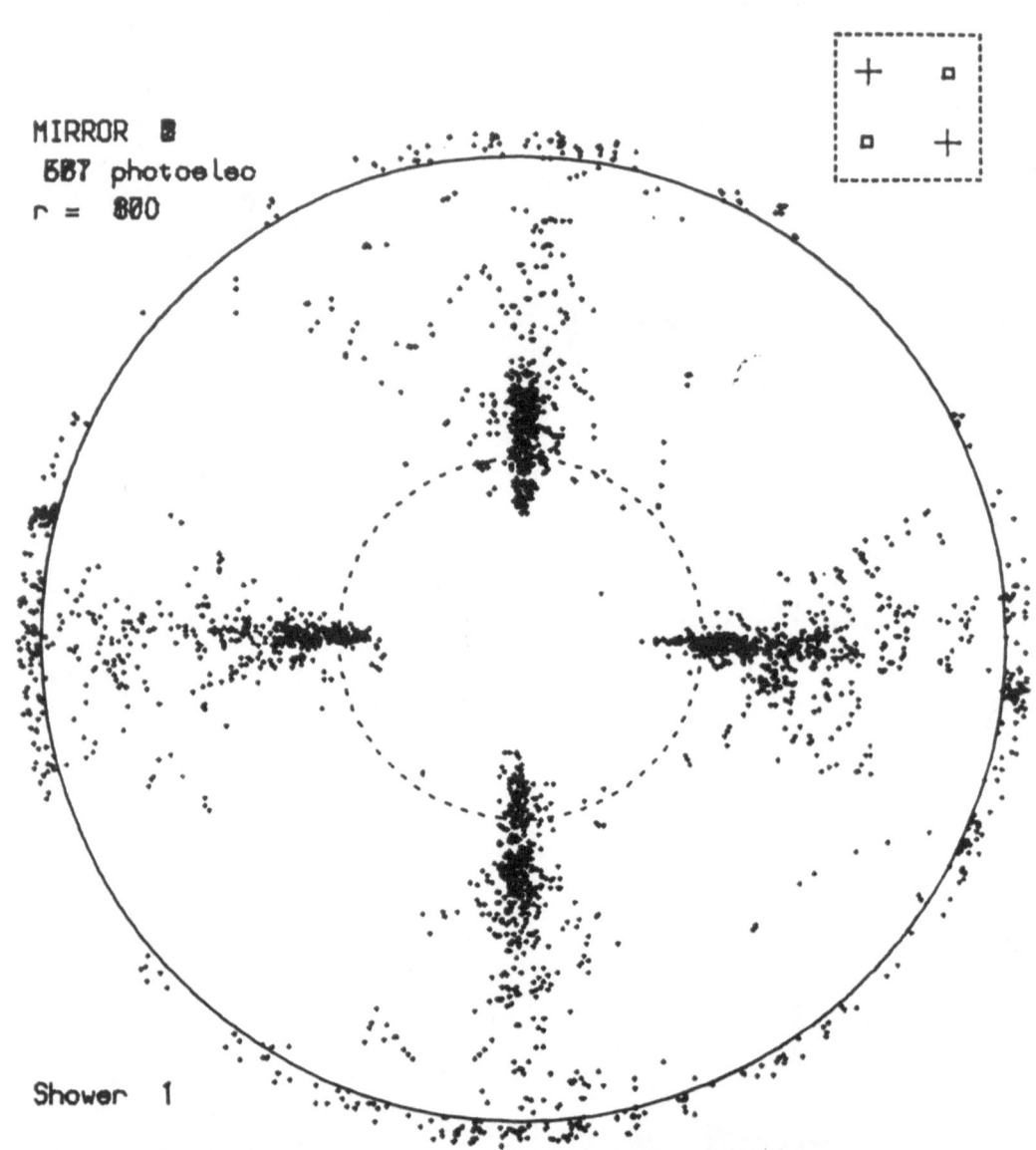

FIGURE C

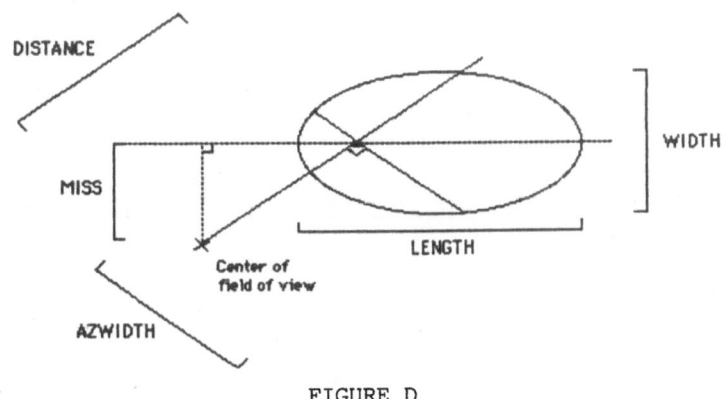

FIGURE D

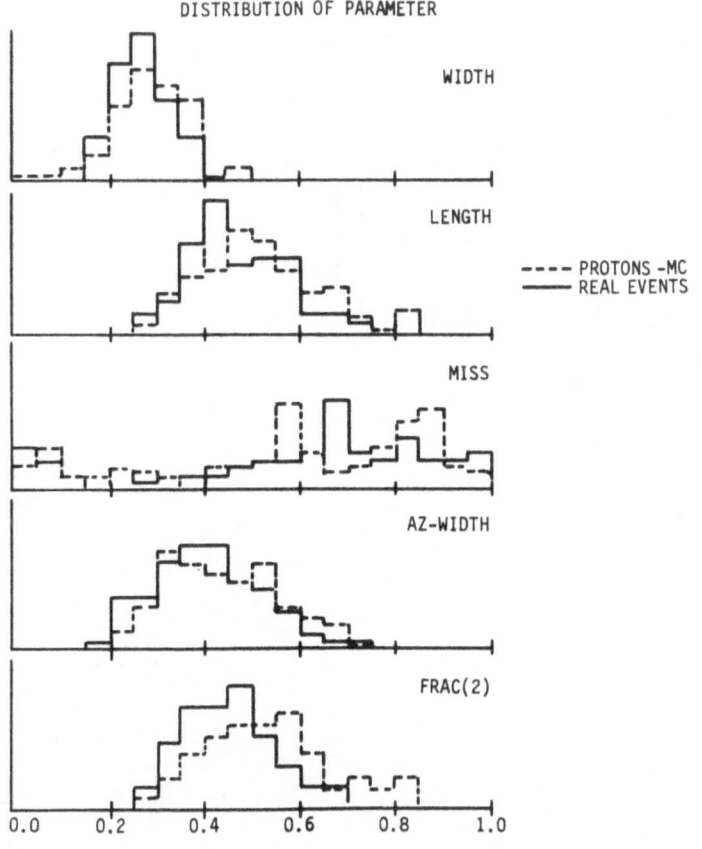

FIGURE E

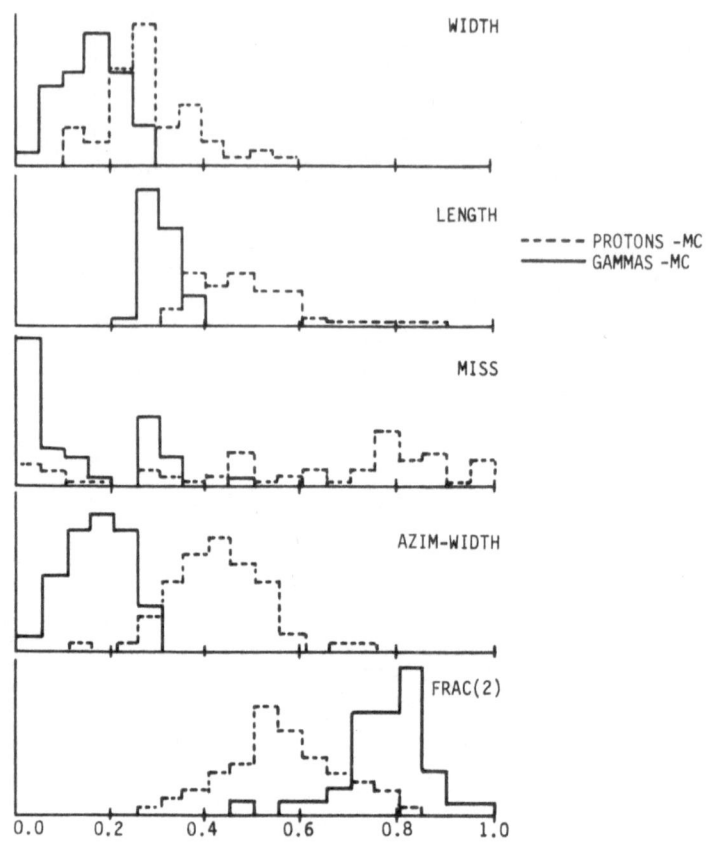

DISTRIBUTION OF PARAMETER

FIGURE F

ple group have only reported results based on their raw data analysis. Improvements in the camera, in particular the use of finer pixels, should allow better isolation of the gamma–ray signal.

III. Studies of Air Shower Directions from Fast Timing Reported by the Durham Group.

This group did extensive studies utilizing the Crab pulsar as a source. In principle, the direction of an air shower can be obtained by the relative timing from two separated telescopes to an accuracy greater that the acceptance angle of these telescopes. Unfortunately the workable separation is limited by the width of the photon pancake. In principle the technique worked. However, by using four telescopes where each unit was seperated by 60 meters, the coincidence rate was reduced to such a low rate that the results were not useful.

Prof. Murthy reported that their group had at one time used 18 detectors with a significant spacing and achieved an angular resolution of 0.3° H.W.H.M. Because of the low data rate they also decided to move the micros back together to form a compact array.

IV. Utilization of the Solar Facility in the Pyrenees

The French group described the results of preliminary studies of the use of the solar mirrors as collectors for a gamma ray Cherenkov telescope. During the Spring 1985, 2 mirrors 7 meters in diameter with a seperation of 35 meters were used in coincidence. The field of view for each mirror was about 3° to 3.5°. About 100 meter long cables were used which deteriated the pulse shape. The pulse rise time was about 4 n.s. The width of the gate for two fold coincidence was 12 n.s. The coincidence rate for cosmic rays was estimated to be about 20 Hertz clearly indicative that the energy threshold was reasonably low.

V. Mirrors and Electronic Advancements.

The Durham group described the construction and testing of their light weight and inexpensive   aluminum mirrors that will be used in their new telescope. Each mirror is estimated to cost only about 30 £. The solid aluminum surface has many advantages over coatings from the stand point of durability and reflectivity both in the visible and near ultra violet. Although the surface does show irregularities their optical quality is adequate.

Up to the present time, many of the large Cherenkov telescopes have successfully used mirrors which have had abberations corresponding to angles of at least 0.25 degrees. Since the acceptance angle may be typically 0.75° to 1° this poor angular resolution of the mirrors is of little consequence. It should be pointed out that there can be cases where a higher resolution is very desirable. The Cygnus X-1 black hole candidate source has a 3.8 magnitude star less than 0.5° away from the source. If the mirror abberations are kept small, a stop can be placed in the focal plane of the mirror to black out the star while accepting photons from gamma showers which come from an angular cone around the stop.

VI. Summary

Although real progress has been made during the last year as evidenced by the discovery of high energy gamma rays from several new sources and the more detailed time stucture information, no one had a proposal for a new technique which would improve the signal to noise ratio by more than one order of magnitude.

It was emphasized that by going to a lower energy threshold not only would the statistics be improved because of the energy spectrum but also because the gamma ray to proton induced shower ratio should be more favorable. A substantially larger fraction of the energy from proton induced showers is expected to go into the charged component of the shower instead of gamma rays.

On the other hand, by going to higher energies (PeV range) new physics may be discovered. It would be possible to test several theories which attempt to explain the production processes for gamma rays. From the discussions presented at this conference I would conclude that major efforts should be made in both directions.

STATISTICAL ANALYSIS IN HIGH ENERGY GAMMA RAY ASTRONOMY

R.J. Protheroe
Department of Physics, University of Adelaide
Adelaide, South Australia 5001

ABSTRACT. In this rapporteur paper I summarize and discuss aspects of the statistical analysis addressed by participants during the NATO Advanced Research Workshop on High Energy Gamma Ray Astronomy held at Durham in August 1986. Topics covered include testing for a D.C. excess, testing for periodic emission at a known period, choice of statistic, searching for an unknown period, Fourier analysis, source variability, and estimating the light curve.

1. INTRODUCTION.

During the NATO Advanced Research Workshop on High Energy Gamma Ray Astronomy a workshop session was held on statistics. The chairman was R. Morse (Wisconsin) and the discussion was led by K.J. Orford (Durham) and A.M. Hillas (Leeds). During the short time allocated for the workshop many important points were raised but time was too short for an adequate treatment of each. In this rapporteur paper, I have tried to identify the main areas of interest and discuss these more fully. In doing this it has not been practical to give credit to everyone who spoke and in some cases I may have unintentionally misrepresented some of the views expressed — if this is indeed the case, I apologize in advance.

2. D.C. SOURCES.

When testing the significance of a D.C. excess it is important to take account of the uncertainty in the background estimate. Consider a region of sky subtending solid angle Ω_{ON} around a suspected source direction, and a background region (usually chosen at the same declination) of solid angle Ω_{OFF}. If N_{ON} and N_{OFF} events were recorded from the source and background regions in times t_{ON} and t_{OFF}, respectively, the problem is to assess the significance of any excess of events from the source region over that expected based on the off-source observation.

This problem may be treated adequately using standard statistical techniques (a similar problem concerning particles emitted in opposite directions by a radioactive source is dealt with by Eadie *et al.* [12]). The present problem has been dealt with in detail by

91

Gibson *et al.* [14], Dowthwaite *et al.* [11] and Li and Ma [19].

Following the notation of Li and Ma [19], and defining the ratio of the on-source and off-source exposures,

$$\alpha = \Omega_{ON} t_{ON} / \Omega_{OFF} t_{OFF},$$

the best estimate of the background during the on-source observation is

$$\hat{N}_B = \alpha N_{OFF}.$$

Then the best estimate of the source count is

$$\hat{N}_S = N_{ON} - \alpha N_{OFF}.$$

$\hat{N}_S$ and $\hat{N}_B$ are the maximum likelihood estimates of the source and background counts.

The significance of any excess can be tested using the maximum likelihood statistic, λ, defined as the ratio of the likelihood of the observation given zero expectaion value of the source count divided by the likelihood for the observation given the maximum likelihood values,

$$\lambda = \left(\frac{Pr. \left(N_{ON}, N_{OFF} | < N_S >= 0, < N_B >= \alpha \left(N_{ON} + N_{OFF} \right) / \left(1 + \alpha \right) \right)}{Pr. \left(N_{ON}, N_{OFF} | < N_S >= \hat{N}_S, < N_B >= \hat{N}_B \right)} \right).$$

Both the numerator and denominator can be obtained as the product of two Poisson laws, so

$$\lambda = \left[\frac{\alpha}{1 + \alpha} \left(\frac{N_{ON} + N_{OFF}}{N_{ON}} \right) \right]^{N_{ON}} \left[\frac{1}{1 + \alpha} \left(\frac{N_{ON} + N_{OFF}}{N_{OFF}} \right) \right]^{N_{OFF}}.$$

It can be shown that $-2 \ln \lambda$ is distributed as $\chi^2(1)$ asymptotically, and hence the significance of the observation is $\sim \sqrt{-2 \ln \lambda}$ standard deviations.

3. SOURCES WITH KNOWN PERIOD.

For periodic sources for which the period is well known, for example from radio or X-ray data, the arrival times of events, t_i, may be readily converted to phases, ϕ_i. This will require a knowledge of P, $\dot{P}$ and (where necessary) $\ddot{P}$, as well as timing corrections to the barycentre of the solar system and, in the case of pulse periods of binary sources, a correction to the barycentre of the binary system. Of course the epoch of zero phase must be known as well if absolute phases are required.

Having obtained phases, the conventional approach has been to bin the data in a histogram. There are many problems associated with doing this, particularly if one wishes to evaluate the statistical significance of any effect found. In particular: (1) bin width must be chosen; (2) bin origin must be chosen; (3) if no effect found, and the data are re-binned this must be taken into account in assigning the significance of any effect found subsequently; (4) to some extent the significance assigned to an effect depends on luck (e.g. happening to choose bins of optimum width that didn't split any phase peak); (5) the histogram is not the best estimate of the light curve. Because of this, the current consensus is that it is best *not* to bin the data.

Alternative techniques exist for testing phase distributions for uniformity and it is recommended that these be used instead to test for evidence of periodic emission. Some of these are described in the next section.

3.1. Circular Statistics.

Statistics for dealing with circular data, such as phases, are referred to as circular statistics. The value of a circular statistic representing the data does not depend on the choice of phase origin (i.e. it is invariant under a change of phase origin). A number of circular statistics are available and many have been described by Mardia [20] and Batschelet [1].

The most commonly used circular statistic is the Rayleigh statistic, $\bar{R}$. This can be visualised by representing the N phases, $\phi_i, i = 1, \ldots N$, by N unit vectors, each oriented at an angle $2\pi\phi_i$ with respect to a particular direction. The Rayleigh statistic is then the length of the resultant vector divided by N.

Other circular statistics commonly used include Kuiper's V_N statistic which is based on a generalization of the Kolmogorov-Smirnov test to circular data, Watson's U^2 statistic, and Hodges-Ajne's statistic. A statistic has also been recently suggested by Protheroe [21], Υ_N, and generalizations of the Rayleigh statistic to include 2nd and higher harmonics have also been suggested (e.g. Buccheri [5]).

3.2. Choice of Statistic.

In most cases, the phase distribution of events due to cosmic rays is expected to be uniform and so it is only necessary to test the null hypothesis, H_0, and see if the observed phase distribution is significantly different from uniform. As noted above, many statistics could be used to do this. However, some statistics are more powerful than others and it is sensible to choose a statistic which is powerful against alternatives to H_0 of the form we may expect in high energy γ-ray astronomy, based on past experience.

Consider a particular statistic, S. At a level of significance α (e.g. 5%), if H_0 is true (i.e. uniform distribution) S will exceed its critical value S_α in a fraction α of all trials, i.e.

$$Pr.(S > S_\alpha | H_0) = \alpha.$$

When used at the same level of significance then, *all* statistics are equally safe (i.e. all statistics will incorrectly reject H_0 a fraction α of the time).

In cases where H_0 is *not* true, we would like S to exceed S_α as often as possible. If a particular alternative hypothesis H_1 were true, we would therefore like the probability of incorrctly accepting H_0,

$$\beta = Pr.(S < S_\alpha | H_1),$$

to be as small as possible. The 'power' of the test is defined as $(1 - \beta)$ and is the probability of correctly rejecting H_0. The power of a statistical test at a particular level of significance is a function of the true distribution and some statistics are more powerful than others against particular alternatives to H_0.

Recently, Kobayakawa *et al.* [17] have attempted to compare the 'safety' of several statistics. They have done this by generating a large number of artificial phase data sets containing fifty or a hundred randomly sampled phases. They then binned the data and used for the safety test only those 5% of the data sets which had the highest χ^2 values. These 5% of phase data sets were then tested for uniformity by using various circular statistics and those tests that rejected H_0 in a large fraction of the data sets were considered less 'safe'. This is a quite unreasonable way to compare the safety of statistical tests as the data sets being tested are *not* drawn from a uniform distribution, but rather from a distribution biased by selecting those fluctuations that lead to high χ^2 values (such as a large excess in one bin — i.e. a narrow phase peak). It is not surprising then that the test that they found to be the most powerful, Υ_N, also rejected H_0 more often for these data sets. In a more reasonable safety test, in which data sets are truly drawn from a uniform distribution, they would have found (as noted above) that *all* the statistics tested would have rejected H_0 on average a fraction α of the time, and are therefore equally safe.

If we expect from past experience, or theory, that the phase distribution may have a particular distribution then it is sensible to use the statistical test that has the highest power for this distribution, i.e. the highest value of $(1 - \beta)$. For example, if for a particular source, from observations at other wavelengths you suspect that the γ-ray light curve may have a double-peaked structure (Fig. 1(a)) it may not be sensible to use the Rayleigh test since the resultant vector (see section 3.1) is likely to be small (see Fig. 1(b)) and one should instead use a statistic powerful against this case. Similarly, if you suspect that the phase distribution may contain a narrow peak, it may be more useful to use a statistic which is more powerful against a narrow peak than the Rayleigh statistic. If one has no information or suspicions about the nature of the phase distribution then perhaps one should be conservative and use the Rayleigh test.

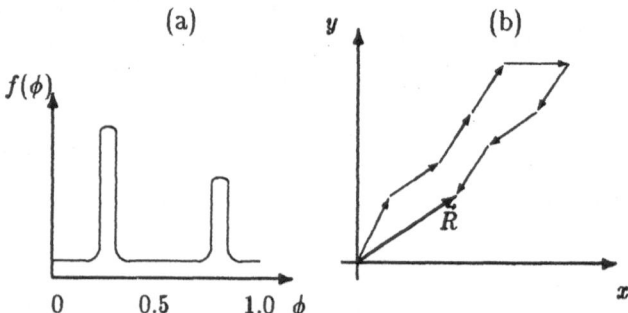

Figure 1. (a) Typical double-peaked phase distribution. (b) Unit vector representation of phases sampled from a distribution such as that shown in Fig. 1(a) (for clarity, far fewer phases are plotted than in a typical observation). The length of the resultant vector, $\vec{R}$, is related to the Rayleigh statistic.

Since phase distributions in γ-ray astronomy have included peaks that are significantly narrower than half a period and, in some cases, a double-peaked structure it may be more sensible to use the Z_n^2 statistic suggested by Buccheri [5]. This statistic is defined,

$$Z_n^2 = \frac{2}{N} \sum_{k=1}^{n} \left[\left(\sum_{j=1}^{N} \cos 2k\pi\phi_j \right)^2 + \left(\sum_{j=1}^{N} \sin 2k\pi\phi_j \right)^2 \right],$$

which is equivalent to using the Rayleigh statistic if $n = 1$. The statistic is thus effectively the Rayleigh statistic modified to include n harmonics. For a uniform phase distribution, Z_n^2 is distributed asymptotically as $\chi^2(2n)$.

4. SEARCHING FOR UNKNOWN PERIODICITY.

When the period of the γ-ray source candidate, P, is unknown, or known only to be within some range $P_1 - P_2$, then a period search is made. This may be done by calculating the value of a circular statistic as a function of period in the range $P_1 - P_2$, and looking for those periods for which the value of the statistic is high. It is then necessary to assess the significance of any periods found. First, the chance probability, p, is calculated as if only one period had been examined as described in section 3. If there are n_I independent periods in the range $P_1 - P_2$ then the chance probability for the statistic having a value as high as that found is $(1 - [1 - p]^{n_I})$. The statistics most often used for these searches are the Rayleigh statistic, and statistics based on the power spectrum or Fourier Transform. The latter are often used for the study of time series and are discussed in the next section.

The number of independent periods depends primarily on the length of the time interval over which the observations were made, T, and the period, P. The spacing in period, ΔP, between these periods may be crudely estimated as the error in period required to give an error in phase of ~ 1 over the observing time,

$$\Delta P \simeq \left(\frac{T}{n} - \frac{T}{n+1} \right),$$

where n is the number of cycles, $n = T/P$. Hence,

$$\Delta P \simeq P^2/T,$$

which is often referred to as the spacing of 'independent Fourier periods'. It should be emphasised that this only gives an approximation to the number of independent periods searched, and that the actual number of independent periods depends on the circular statistic chosen. For example, it was pointed out by O.C. De Jager (Potchefstroom), R.C. Lamb (Iowa) and K.J. Orford (Durham) that in the case of the Rayleigh test this underestimates n_I by a factor of ~ 3. It is particularly important to estimate n_I either by Monte Carlo methods or, preferably, by using experimental data similar to that being analysed.

Other problems such as aliasing, interference, etc., will also arise due to the finite observing time and the digitization of event times as discussed in the next section.

4.1 Fourier Analysis.

The analysis of time variability has often been carried out using Fourier methods. Consider some quantity, $X(t)$, which varies with time, t. This quantity, X, could, for example, be anything from the apparent magnitude of a star, to the price of γ-ray telescopes in Durham. Information about periodicities in this quantity could be found from the Fourier transform,

$$F(\omega) = \int_{-\infty}^{\infty} X(t) \exp(i\omega t) dt,$$

or from the 'power' at angular frequency ω, $|F(\omega)|^2$. Note that also, just as there is no standard definition of the Fourier transform (variations of factors of 2π or a minus sign in the exponential occur), definitions of the quantities discussed in this section also vary in the literature.

In the case of a stochastic process, where one may expect to find some correlations in the data on certain timescales, τ, this information could be obtained from the 'power spectrum' (see, e.g. refs. [4] and [8]),

$$P(\omega) = \int_{-\infty}^{\infty} r_X(\tau) \exp(i\omega\tau) d\tau,$$

where r_X is the autocorrelation function,

$$r_X(\tau) = \int_{-\infty}^{\infty} X(t) X(t + \tau) dt.$$

The autocorrelation theorem (see e.g., ref. [3]) can be used to show that $P(\omega) \propto |F(\omega)|^2$, i.e. that the power spectrum and Fourier power are identical.

In practice, the data are not taken continuously over an infinite time, but are often taken at discrete time intervals, forming a 'time series', i.e. a series of measurements of some quantity, $X(t_i)$, made at a series of times, $t_i, i = 1, \ldots, N$. For such data, there is a considerable amount of literature (see e.g., refs. [2], [4]) dealing with techniques for estimating $F(\omega)$ and $P(\omega)$, and some of these techniques have been applied to arrival time data in γ-ray astronomy. For this reason, I will discuss some of these techniques below.

In the case of a continuous observation over a time, T, one can obtain the 'finite Fourier transform',

$$F_T(\omega) = \int_{-T/2}^{T/2} X(t) \exp(i\omega t) dt,$$

and in the case of a time series one can obtain the 'discrete Fourier transform',

$$F_N(\omega) = \sum_{j=1}^{N} X(t_j) \exp(i\omega t_j).$$

Fast Fourier Transform algorithms (e.g. ref. [13]) are often used to compute $F_N(\omega)$. The discrete Fourier transform is in fact proportional to the convolution of the true Fourier transform with the 'spectral window' [8],

$$\delta_N(\omega) = \sum_{j=1}^{N} \exp(i\omega t_j).$$

This spectral window contains all the information about the times of the observation. The finite data spacing gives rise to spurious peaks at multiples of the sampling frequency (i.e.

multiples of $2\pi N/T$ in equally spaced data), and interference between these peaks and any real periodicity in the data will give rise to 'aliasing'. For example if the spectral window contains a peak at a frequency ω_W and the true Fourier transform contains a peak at ω_0 then there will also be a peak at $\omega_W - \omega_0$ in the discrete Fourier transform.

A quantity closely related to the power spectrum is the 'periodogram',

$$P_X(\omega) = \frac{1}{N}|F_N(\omega)|^2,$$

which is proportional to the convolution of the power spectrum with the 'power spectral window', $|\delta_N(\omega)|^2$ (e.g. Deeming [8]). An alternative definition of the periodogram was introduced by Scargle [23] and has been discussed by Horne and Baliunas [16]. For the case of evenly spaced data one would expect, based on section 4, that there would be $\sim N/2$ independent periods up to the Nyquist frequency, $\pi N/T$, whereas Horne and Baliunas [16] found that $N_I \sim N$ for $N = 10$, and $N_I \sim 1.5N$ for $N = 400$. Applications of the periodogram in astrophysics have been discussed by Deeming [8], Scargle [23], Horne and Baliunas [16], and Leahy et al. [18].

Data in γ-ray astronomy are not usually a typical time series. Data typically consist of a series of arrival times, t_i, of air showers mostly due to cosmic rays but hopefully containing some due to γ-rays. The techniques above are often applied by setting $X(t_i) = 1$. This, clearly, is not strictly correct however since during an observing run, apart from dead time, the telescope effectively samples X at all times which are spaced by an amount equal to the digitization time of the experiment (e.g. 1 μs), and at most of these times $X = 0$. This arises due to the very low rate of receiving photons compared to, say, in optical astronomy. Nevertheless, $F_N(\omega)$ and $P_X(\omega)$ obtained in this way provide circular statistics for analysing these data. For example, in the case of the γ-ray data described above one can easily show that $P_X(\omega) = N\bar{R}^2$ where $\bar{R}$ is the Rayleigh statistic. As with the Rayleigh statistic, the power spectrum (when suitably normalized) has an exponential probability distribution, $p(P)dP \simeq \exp(-P)dP$, which makes it relatively easy to estimate the significance of an observed high power level. As an example of this, Fig. 2 reproduces the distribution of power for all independent periods searched by Gorham et al. [15] during their June 1985 observation of Her X-1.

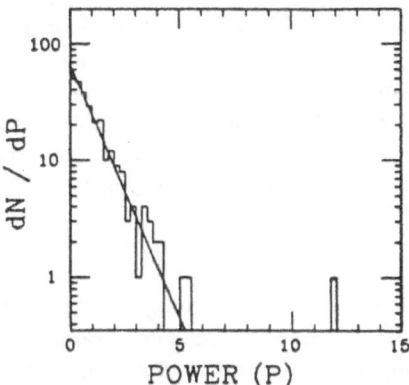

Figure 2.   Observed distribution of power, P, obtained by Gorham et al. [15] during an observation of Her X-1.

The value at $P \sim 11.9$ is attributed to signal.

It is not clear to me what, if any, advantage the Fourier methods have over the use of other circular statistics, however it is useful to be aware of them as many of the problems in periodic analysis mentioned above (e.g. aliasing) which apply here will apply also to cases where other circular statistics are used.

4.2 Source Variability.

It is possible to obtain an estimate of the signal strength from the power spectrum, $P(\omega)$, (or from the periodogram, $P_X(\omega)$). In the case where the signal is dominated by noise (the case with present VHE γ-ray observations) a constant signal strength can give rise to very large variations in the observed power or periodogram. M.F. Cawley (Whipple Observatory) reported that he had carried out Monte Carlo simulations for such situations and found that when when he injected a constant signal strength corresponding to $< P >= 10$, fluctuations gave rise to P in the range 4 – 20. In real data such variations in P may have been considered evidence for time variability. Clearly, it is important to quote realistic errors on flux estimates which take account of such fluctuations.

5. LIGHT CURVES.

As mentioned earlier, the phase histogram is certainly not the best estimate of the light curve — it might even be the worst. In particular, it is difficult to answer questions like "How wide is that phase peak?" and "Is that secondary peak in the phase distribution real?" in an unbiassed way. To better describe the true phase distribution, De Jager et al. [9] have proposed the use of 'kernel density estimators'. In these techniques, the "true" light curve, $f(\phi)$, as a function of phase, ϕ, is estimated by

$$\hat{f}(\phi) = \frac{1}{Nh} \sum_{j=1}^{N} K\left(\frac{\phi - \phi_j}{h}\right)$$

where ϕ_j are the phases corresponding to the observed times, t_j, of arrival of air showers, K is the 'kernel' or weighting function, and h is the smoothing parameter. $\hat{f}(\phi)$ is thus just a smooth function which results from convolving a 'comb' of delta functions at the observed phases with a smoothing function. The simplest smoothing, with K =constant over the width $2h$, is equivalent to plotting values of the histogram at the centres of the bins of width $2h$ for all possible bin origins. This is referred to as the 'naive estimator'.

De Jager et al. [10] suggest the use of the normal kernel,

$$K(x) = \exp(-x^2/2)/\sqrt{2\pi},$$

which gives a smoother result and is less time consuming to compute than more complicated estimators such as the Fourier series estimator. The key to obtaining a reliable estimate of $f(\phi)$ lies in the correct choice of h, i.e. not so wide so that real, narrow peaks are eliminated, but wide enough to smooth out statistical fluctuations of the background. They show how to obtain a value of h which minimizes both the variance and bias in a consistent way and have applied this technique to the Potchefstroom data [22] on PSR 1802–23 (Fig. 3) to obtain an estimate of the light curve and confidence bands on this.

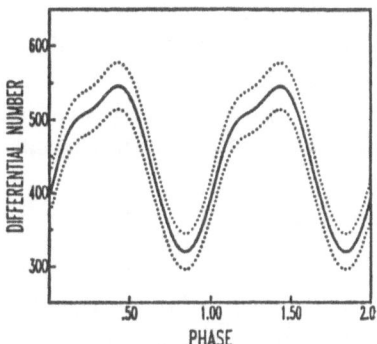

Figure 3. Light curve of PSR 1802-23 obtained by De Jager *et al.* [10] by applying the normal kernel density technique.

At present, $\hat{f}(\phi)$ itself is not used to obtain the significance of the source detection, but the maximum value of $\hat{f}$ or the area of $\hat{f}$ above the background level are perfectly valid circular statistics which could be used for this purpose. Whether this would be worthwhile would depend on how difficult it is to work out their probability distributions given the null hypothesis, i.e. a uniform phase distribution.

6. CONCLUDING REMARKS.

Finally, I should mention the worry expressed by P.V. Ramana Murthy (Tata Institute) concerning the reputation of the field of VHE and UHE γ-ray astronomy. His concern particularly was that we have at present detected sources mainly at the 3σ – 5σ levels, and that astronomers in other fields may doubt the authenticity of the γ-ray results. It was generally agreed that this was an unfortunate situation to be in and that perhaps the reason for this was that, unlike in X-ray astronomy which has Sco X-1, there are no anomalously nearby powerful emitters of VHE or UHE γ-rays. In X-ray astronomy, very early in the field it was possible to obtain extremely significant results because of Sco X-1 and this led to a very rapid development of the field. It was hoped that with the improvements in sensitivity in VHE γ-ray telescopes which are presently taking place (according to T.C. Weekes (Whipple Observatory) until recently there has been very little improvement in sensitivity over the last two decades), the statistical significance of results in VHE γ-ray astronomy should soon increase. As an example of this, it is estimated that with the new Durham Mark III telescope [7] it would have been possible to detect pulsed emission from the Crab Pulsar at the 3σ level in ~ 1 hour of observing, or a burst of pulsed VHE emission from Cyg X-3 such as that reported by Chadwick *et al.* [6] with a chance probability of $\sim 10^{-37}$. There was optimism that current and new observations by more than one group of the 12 ms periodicity during bursts from Cyg X-3 may soon lead to *extremely* significant results.

In conclusion, in these early days of VHE and UHE γ-ray astronomy, a proper statistical analysis of data is crucial. Unfortunately this subject is often given far less time than other important aspects of an experiment, e.g. building and testing electronics, designing new telescope systems, and spending many long hours taking data. We must be careful that it does not become the "weak link in the chain" and mar an otherwise excellent experiment. Many problems in statistical analysis in γ-ray astronomy were raised in the workshop and I have attempted here to reproduce some of the ideas and results discussed and add some comments of my own.

REFERENCES.

1. Batschelet, E., 1981: *Circular Statistics in Biology*, (London: Academic Press).
2. Bloomfield, P., 1976: *Fourier Analysis of Time Series: An Introduction*, (New York: J. Wiley and Sons).
3. Bracewell, R.N., 1978: *The Fourier Transform and Its Applications*, (New York: McGraw-Hill), p. 115.
4. Brillinger, D.R., 1975: *Time Series, Data Analysis and Theory*, (New York: Holt, Rinehart and Winston).
5. Buccheri, R., 1985: in *Proc. Workshop on Techniques in Ultra High Energy Gamma Ray Astronomy, La Jolla*, eds. R.J. Protheroe and S.A. Stephens, (Adelaide: Univ. of Adelaide), p. 98.
6. Chadwick, P.M., *et al.*, 1985: Nature, **318**, 642.
7. Chadwick, P.M., *et al.*, 1987: these proceedings.
8. Deeming, T.J., 1975: Astrophys. and Space Sci., **36**, 137.
9. De Jager, O.C., *et al.*, 1986: Astron. Astrophys., in press.
10. De Jager, O.C., *et al.*, 1987: these proceedings.
11. Dowthwaite, J.C., *et al.*, 1983: Astron. Astrophys., **126**, 1.
12. Eadie, W.T., *et al.*, 1971: *Statistical Methods in Experimental Physics*, (Amsterdam: North Holland), p. 51.
13. Elliott, D.F. and Rao, K.R., 1982: *Fast Transforms*, (New York: Academic Press).
14. Gibson, A.I., *et al.*, 1982: in *Proc. Int. Workshop on Very High Energy Gamma Ray Astronomy, Ootacamund*, eds. P.V. Ramana Murthy and T.C. Weekes, (Bombay: Tata Institute), p. 97.
15. Gorham, P.W., *et al.*, 1986: Astrophys. J. (Lett.), in press.
16. Horne, J.H., and Baliunas, S.L., 1986: Astrophys. J., , **302**, 757.
17. Kobayakawa, K., *et al.*, 1986: in *Proc. Japan-USA Joint Seminar on Cosmic Ray Physics/Astrophysics, Tokyo*, in press.
18. Leahy, D.A., *et al.*, 1983: Astrophys. J., **266**, 160.
19. Li, T.P., and Ma, Y.Q., 1983: Astrophys. J., **272**, 317.
20. Mardia, K.V., 1972: *Statistics of Directional Data*, (London: Academic Press).
21. Protheroe, R.J., 1985: Astron. Express, **1**, 137.
22. Raubenheimer, B.C., *et al.*, 1986: Astrophys. J. (Lett.), in press.
23. Scargle, J.D., 1982: Astrophys. J., **263**, 835.

A SURVEY OF GAMMA RAY FACILITIES - CURRENT AND PLANNED

K. E. Turver

Department of Physics
University of Durham
Durham DH1 3LE
United Kingdom

This survey is based on the information provided by participants at a Workshop session under the Chairmanship of Professor J Gaidos. The workshop was held to summarize the specifications of detectors for gamma rays of TeV and PeV energy which are currently operating or for which firm plans exist.

Table I describes the detector systems employing the atmospheric Cerenkov light technique for the detection of TeV gamma rays. Table II describes the air shower arrays for the recording of PeV gamma rays.

The response of participants in providing details of their equipment and future plans is gratefully acknowledged.

K. E. Turver (ed.), Very High Energy Gamma Ray Astronomy, 101–103.
© *1987 by D. Reidel Publishing Company.*

TABLE I   TeV Gamma Ray Projects

PROJECT (Institution)	LOCATION				MIRRORS				NO. OF MOUNTS (Separation)	THRESHOLD (TeV)	OPERATION	COMMENTS
	PLACE	ALT.	LAT.	LONG	AREA m²	NUMBER	SIZE (m)	F.L.				
ADELAIDE (Un. of Adelaide)	Woomera Aus.	600	32 S	137 E	10	3	3.5	?	1	< 1	Late 1987	Design Incomplete
HALEAKALA (Un. of Athens, Hawaii, Purdue, Wisconsin)	Maui Hawaii U.S.A.	3300	20 N	156 W	1.8	6	1.5	1.5m	1	-	Current	
HERCULES (Smithsonian USA, Leeds Un. U.C. Dublin)	Whipple Obs. AZ USA	2300	31 N	110 W	75	2	10	7.1m	2 (120m)	<0.1	Current 1987? 1988?	(1 camera) (1 high res. camera) (2 cameras)
NARRABRI (Un. of Durham)	Narrabri NSW Aus.	210	30 S	149 E	12	3		2.5	1	0.24	Sept.'86	
NORTHERN HEM. (Un. Durham)	La Palma	240	29 N	18 W	5	3		2.5	1	0.3	June'87	
PACHMARHI (Tata Inst.)	Pachmarhi	1200	22.5 N	784 E°	0.6 1.8	10 8	0.9 1.5	0.42 0.6	18	0.6	Current	
POTCHEFSTROOM (Un. of Potchefstroom)	Potchefstroom	1400	28 S	29 E	1.8	3	1.5	0.66	3 (55m)	1.2	Current	3 telescopes each 3 mirrors
THEMIS (Saclay)	Themis Pyrenees France	1700	42 N	2 E	38	7	7.0	4.8	7 (60m)	0.1	Late'87	(If funded)
WHITE CLIFFS (Un. of Adelaide)	White Cliffs NSW,Aus.	600	32 S	142 E	19	3	5	1.8	3 (50m)	5	March'86	

TABLE II   PeV Air Shower Detectors

PROJECT	LOCATION				ARRAY SIZE (mxm)	ANG RES (deg)	THRESH-HOLD (PeV)	NO. OF ELEMENTS	MUON DE-TECTORS(m²)	DATE OF OPERATION	COMMENTS
	PLACE	ALT.	LAT.	LONG							
AKENO (Un. of Tokyo)	AKENO JAPAN	900	35 N	140 E	210x90	2.5	<1	~100	225	Dec.'85	
BUCKLAND PARK (Un.of Adelaide)	BUCKLAND PARK ADELAIDE	0	35 S	138 E	$3 \times 10^4 m^2$ at 10 PeV	1.5	0.5	27	=	Jul.'85	
CHICAGO (Un. of Chicago)	FLY'S EYE, UTAH U.S.A.	1450	40 N	113 W	10^5	0.25	0.1	1064	1200		Proposal
GREX (Un. of Leeds)	HAVERAH PARK U.K.	210	54 N	0	$4 \times 10^4 m^2$	1.0	0.2	32	40	Feb.'86	
HEGRA (Un. of Kiel,West Germany)	LA PALMA Canary Islands	2400	29 N	18 W	$30000 m^2$	1.0	0.1	37	-	mid.'87	
K.G.F. (TATA Ins.)	KOLAR GOLD FIELDS India	900	12 N	78 E	$1.6 \times 10^4 m^2$	<1.0	0.5	61	200	Current	
SPASE (Bartol,Un. of Leeds)	S.POLE	1300	90 S	0	7.2×10^3	1	0.1	16	-	Dec.'87	24 hour of all sky Objects same zenith
TOP ARRAY	GRAN SASSO (Italy)	2000	42 N	13 E	$10\text{-}100 \times 10^3 m^2$	1.15	0.05	28	200	4 units 14 units 28 units	Late'86 Late'87 Late'88
WHIPPLE (Smithsonian, UCD Dublin,Un. of Hongkong	WHIPPLE OBSERV. Ariz. U.S.A.	2300	31 N	110 W	-	1	0.1	13	-	Current	Col with Hercules (TeV)
CYGNUS(Los Alamos,Nat. Lab.Un.of Maryland	LCS ALAMOS, NM,U.S.A.	2100	41 N	103 W		1	0.2			1986 Current	

VHE GAMMA RAYS FROM CYGNUS X-3

L. Resvanis
Physics Laboratory, The University of Athens
10680 Athens, Greece

J. Learned, V. Stenger, and D. Weeks
Department of Physics, The University of Hawaii
Honolulu, HI 96822, USA

J. Gaidos, F. Loeffler, J. Olson, T. Palfrey, G. Sembroski,
and C. Wilson
Department of Physics, Purdue University
West Lafayette, IN 47907, USA

U. Camerini, J. Finley, W. Fry, M. Jaworski, J. Jennings,
A. Kenter, R. Koepsel, M. Lomperski, R. Loveless,[*] R. March,
J. Matthews, R. Morse, D. Reeder, P. Sandler, P. Slane,
and A. Szentgyorgyi
Department of Physics, The University of Wisconsin
Madison, WI 53706, USA

ABSTRACT. The Haleakala 10 m² Cherenkov light telescope observed Cyg X-3 for 113 hours during the summer and fall of 1985. A 60-sec. burst of gamma rays was observed during the radio flare of early Oct. at a phase of $\phi_{BB} = 0.74$. No evidence for pulsar periodicity was found during the burst. Searches for periodicity outside the burst have not yet produced positive results.

1. INTRODUCTION.

The Haleakala telescope has been described in another contribution to this conference(1). Briefly it consists of two apertures, each with 18 photomultiplier tubes, which trigger on a coincidence of 7 PMT hits within 10 ns.

During the summer and fall 1985, 113 hours of data were taken with this telescope pointing at Cyg X-3. The A and B apertures were alternately pointed at Cyg X-3, shifting every 720 sec. The total trigger rate was ~2 Hz with each aperture accounting for half the rate. Triggers due to ambient light (skyshine, starlight, etc.) were eliminated by requiring at least 7 TDC hits within 6 ns. The slope of the multiplicity distribution (shown in Fig. 4 of Ref. 1) changes abruptly

---

[*] Presenter of this paper.

K. E. Turver (ed.), Very High Energy Gamma Ray Astronomy, 105–110.

at this point; events with greater than 7 hits are due primarily to cosmic ray showers with some admixture of gamma ray showers. The shower rate was ~0.35 Hz for both apertures.

2. ANALYSIS.

One method of searching for gamma rays signals is the observation of an excess rate, either simultaneously between the two apertures, or sequentially using either of the apertures. We looked for such rate effects using 100 sec. bins with both normal and offset (by 50 sec.) histograms. For the Oct. 1985 data this results in 784 bins, roughly half of which occur while the telescope is pointed at Cyg X-3. From random statistics we would expect 1.1 bins with a fluctuation in excess of 3σ. However, as shown in Fig. 1, we observe 3 such bins, all of which point at Cyg X-3. Fig. 2 shows the bin with the largest excess ($\sim 3.9\sigma$), which occurs in the A aperture while it is pointing at Cyg X-3. Fig. 3 shows an expanded version with 40 sec. bins, and a peak of 60-80 sec. is visible. Taking the three largest contiguous bins (1320-1380 sec.), we have 19 gamma showers with a background of ~20 showers due to cosmic rays. The significance of this peak is 4.4σ, and the signal to noise ratio is 0.95.

More evidence of the anomalous behavior of this burst comes from the pulse height spectrum observed by the phototubes. Within the burst the fraction of events with $\sum \text{ADC} \geq 7500$ channels is 20.8% whereas normal events have only 4.9%. Fig. 4 shows an integral pulse height spectrum displaying this difference. Looking only at the high energy showers ($\sum \text{ADC} \geq 10000$ channels), the hypothesis that the two spectra come from the same event distribution has $\chi^2 = 3.2$ for 1 dof.

Thus both rate and pulse height provide evidence that the peak is anomalous and consistent with gamma ray showers. In addition, there was no change in the off-source shower rate during the period of the burst. No evidence for anomalies in the raw trigger rates or in the tube singles rates has been observed.

To put this observation into a wider perspective, we quote the time of the center of the peak (12 Oct. 1985; 6 hours, 4 minutes, 24 seconds UT) and the phase relative to the 4.79 hour period of Cyg X-3 (ϕ_{BB}=0.74) using the ephemeris of Bonnet-Bidaud(2). The burst is coincident with the large radio flare observed in early Oct. 1986 and with the large flux increase of underground muons observed at the Soudan mine(3). In particular, the muon flux occurs at a phase $\phi_{BB} = 0.74$, in excellent agreement with the phase of our burst. Using the half-angle acceptance ($\sim \frac{1}{4}^\circ$) and a typical distance of 7 km to shower maximum, we compute the flux of gammas within the peak to be $2.1 \times 10^{-9} \text{cm}^{-2}\text{sec}^{-1}$. It is, of course, tantalizing to note the correspondence of phase between the two experiments, but it is difficult to reconcile the fluxes.

In a recent paper Chadwick et al.(4) have reported a pulsar period of 12.5908 ms for Cyg X-3. This period was observed by searching data which exhibits an excess rate near a phase of 0.625. Although our burst occurs at a different phase, it would be exciting if we also observed a similar pulsar period. Hence we have carefully searched the 100 seconds which encompasses the burst.

Studies of the Crab pulsar indicate that the cuts necessary to search for rate excesses (7 hits in 6 nsec.) are unduly restrictive in pulsar searches because the majority of rejected events are 5 or 6 hit showers with a few random hits to cause a 7-fold trigger. Of course, these 5 or 6 hit showers can also be gamma showers either with lower energy or larger impact parameter. Each gamma shower is important in establishing the significance of a signal at a particular phase whereas the background adds randomly in the pulsar phase. As a result we have relaxed the shower threshold in the pulsar searches. We required only 5 hits in 10 ns, a less restrictive cut than the hardware trigger, which requires 7 hits in 10ns.

We have searched for pulsar periods from 10 ms to 2.2 sec in both on-source and off-source apertures. Once again, the off-source data provides an illuminating measure of the statistical probability of any single candidate. In particular, from 10 - 62 ms there are 5 peaks with $nR^2 \geq 9.0$ (probability $\leq 1.2 \times 10^{-4}$). Three come from off-source data and two from on-source. Fig. 5 shows the log of the Rayleigh probability for the scan from 10-15 ms which was made at intervals of 0.4 μs. Both on-source and off-source curves look very similar and exactly what one would expect from random background (mean = 1.0, RMS deviation = 1.0). There is no evidence for any pulsar periodicity within the burst, contrary to tranparencies presented at Durham.

Searches have also been made throughout the 113 hours of Cyg X-3 data taken during the summer and fall of 1985. Following the prescription of Chadwick et al. data taken at phase 0.625±0.058 (1000 sec on either side of 0.625) were searched for pulsar periods near 12.59 ms. No evidence for pulsar activity was observed(5).

Preliminary evidence for a signal occurring shortly after the burst (phase = 0.82) was presented at the conference showing a Rayleigh probability $nR^2 = 10.5$ (probability = 2.8×10^{-5}) with p= 12.5903 ms and $\dot{p} - -8 \times 10^{-10}$ s/s. Without the $\dot{p}$ term this peak does not appear and has not been observed in other analyses. Furthermore, no excess rate is observed during this time period of ~ 720 sec. An analysis of the complete data run (3.3 hours) looking at a Rayleigh analysis every 720 seconds (the period of switching on and off source apertures) shows off-source peaks with $nR^2 \sim 10$. It is, of course, possible that the observed on-source Rayleigh peak represents a gamma signal although we cannot at present provide any additional supporting evidence.

In conclusion, we have observed a 60 second burst from Cyg X-3 during the peak of the radio emission of early Oct 1985. No conclusive evidence for any pulsar activity has yet been demonstrated.

REFERENCES

1. R. Morse, *The Haleakala Gamma Observatory*, presented at "High Energy Gamma Ray Astronomy and Related Topics", Durham, England, Aug. '86.

2. J. M. Bonnet-Bidaud and M. van der Klis, Astr. Astrophys. Lett., **97**, L5-L7 (1981).

3. D. S. Ayres, *"New Evidence from Soudan I for Underground Muons associated with Cygnus X-3"*, presented at 2nd Conference on the Interactions Between Particle and Nuclear Physics, Lake Louise, Canada, May, 1986.

4. P. M. Chadwick *et al.*, Nature, **318**, 642 (1985).

5. A. Szentgyorgyi, thesis, University of Wisconsin.

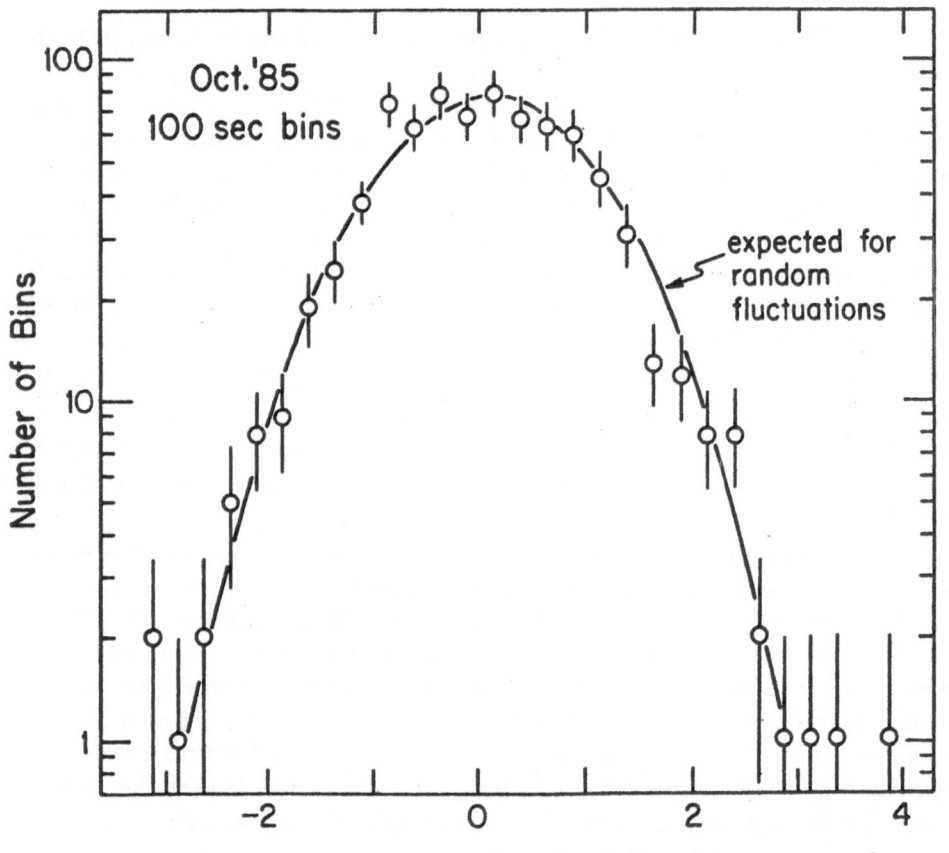

Fig. 1

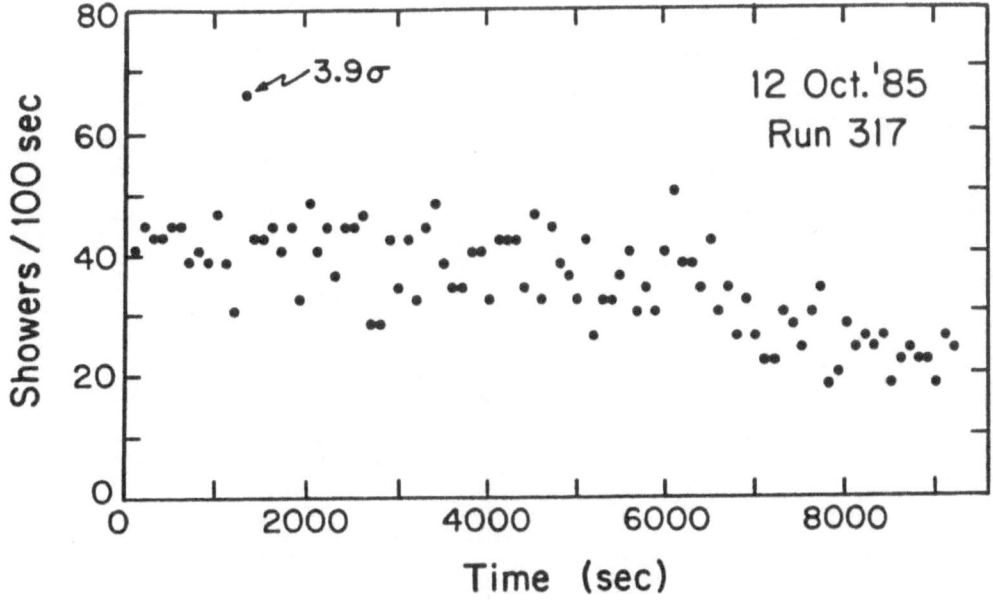

Fig. 2

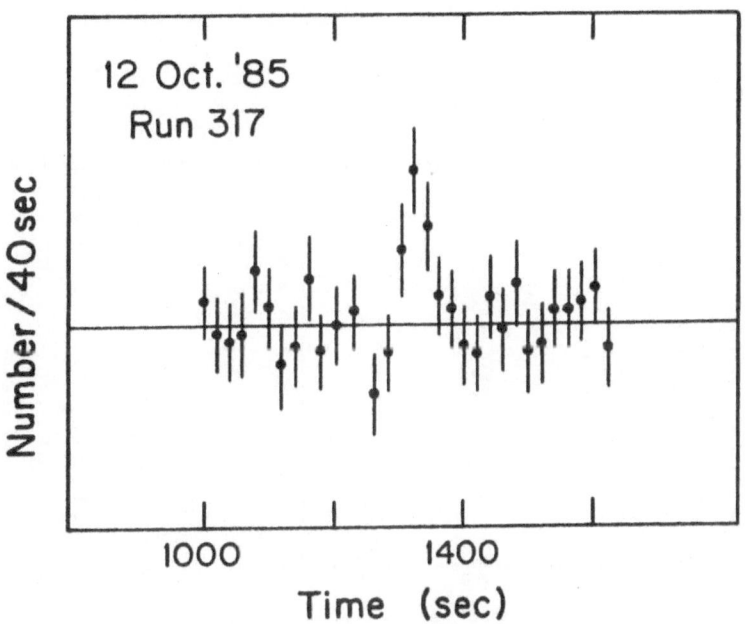

Fig. 3

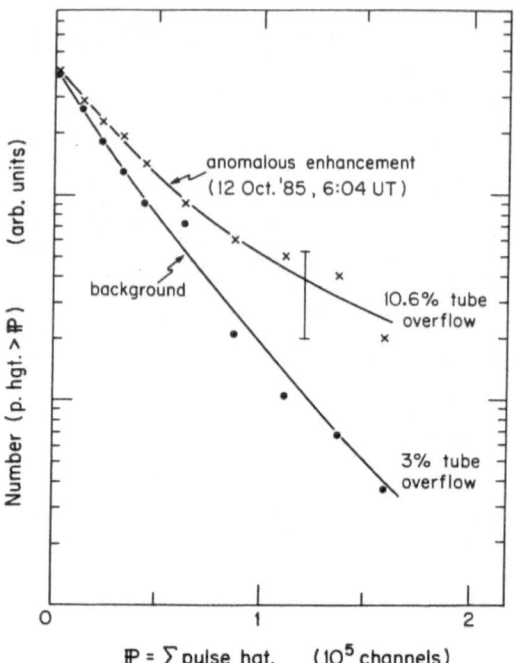

Fig. 4

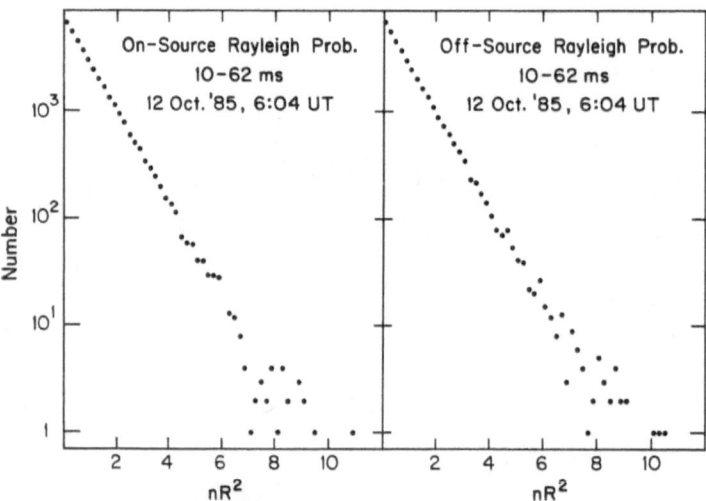

Fig. 5

SEARCH FOR A 12.59 ms. PULSAR IN CYGNUS X-3 AT E > 400 GeV

D.J.Fegan[1], M.F.Cawley[2], K.Gibbs[3], P.W.Gorham[4], R.C.Lamb[2], N.A.Porter[1], P.T.Reynolds[1], V.J.Stenger[4] and T.C.Weekes[3].

1. University College Dublin.
2. Iowa State University, Ames, Iowa.
3. Harvard-Smithsonian Center for Astrophysics.
4. University of Hawaii.

ABSTRACT.
  A high energy (E > 400 GeV) gamma ray database on Cygnus X-3 covering the observational interval 1983-1985 has been examined for possible evidence of a 12.5908 ms. pulsar, as recently reported by Chadwick et al. (1). The results of our period search fail to substantiate the original observation. Reasons why the two results may not be incompatible are discussed.

1. The Cygnus X-3 database and the analysis procedure.

  The binary X-ray source Cygnus X-3 has been observed on many occasions to emit U.H.E. gamma rays (2). Recently Chadwick et al.(1) have reported the detection of a 12.5908 +/- 0.0003 ms. pulsar in data recorded on September 12th 1983 at phase .625 in the 4.8hr. cycle of the source, with a chance probability of less than one in a million.
  We report here on observations made on Cygnus X-3 using the Whipple Observatory 10m. imaging telescope(3). The database analysed consists mainly of 28 minute tracking scans taken between April 1983 and November 1985, where at least part of each scan encompasses the .55 to .70 orbital phase. This interval covers the range of previously successful Tev gamma ray observations (4) and also includes the .58 phase interval where Chadwick et al. have also claimed to have observed 12.5908 ms. emission in their 1982 data. A total of 44 scans corresponding to 1205 minutes of observation constitutes the data base.
  Relative times of arrival of each event were recorded to a precision of 1 microsecond, with an absolute accuracy of 0.5 milliseconds. Each event time was corrected to the solar system barycenter using the M.I.T. ephemeris. In order to reduce the effects of Doppler orbital smearing of any periodic signal each scan was split into six individual 8-minute segments with 50% overlap between adjacent segments. While this generates some redundancy it also minimizes the possibility of losing any effect between adjacent segments.
  Three independent analyses were performed. Analysis A involved

111

K. E. Turver (ed.), Very High Energy Gamma Ray Astronomy, 111–114.
© *1987 by D. Reidel Publishing Company.*

testing 31 independent trial periods over the period range 12.586
to 12.596 ms. with a period derivative of zero. Analysis B was
similar but with a period derivative of $-0.75 \times 10^{-9} \text{s.s}^{-1}$. Analysis
C was a control periodicity search in which the database was
subjected to 31 independent trial periods over the range 14.5459 to
14.5593 ms., with the same period derivative as used in B. Clearly,
no significant periodic component should be present here. For each
analysis, we have calculated the Rayleigh power associated with each
8-minute data segment, for each trial period. We have taken the peak
power values and combined them into 3 independent ensembles, one per
analysis.

2.Results.

In Fig.1(a,b,c) we show individual differential power spectral
plots of the form $-dn/dz = k \, \text{Exp}(z(w))$ where $z(w)$ represents the
Rayleigh power at some sample frequency w. All three distributions
show the exponential behavior characteristic of a noise source.
In analysis A there are four occasions where the power exceeds the
arbitrary level of 10.0 while in both the B and C analyses the
figure is three. The database consists of 44x6=264 8-minute segments
each of which corresponds to 31 independent trial periods over the
search ranges in question. Introducing a factor of approximately 3 to
allow for oversampling (de Jager private communication), the
probability of finding four power values greater than 10.0 is .02
while the probability of finding three is .07.
   The details of the scans which contribute to the tail of the
distributions of Fig.1(a,b) are as follows.

Method of Analysis	Date D M Yr	4.8hr Phase	Power	Prob.	Period (ms.)
A	01/12/83	.68	11.62	.0008	12.5955
B			10.87	.0017	12.5956
A	25/05/85	.61	11.12	.0013	12.5938
B			11.14	.0013	12.5940
A	12/10/85	.75	10.41	.0028	12.5876
B			9.13	.0098	12.5877
A	15/11/85	.65	10.23	.0033	12.5919
B			10.09	.0037	12.5921

The probability values quoted above are based upon 31x3 trial
periods. Since 264 segments were analysed these values are not
significant. We would conclude that there is no evidence in our
database for any statistically significant periodic emission from
Cygnus X-3 for periods between 12.586 and 12.596 ms. for scans which
correspond to the 4.8hr. phase range 0.5 to 0.7. These conclusions
are substantiated by our C analysis which produces a similar ensemble
to that obtained in both the A and B analyses.

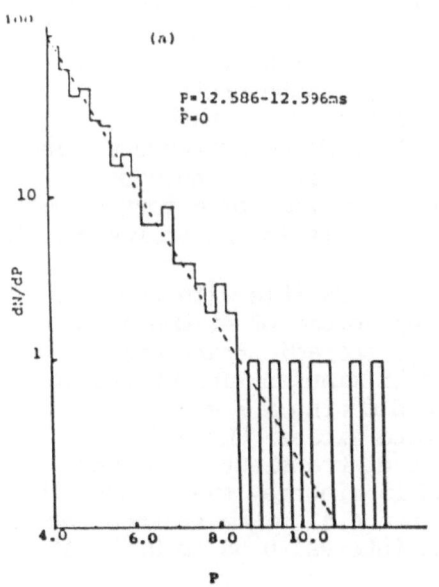

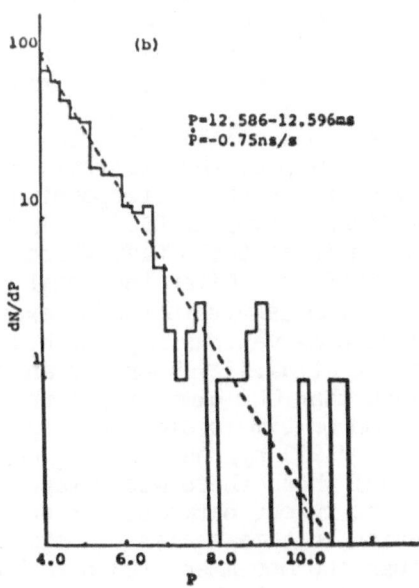

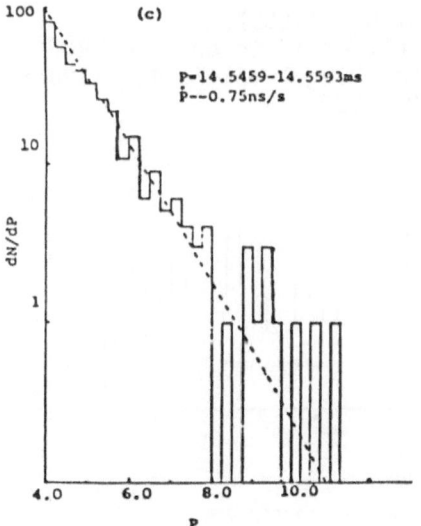

a) Period derivative = 0.0
b) Period derivative = -0.75 ns. / s.
c) Control analysis.

Fig.1 Power spectrum ensembles.

Of the scans listed above, that for 12/10/1985 is of interest since the Haleakala group have contemporaneous observations of Cygnus X-3 and report (this workshop) a possible one minute transient at 06:04 U.T. with a periodic component close to the period reported by Chadwick et al.(1). Our scan is from 05:51 to 06:18 U.T. We have analysed this particular scan on a minute by minute basis and in Fig.2(a) we plot the statistic -Log(Power) for each minute. Fig.2(b) shows the corresponding count rates. We do not observe any significant power at 06:04 U.T. The most significant minute of emission is at 06:13 U.T. where -Log(Power) = 4.12 for a test period of 12.5869 ms.   Given the total number of trials in the analysis such a deviation is within random expectation.

These observations may not however be incompatible with those of Chadwick et al.(1). There is an increasing volume of evidence to suggest that UHE gamma ray sources undergo sporadic transient emission on time scales of between 1 and 15 minutes. The obseration of 12/09/1983 by the Durham group occurred during a 7 minute interval where there was a large count rate excess. There is no evidence in our database for any significant transient gamma ray emission on time scales of between 1 and 15 minutes, even though the October and November 1983 data have shown a 4.41 sigma enhancement in the gamma ray light curve over the phase interval 0.58 to 0.67 (4).

This work is supported by U.S.D.O.E. and the N.B.S.T. Ireland.

REFERENCES.

1. Chadwick,P.M. et al., Nature, 318, 642-644 (1985).
2. Weekes,T.C. New Particles 1985, 288-301. World Scientific(1986).
3. Cawley,M.F. et al.,Proc.19th. ICCR (San Diego) V3,453-456 (1985).
4. Cawley,M.F. et al.,Ap.J. 296, 185-189 (1985).

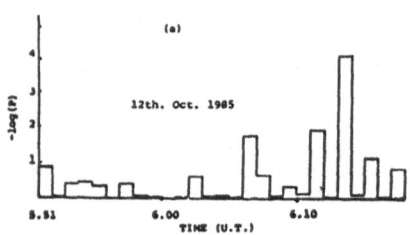 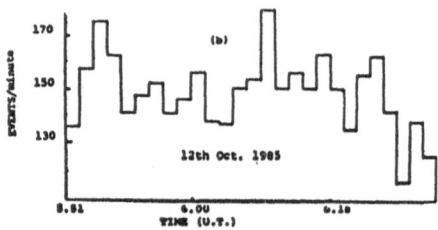

Fig.2a) Event rates per minute for scan of 12/10/1985.

Fig.2b) -Log(Power) values per minute for scan of 12/10/1985.

FURTHER EVIDENCE FOR THE EMISSION OF 1000 GeV GAMMA RAYS WITH 12 ms PERIODICITY FROM CYGNUS X-3

P. M. Chadwick, N. A. Dipper, J. C. Dowthwaite,
T. J. L. McComb, K. J. Orford, and K. E. Turver

Department of Physics
University of Durham
Durham DH1 3LE
United Kingdom

ABSTRACT. The X-ray binary Cygnus X-3 has long been suspected of containing a young fast pulsar and the first evidence for this, suggesting a period of 12.59 ms, was reported by us in 1985. We here report further evidence for the reality of the 12.59 ms periodicity and indications of value for the secular derivative for the period.

In an earlier paper (Chadwick et al 1985) we have reported evidence from observations in 1983 for the emission of 1000 GeV gamma rays with a periodicity of 12.5908 ms. On 1983 September 12 0519 hrs UT an excess of counts close to the time of X-ray maximum occurred - see Figure 1. In an interval of 420 secs an excess of counts was recorded, being about 20% of the cosmic ray background. A subsequent period search of this potentially rich dataset yielded evidence for periodicity at 12.5908 ms - see Figure 2.

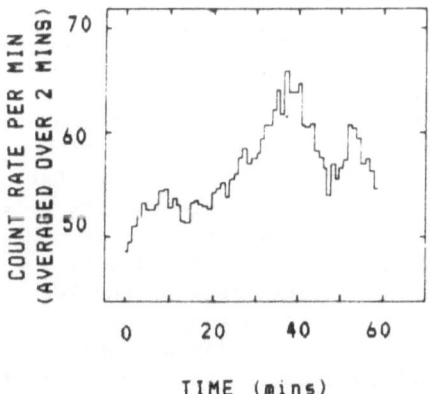

TIME (mins)

Figure 1: The count rate during the 1983 Sept 12 observation.

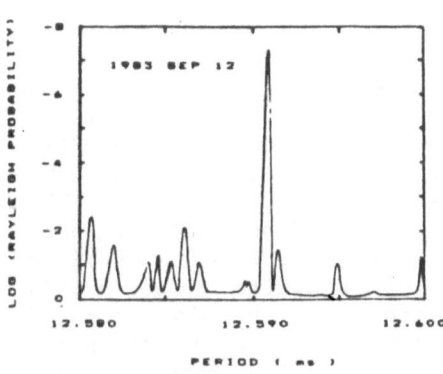

Figure 2: The Rayleigh probability of uniformity in phase for the events in the burst as a function of trial period.

115

K. E. Turver (ed.), Very High Energy Gamma Ray Astronomy, 115–119.
© *1987 by D. Reidel Publishing Company.*

On 1983 October 2 during a 420 sec interval precisely 100 4.8 hr cycles
later (but under non-ideal viewing conditions) a similar occurrence of
periodicity was noted - see Figure 3.

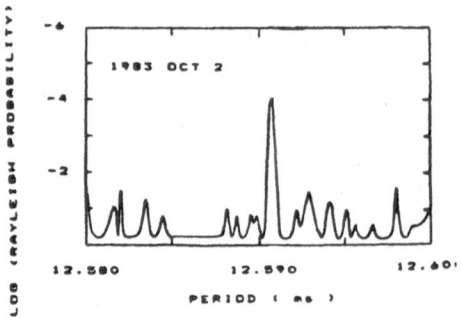

Figure 3: Rayleigh probability of uniformity in phase as a function of
trial period for data for 1983 Oct 2 selected precisely 100 4.8 hr
cycles after the burst shown in Figure 1.

We here report analyses of more data taken in Utah prior to 1983
when the sensitivity of the telescopes was less and so the significance
of the periodicities, if any, would also be less. In addition a limited
series of observations were taken in Durham in 1985 October and November
using a single, more sensitive telescope incorporating new high
reflectivity mirrors of longer focal length. The first requirement in
searching data taken a considerable time before or after 1983 is to
establish as accurately as possible the short intervals in the 4.8 hr
cycle corresponding to the 1983 7 min intervals of activity. A
limitation here is the accuracy of and indeed the choice of 4.8 hr
ephemeris, since recently the preferred emphemeris has been questioned.
We here adopt the convention that the ephemeris published by Van der
Klis and Bonnet-Bidaud (1981) (referred to as VdK) is the "reference
ephemeris". According to EXOSAT data taken in 1985 July the time of
X-ray maximum as defined by VdK was 0.035 of a 4.8 hr cycle later than
actually observed (Molnar 1985). On this basis we have selected the data
from the 1985 observations (potentially the most sensitive) for an
interval of 1000 sec (thus allowing for uncertainties in the 4.8 hr
cycle) and centred on a time 600 sec before that predicted by the VdK
ephemeris. On the assumption that the pulsar in Cygnus X-3 is spinning
down at a rate typical of that of other young pulsars (say as rapidly as
10^{-13} s/s) we have tested the 40 independent periods in the range 12.5908
- 12.5950 ms using the Rayleigh test for uniformity during observations
on 6 nights. We show in Figure 4 the periodogram for data for all 6
nights where the chance probabilities have been combined. The
periodicity at 12.5928 ms arises mainly from the observation on the
night 1985 October 12 - see Figure 4. Allowing for the number of periods
searched, the probability of the effect arising by chance is about 0.01.

        If it is true that the ms periodicity occurs at a time in 1985
which is before that according to the VdK reference ephemeris and that
the pulsar period is longer than in 1983, then we may expect that prior
to 1983 the activity will occur at times later than the VdK prediction
and with a shorter pulsar period. If we assume that the gamma ray
emission occurs at fixed X-ray phase and making no allowance for second
derivitives in the 4.8 hr cycle or the pulsar period (viz. allowing for
linear variation with time), we predict that in 1982 the activity should
be detectable in a 1000 sec interval centred at a time specified by VdK
plus 200sec and the pulsar period should be in the range 12.5885 to
12.5905 ms. Data from observations on 29 nights in 1982 July - November
have been analysed and the periodogram showing the combined chance
probability is shown in Figure 5.

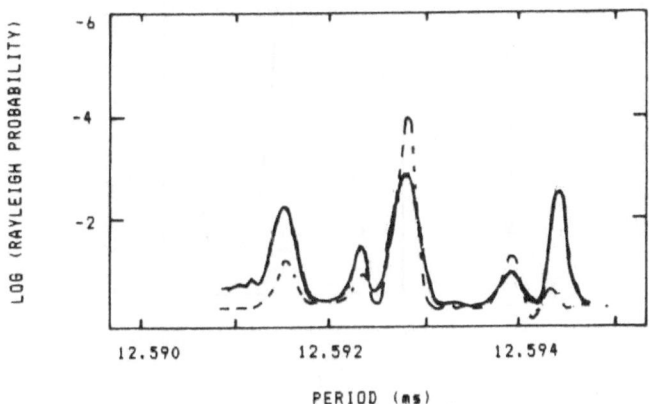

Figure 4: Rayleigh probability for uniformity in phase as a function
of trial period for data in 1985 (solid line all 1985 data, broken
line 1985 Oct 12 only).

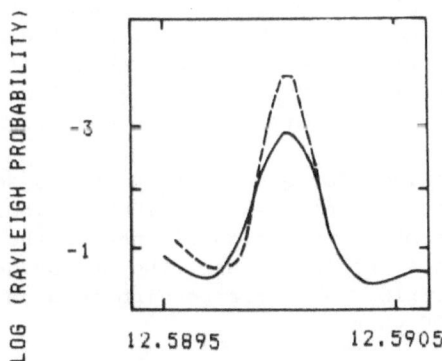

Figure 5: Rayleigh probabilty for uniformity in phase as a function of
trial period for data in 1982 (solid line all 1982 data, broken line
1982 July only).

Since the assumption of the constancy of the period over an elapsed
interval of 5 months may not be valid, we have also analysed the data
month-by-month. This shows that there is a strong effect during the 5
observations in 1982 July - see Figure 5. Similarly for data recorded on
3 nights in 1981 October, we have assumed that the activity will be
about 1000 sec later than Vdk would predict and at a period in the range
12.5875 to 12.5895 ms. The combined probability for non-uniformity for
these data is shown in Figure 6.

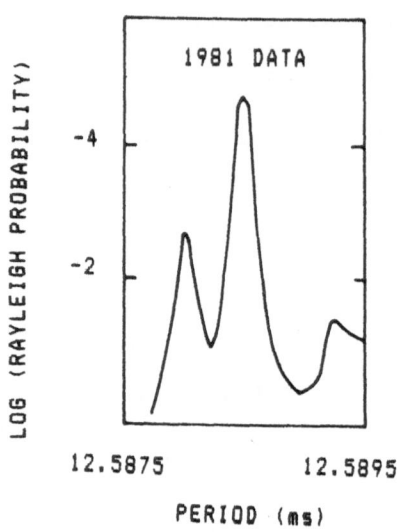

Figure 6: Rayleigh probability for uniformity in phase as a function
of trial period for data in 1981.

We note that the differences between the times at which VHE gamma
ray activity has been noted and the times predicted by the VdK reference
ephemeris for X-ray maximum are similar to those differences reported
recently by Mason (private communication, 1986) for the time of X-ray
maximum as a result of a new interpretation of a consistent X-ray
dataset.
    The candidate periods for each year suggest a secular variation of
the pulsar period over the interval 1981-1985 as indicated in Figure 7.
We also show the estimate of the significance of the detection of
periodicity in the data from each year. The period derivative derived
from these data is 2.8×10^{-14} s/s, corresponding to a characteristic age
for the pulsar of 7000 years.

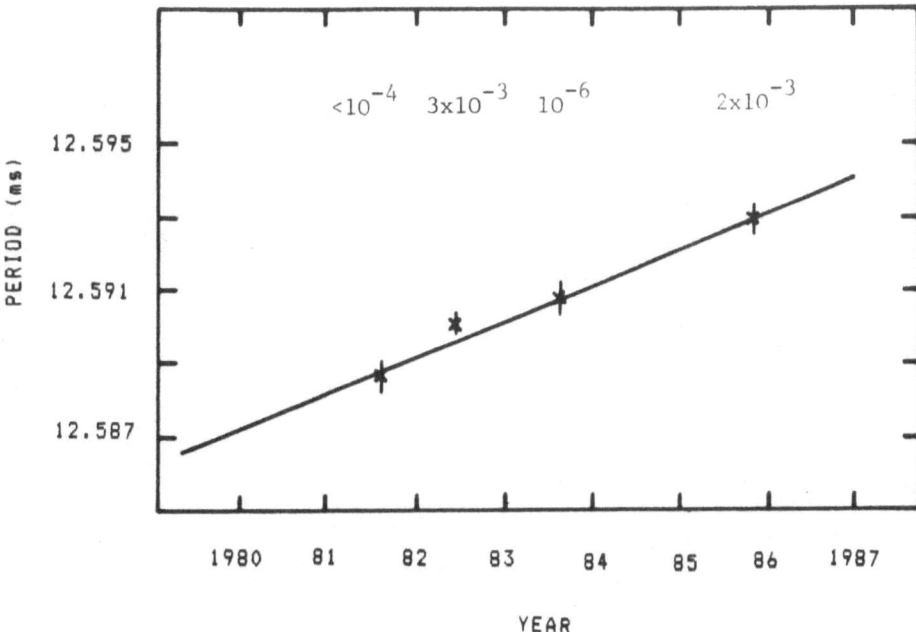

Figure 7: The secular variation of the pulsar period.

REFERENCES

(1) Chadwick,P.M. et al., Nature, (1985), 318, 642.
(2) Molnar, L.,(1985), Ph D Thesis, Harvard University.
(3) Van der Klis, M and Bonnet-Bidaud,J.M., (1981), Astro.Astrophys.,95, L5.

SIMULTANEOUS MEASUREMENTS OF VHE GAMMA RAYS FROM HERCULES X-1

P. M. Chadwick, N. A. Dipper, I. W. Kirkman,
T. J. L. McComb, K. J. Orford and K. E. Turver.

Department of Physics
University of Durham
Durham DH1 3LE
United Kingdom

ABSTRACT. We report the results of a measurement of VHE gamma rays
from Her X-1 made using the Dugway telescopes which coincided with a
similar measurement at the Whipple Observatory reported by Gorham
(1986). In both observations evidence for emission of VHE gamma rays
showing the same periodicity and significant at about the 3 sd level
was obtained. This represents the first simultaneous detection of a
VHE gamma ray source by two independent telescopes.

Hercules X-1 was first observed as a source of 1000 GeV gamma rays by
us in 1983 (Dowthwaite et al, 1984). A burst of additional Cerenkov
light events lasting for 3 min was noted and subsequently shown to
have the 1.24 s timing signature typical of the X-ray pulsar. More
recently pulsed VHE gamma ray emission from Her X-1 has been detected
on numerous occasions by the Whipple collaboration (reported by
Gorham, 1986) and by the Haleakala collaboration (private
communication). Hercules X-1 must therefore rate as probably the best
measured VHE gamma ray emitter at present. However, to date no
instance has been recorded when two or more independent and separated
telescopes have observed one of these bouts of pulsed emission from
Hercules X-1 (or indeed any source) at precisely the same time. Such a
simultaneous detection is considered to be an important part of the
establishment of the field of VHE gamma ray astronomy.

We here report the results of a measurement of Her X-1 made from
1984 April 4 0930 - 1030 hrs UT at the University of Durham TeV gamma
ray system at Dugway, Utah. At the same time a successful 28 min
observation was made with the Whipple telescope 800 miles to the South
in Amado, Arizona. The threshold of the Dugway telescopes was about 1
TeV and that of the Whipple telescope was about 250 GeV. In both cases
Her X-1 was being tracked continuously during the observation and the
observing conditions in both localities were good. For these
measurements Her X-1 was in the HIGH ON state in its 35 d cycle (phase
0.2) and was at phase 0.4 in the 1.7 d orbital cycle.

K. E. Turver (ed.), Very High Energy Gamma Ray Astronomy, 121–123.
© 1987 by D. Reidel Publishing Company.

The result from the Whipple telescope, reproduced from Gorham (1986), is shown in Figure 1 and the data from Dugway for the same epoch are also shown.

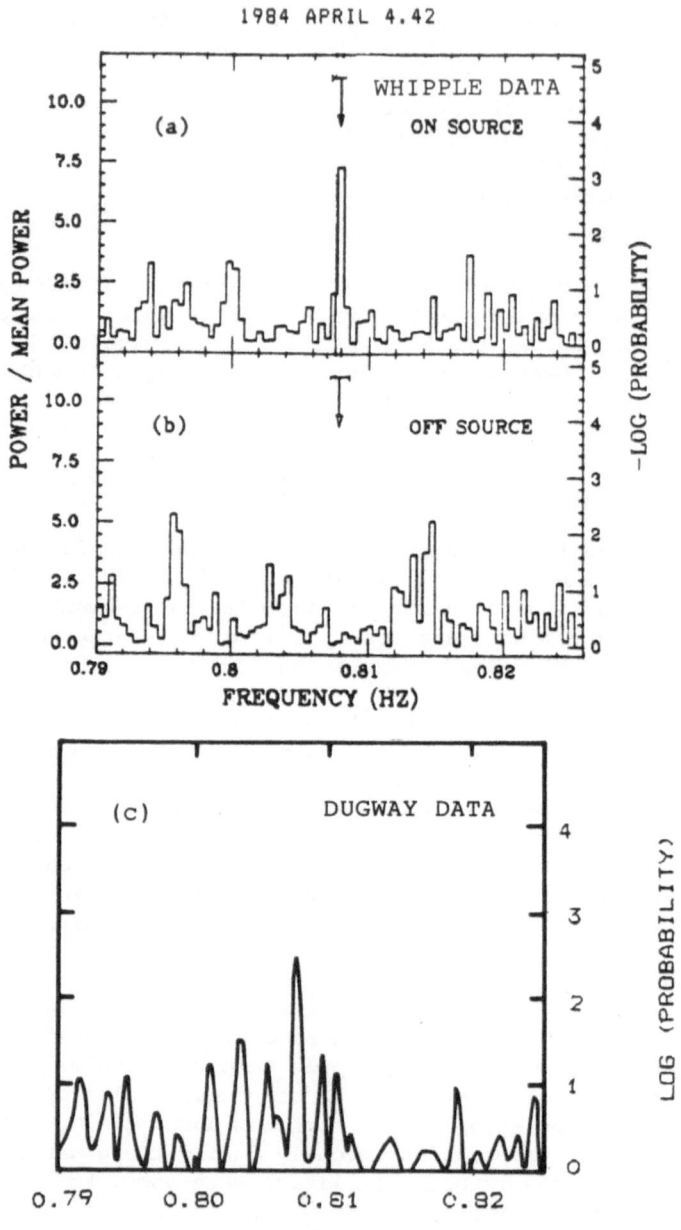

1984 APRIL 4.42

Both the Whipple and Dugway measurements indicate a periodicity of 1.2376 ± 0.0004 s, with chance probabilities of 0.0008 for the Whipple measurement and 0.003 for the Dugway measurement. The two measurements indicate similar values for an approximate flux of 3×10^{-10} $cm^{-2} s^{-1}$ despite the differences in energy threshold. At this stage in the understanding of the thresholds of the two detectors this does not, in our view, represent a serious problem.

REFERENCES

(1) Dowthwaite J.C., Harrison, A.B., Kirkman,I.W., Macrae, H.J., Orford, K.J., Turver, K.E. and Walmsley,M., (1984), Nature, 309, 691.

(2) Gorham, P.W., (1986), Ph D Thesis, University of Hawaii.

TeV Observations of Her X-1 at the Whipple Observatory: 1984-1985

P.W.Gorham[1], M.F.Cawley[2], D.J.Fegan[3], K.G.Gibbs[4], R.C.Lamb[2],
N.A.Porter[3], V.J.Stenger[1], T.C.Weekes[4].

1. University of Hawaii, Honolulu, Hawaii, U.S.A.
2. Iowa State University, Ames, Iowa, U.S.A.
3. University College, Dublin, Ireland
4. Harvard-Smithsonian Center for Astrophysics, Amado,
   Arizona, U.S.A.

ABSTRACT. Her X-1 has been observed for 73 hours with the Whipple
Observatory Cerenkov imaging camera on the 10m reflector between April
1984 and June 1985. Seven distinct episodes of emission were seen,
ranging in duration from 25 to 80 minutes, totalling 7% of all
observations. Evidence is presented which suggests that the gamma-ray
source is not always coincident with the X-ray source; in particular,
the strongest effect was observed after the pulsar entered eclipse.
All effects were observed to occur during the active states of the 35
day X-ray cycle.

1. INTRODUCTION.

Her X-1 was first observed as a TeV gamma-ray source by the Durham
group (Dowthwaite et al. 1984). A 3 minute outburst was seen in April
1983, close to the expected turn-on of the long-term 35 day X-ray
cycle of the Her X-1 system, and at a phase of 0.76 in the 1.7 day
orbital cycle. The $33\pm10\%$ excess was found to be approximately 100%
modulated at the 1.237s pulsar period, with a light curve displaying a
broad single peak, somewhat similar to that seen in X-rays. The chance
probability for this effect was 7×10^{-5}. Emission above 500TeV from Her
X-1 was claimed by the Fly's Eye group for a 40 minute observation
taken in July 1983 (Baltrusaitis et al. 1985). Again, the signal was
modulated at the pulsar period, with a chance probability of
2×10^{-4}. The 35 day phase was 0.63 in this case, corresponding to a
position when an intermediate state of activity is usually seen in
X-rays. The orbital phase was 0.66. A relatively narrow light curve
was observed (10% duty cycle). The observation was not confirmed at
TeV energies by the Durham group who were simultaneously observing the
source (Chadwick et al. 1985), suggesting a very flat source spectrum.
   X-ray observations taken around this time indicated that the
source entered an extended low state shortly before the Fly's Eye
observation. However, reprocessed optical and infrared pulses were
still detectable (Middleditch et al. 1985). Normal X-ray behaviour was
seen again in March 1984 (Ogelman et al. 1985).
   The Whipple Observatory TeV observations of Her X-1 were
initiated shortly after the cessation of the extended X-ray low state
in the spring of 1984 (Gorham et al. 1986a). We report here on
observations taken up to the end of June 1985. We are continuing to
monitor the source in 1986. Seven separate episodes of emission were

K. E. Turver (ed.), Very High Energy Gamma Ray Astronomy, 125–130.
© *1987 by D. Reidel Publishing Company.*

observed between April 1984 and June 1985, lasting a total of 5
hours. The source was monitored for a total of 73 hours, implying a 7%
duty factor for detection of pulsed emission. All the 35 day phases of
the detections lie either in the high-on state or low-on (intermediate
activity) state; we do not find any preferred orbital phase for TeV
emission, with observed phases ranging from 0.2 to > 0.9 of the 1.7
day cycle. The strongest emission was observed to occur after the
source has entered X-ray eclipse, indicating that the source of
gamma-ray emission is not coincident with the X-ray source (Gorham et
al. 1986b).

2. OBSERVATIONS.

Her X-1 was observed with the Whipple Observatory 10m gamma-ray
telescope during three successive moonless periods from March-May 1984
and again for 4 darkruns from March-June 1985, for a total of
approximately 73 hours observing time over 43 nights. The times of the
Cerenkov flashes were digitized to 1 microsecond precision, with a
WWVB clock providing an absolute reference to an accuracy of about
0.5mSec. The times were corrected to the solar system barycenter and
to the center of mass of the Her X-1/Hz Her system, using the
ephemerides of Deeter et al. (1981), Joss et al. (1980), and Ogelman
et al. (1985). The data base was first tested in half-hour segments
for the presence of periodic emission near the pulsar period, using
both the Rayleigh test and Fourier transform techniques. A frequency
band of width 1.34mHz centred around the expected X-ray frequency was
chosen as a search window in the power spectrum; this is slightly
larger than ± 1 independent Fourier frequency about the central
frequency for a half-hour segment of data, but corresponds to
± 200km/sec in terms of the Doppler shifts due to radial velocity in
the source. A similar window was used by Groth (1974) and Middleditch
and Nelson (1976) in their detections of optical pulsations from the
Her X-1/Hz Her system. In addition to searching for periodicity near
the fundamental of the pulsar period, we searched near the second
harmonic of this period (ie. near 0.62sec, or 1.616Hz), since Her X-1
has frequently been observed in X-rays to exhibit pulsations with a
strong double-pulsed content, particularly during the intermediate
state of activity in the 35 day cycle (see, for example, Ogelman et
al. 1985). Symmetric double pulses may in fact show little or no power
at the fundamental; this procedure was necessary to the detection of
one of the outbursts reported here.

In the first search through the data, a power threshold of 6 was
used to reduce the accidental rate to near 1% within a given half-hour
segement, allowing for 6 independent test frequencies (3 at the
fundamental, and 3 at the second harmonic). This threshold was
exceeded in 8 instances, where 2.2 were expected for a total of 876
trials. Using the binomial distribution, this gives a probability of >
99.85% that we have observed pulsations attributable to Her X-1. A
similar analysis was applied to a smaller amount of background data
which was found to behave in accordance with expectation. These 8

instances of suspected emission were then examined in further detail, and combined with neighbouring segments to test for the presence of extended emission at lower power. This resulted in a total of 7 distinct episodes of emission, ranging in duration from 25 to 80 minutes, as summarised in table I. The chance probabilities of these episodes range from 10^{-3} to 7×10^{-6}.

In fig. 1 we present a scatter plot of the 1.7 and 35 day phases for all observations. The detections are indicated by diamonds whose size is scaled according to the relative power in the power spectrum. The crosses indicate observations with no detected pulsations. The positions of the Durham effect (solid circle) and Fly's Eye effect (open circle) are also indicated. Although the total number of detections is small (9) in this plot, it is apparent that the 35 day X-ray ON states are almost exclusively preferred for detections of gamma-ray pulsations. There is no clear correlation with the 1.7 day orbital cycle, although two strong effects which occur at the 35 day turn-on also occur in the orbital phase range 0.6 - 0.8; this may be significant, as the X-ray turn-on of the 35 day cycle occurs almost exclusively at orbital phases 0.23 or 0.68 (Middleditch and Nelson 1976). It is interesting to note in this context that gamma-ray emission from Cyg X-3 is also seen predominantly at orbital phases 0.2 and 0.6 (Weekes 1985). The orbital phase of the 1984 May 5 effect may also be significant in that the phase region 0.55 - 0.6 is known for the appearance of "anomalous dips" in the X-ray intensity, occurring usually one or two orbits after the X-ray turn-on in the 35 day cycle (Crosa and Boynton 1980). Our effect at this phase most likely occurred during the second orbit after the 35 day turn-on. Finally, the strongest effect observed occurred at an orbital phase corresponding to transit into source eclipse. This effect is reported in detail elsewhere (Gorham et al. 1986b) and will be but briefly discussed here. The observation of 1985 June 16 lasted for ~2 hours, with source eclipse occurring ~42 minutes after the start of the observation. In fig. 2 the power spectra around the expected pulsar frequency is shown, for both the data taken before and during eclipse. No significant pulsations were detected before eclipse, but a signal near the pulsar frequency was seen with a chance probability of 7×10^{-6} in the data taken during eclipse (a probability of 4×10^{-5} after all trials have been considered). The signal was found to be present at roughly equal strength throughout the data taken during eclipse.

3. DISCUSSION.

We have evidence from the observation of strong pulsations after eclipse that the source of X-rays and gamma rays are not always coincident. In the case of the 1985 June 16 effect, the pulsar should have moved 20% of its way across the diameter of the intervening companion star by the end of the observation. For reasonable source models, this would present prohibitively large column densities to both TeV gamma rays and higher energy protons. Models invoking the 'beam dump' process for generation of gamma rays in the limb of the

Table I. Summary of Observed Effects from Her X-1.

(1)	(2)	(3)	(4)	(5)	(6)	(7)	(8)	(9)	(10)
1984 April 4.42	1670	0.41-0.42	0.2 High On	0.80789037(10)	0.80795(9) +0.06mHz	10^{-3}	0.25	3.1±1.5 2.3×10^{35}	broad + sharp features.
1984 May 5.26	3360 (c)	0.55-0.61	0.08 High On	0.80789070(27)	0.80824(4) +0.35mHz	2×10^{-5}	0.25	1.8+1.1 1.3×10^{35}	sharp feature at phase 0.6-0.65
1984 May 23.33	1500	0.18-0.19	0.6 Low On	0.80789090(35)	0.80827(9) +0.38mHz	3×10^{-4}	0.25	3.7+2.0 2.7×10^{35}	double peak, power entirely in second harmonic
1985 April 23.35	1680	0.23-0.25	0.19 High On	0.807894(2)	0.80737(8) -0.52mHz	3.7×10^{-4}	0.8	0.8+0.7 1.8×10^{35}	broad + sharp features.
1985 April 24.34	1680	0.81-0.83	0.22 High On	0.807894(2)	0.80771(9) -0.18mHz	5.5×10^{-4}	0.9	0.4+0.7 1.0×10^{35}	double peak, power in both 1st & 2nd har.
1985 May 21.22	3600 (d)	0.63-0.66	0.99 turn-on of High On	0.807895(2)	0.80783(2) -0.06mHz	9.2×10^{-6}	1.3	1.1+0.7 4.1×10^{35}	broad, with sharp feature near peak (0.5 phase).
1985 June 16.24	4680	0.91-0.96	0.73 turn-off of Low On	0.807896(3)	0.80858(3) +0.69mHz	6.8×10^{-6}	0.6	1.5+0.9 2.3×10^{35}	broad, with sharp feature near peak (0.6 phase).

KEY: (1) UT date;   (2) Duration (sec.)   (3) 1.7 day phase   (4) 35 day phase(a)   (5) Pulsar freq. at CM (Hz) (b)   (6) Observed freq. (Hz), and shift relative to pulsar freq.   (7) chance probability   (8) threshold (TeV)   (9) flux (10^{-10} cm^{-2} s^{-1}), and luminosity (erg s^{-1})   (10) shape of light curve.

NOTES: (a) Phase 0.0 corresponds to X-ray turn-on.
(b) Pulsar frequency reduced to rest frame, extrapolated from X-ray results of Ogelman et al. (1985), with standard error on last digits in parentheses.
(c) 1984 May 5: 30 minute gap in data due to comparison run on OFF-source region.
(d) 1985 May 21: Eight 30-minute observations were taken consecutively on Her X-1. Coherent emission was found when the second and fourth runs were combined; the intervening 30-minute run did not contribute to the effect and was omitted.

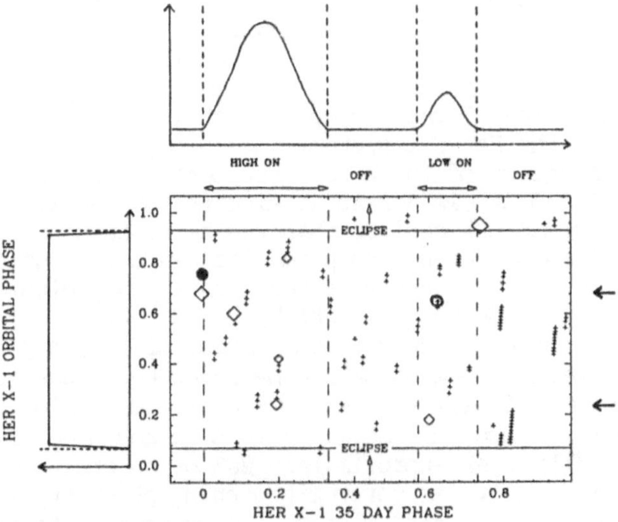

FIGURE 1: Scatter plot of orbital phase against 35 day phase.
Key: ◇ = Whipple Observatory detections (size scaled with power).
+ = Other Whipple Observatory observations (no detected pulsations).
● = Durham detection.
o = Fly's Eye detection.
← = Preferred orbital phases for turn-on of 35 day cycle.

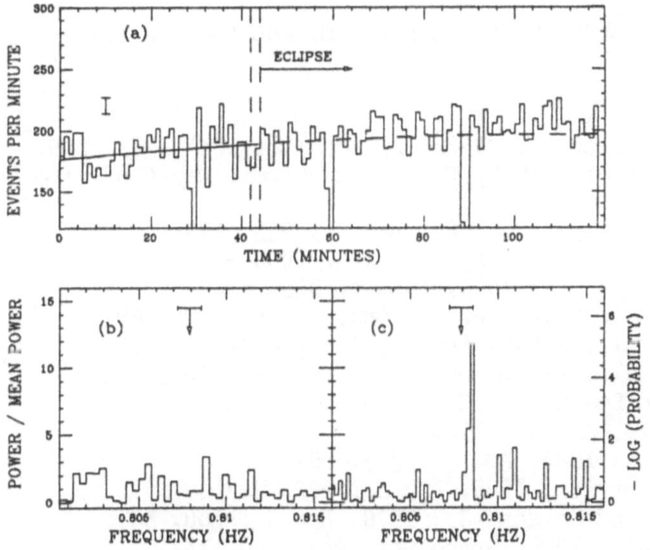

FIGURE 2: (a). Time series for observation of 1985 June 16. The gaps at ~30min intervals are artifacts of the data acquisition. Transition into eclipse is indicated by vertical dashed lines.
(b). Power spectrum for the data preceding eclipse, with the expected X-ray frequency of Her X-1 marked by the arrow, and the horizontal bar indicating a ± 200km s^{-1} band about this frequency.
(c). Similar power spectrum for the data during eclipse, showing strong evidence for pulsations within ~0.6mHz of the expected frequency.

companion star (Eichler and Vestrand 1985) become untenable under such circumstances unless some mechanism for bending or steering the proton beam can be invoked (Gorham and Learned 1986).

The second most significant observed effect (1985 May 21), occurring very close to the expected pulsar frequency, may provide evidence for the interplay of the accretion disk in the production of TeV gamma rays within the Her X-1 system. The 35 day X-ray turn-on, if it represents the emergence of the X-ray (and proton beam) source from behind the edge of the disk, is a natural time for gamma-ray production, which would occur earlier than the X-ray turn-on due to the greater penetration of the proton primaries. This is consistent with the behaviour of the 1985 May 21 effect, where the emission precedes the nominal turn-on phase (0.68) by 1 - 2 hours.

Similar constraints apply to the observation of 1984 May 5. The absorption dips in the X-ray emission are believed to be due to relatively dense clouds of matter released from the companion star arriving above the outer edge of the accretion disk and gradually relaxing in scale height over approximately 20% of an orbit (Cross and Boynton 1980). The nominal phase for the beginning of absorption dips is 0.55; the strongest pulsations in gamma rays were seen to occur ~3 hours later, consistent with gradually decreasing matter density in the line of sight.

Thus the three most statistically compelling effects all provide circumstantial evidence for the presence of a coherent multi-TeV particle accelerator in the Her X-1/Hz Her system.

This work is supported by the U.S.D.O.E. and the N.B.S.T. of Ireland.

References.

Baltrusaitas, R.M. et al.; 1985, Ap. J. (Letters), 293, L69.
Chadwick, P.M. et al.; 1985, Proc. Int. Cosmic Ray Conf. (La Jolla)., 1, 251.
Crosa, L. and Boynton, P.; 1980, Ap. J., 235, 999.
Deeter, J.E. et al.; 1983, Ap. J., 247, 1003.
Dowthwaite, J.C. et al.; 1984, Nature, 309, 691.
Eichler, D., and Vestrand, W.T.; 1985, Nature, 309, 691.
Gorham, P.W. et al.; 1986a, Ap. J., (in press).
Gorham, P.W. et al.; 1986b, Ap. J. (Letters), (in press).
Gorham, P.W. and Learned, J.G.; 1986, preprint.
Groth, E.J.; 1974, Ap. J., 192, 517.
Joss, P.C. et al.; 1980, Ap. J., 235, 592.
Middleditch, J. et al.; 1985, Ap. J., 292, 267.
Middleditch, J. and Nelson, J.; 1976, Ap. J., 208, 567.
Ogelman, H. et al.; 1985, Space Sci. Rev., 40, 347.
Weekes, T.C.; 1985, Proc. of Conf. on New Particles '85, Madison (in press).

VHE GAMMA RAYS FROM HER X-1 IN JUNE-JULY 1985

L. Resvanis
University of Athens, Physics Laboratory
Solonos Str. 104, 10680 Athens, Greece

J. Learned, V. Stenger, and D. Weeks
Department of Physics, University of Hawaii
Honolulu, HI 96822, USA

J. Gaidos, F. Loeffler, J. Olson, T. Palfrey, G. Sembroski,
and C. Wilson
Department of Physics, Purdue University
W. Lafayette, IN 47907

U. Camerini, J. Finley, W. Fry, M. Jaworski, J. Jennings,
A. Kenter, M. Lomperski, R. Loveless, R. March, J. Matthews,
R. Morse, D. Reeder, P. Slane, and A. Szentgyorgyi
Physics Department, University of Wisconsin
Madison, WI 53706, USA

ABSTRACT. We have observed Her X-1 for 29 hours during June and July,
1985, with the Haleakala 10 m^2 Cherenkov-light gamma-ray telescope,
operated at a threshold below 300 GeV. Special attention was given to
a search for bursts. We report evidence for possible bursts on days
165 and 168 at orbit phases of .87, .93, and .60 of the binary system.

1. INTRODUCTION.

The first operation of the Wisconsin-Hawaii-Athens-Purdue gamma-ray
telescope on Haleakala, Maui, Hawaii, as an essentially complete apparatus, began during the lunar dark period of June, 1985, when we observed Her X-1 for 26.5 hours (and for 2.5 more hours in July). About
half of the June observations coincided with the low-on part of the
Her X-1 35-day cycle. Previous observations at other laboratories have
been summarized by T. C. Weekes (1).
        The essential features of our telescope in 1985 were (1) the
mirror area was 10m^2; (2) simultaneous observations were made on the
source and 3.6° off-source in declination; (3) counting was at the one-
photoelectron level, with removal of chance coincidences in software,
and with the possibility of imposing energy cuts in software; and (4)
the aperture was 0.5°, unfiltered, with the threshold below 300 GeV. A
full description of the telescope is being prepared for publication.

K. E. Turver (ed.), Very High Energy Gamma Ray Astronomy, 131–134.
© 1987 by D. Reidel Publishing Company.

2.   ANALYSIS AND CONCLUSIONS.

The identifying features of a signal from Her X-1 are a counting-rate
excess and the 1.24s period of the pulsar. We carried out two classes
of searches through the data: first, a global search for periodicity,
and, second, a search requiring the presence of a counting-rate excess
before imposing the demand for periodicity.

        In the global search we examined both the on-source and the
off-source data in overlapped time intervals of 3, 6, and 12 minutes,
using the Rayleigh test at the fundamental and first overtone of the
pulsar period. The distribution of Rayleigh powers was as expected
from chance, with no on-source interval having a Rayleigh probability
10 times less probable than that expected from chance, given
the number of trials, and with no significant difference between on-
and off-source Rayleigh power distributions.

        However, in a preliminary analysis of the data, in which we
used only a rough criterion for counting-rate excess, we discovered a
400s interval that showed both a significant counting-rate excess and
an anomalous, statistically very improbable, narrow peak in its light
curve. This find led us to a systematic search in high-counting-rate
time intervals for similar narrow structures in the light curve.

        We first selected intervals longer than 200s that had a sig-
nificant (i.e., >2.5σ) counting-rate excess. We found 26 such intervals, 250s to 500s in length, of which 14 showed on-source excess and 12
showed off-source excess--the latter for comparison. The originally
interesting interval survived this selection.

        The Rayleigh test was applied to each of the 26 intervals at
photomultiplier coincidence multiplicity cuts of 8, 10, and 12 (corresponding to different gamma-ray threshold energies), using periods
1.23776s and 0.61888s. The distribution of probabilities was consistent with chance in this small sample, even for our "interesting"
interval, for which there is some first-harmonic cancellation in the
light curve.

        We used 20-bin light curves to search for the presence of
sharp spikes in the high-counting-rate intervals. We created light
curves at 10 periods from 1.2372 to 1.2381s, and looked for single bins
that were at least 3σ higher than expected. We used as a measure of
peak height the difference between the peak and the average value in
the histogram in units of standard deviations of the average value.
Three of the 26 intervals, all in the on-source data, exhibited unusual
peaks. These are shown as the first three entries of Table I.   The
fourth entry extends the first entry in time, as described below. The
fifth is the light curve from the off-source data that showed the
sharpest -- least probable -- peak. The light curves, arbitrarily
phased to peak at the same phase, for entries 1, 2, 3, and 5 are shown
in Figure 1 (A), (B), (C), and (D), respectively.

        All three of these on-source intervals lie in the "low-on"
portion of the Her X-1 35^d cycle. The phase of the third interval is
just as eclipse of Her X-1 is starting, and is exactly one orbit
earlier than the similarly phased burst reported by Gorham, et al. (2).
As it happens, the first and most significant of these intervals is the

one discovered in our preliminary analysis.

    We have applied binomial probability considerations and a
Monte-Carlo to determine the probability of finding such peaks. The
results agree. The Monte-Carlo calculation additionally permits us to
assess the combinatorial precision penalty for binning the data. The entry in
Table I labelled "Prob" includes corrections for the fact that we
accept the peak in any one of the twenty bins, and that we moved the
binning across the light curves in steps of 1/5 bin.

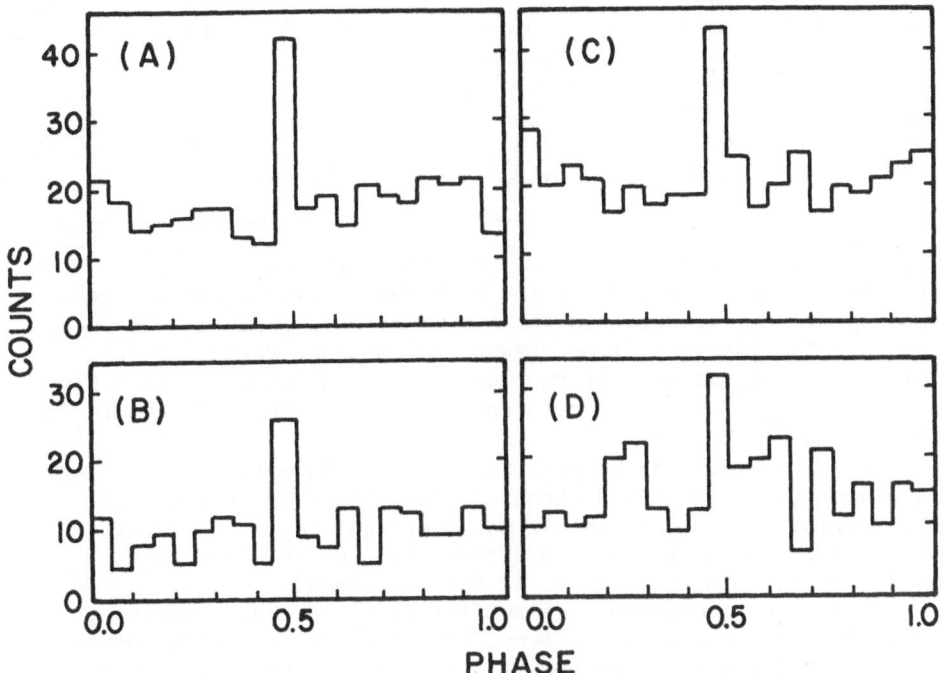

Figure 1. Arbitrarily phased light curves for high-counting-rate
intervals. Curves A, B, C, and D correspond to Table I columns 1, 2, 3,
and 5, the last being the sharpest off-source peak.

    The <u>maximum</u> remaining multiplicative, combinatorial factors for
the systematics of our search are as follows:
  1) For searching over 14 on-source and 12 off-source high-counting-
     rate intervals: 14, and 12, respectively.
  2) For searching over one independent pulsar period: 3.
  3) For adjusting start and stop of high-counting-rate intervals by
     ±100s: 3.
    It appears that the penalty is 126 for on-source and
108 for off-source data. If one applies the calculated penalty to the
best off-source run, one obtains a "probability" of about 4. Therefore
we treat the penalties as overestimated by a factor of 4 to obtain the
relative probabilities in the last row of Table I, a reasonable
procedure since we have treated off-source and on-source data equally

throughout the analysis. From this point of view the weakest of the
first three on-source claims is a factor of 20 better than the best
off-source interval, while the first (on day 165) is 400 times better.

TABLE I

Source	on	on	on	on	off
Day	165	165	168	165	165
Start(TDB)	37949	47007	32992	37949	48882
End(TDB)	38449	47407	33492	40933	49352
Orbit Phase	.87	.93	.60	.87	.95
Period	1.2376	1.2379	1.2372	1.2376	1.2375
Tot. Hits	372	199	435	433	185
Peak Hits	42	26	43	47	21
Prob($\times 10^{+5}$)	9.92	82.8	21.5	7.32	4220
Rel.Prob.	2.4×10^{-3}	2.0×10^{-2}	5.1×10^{-2}	1.7×10^{-3}	1

The first signal on day 165 occurs at the end of a run. When
we attach to it the first 100s of the next run we find a further reduction in probability, as shown in the fourth entry of Table I. However,
there was a 2300s interval of no observation between these two runs.
Therefore the burst, if such it was, was as long as 400s and probably no
longer than 2500s. The strongest signal amounts to about 20 counts in
400s, or 6×10^{-10} photons/cm^2-s.

We have examined the data in each peak in more detail. There
is no obvious structure within the high bin. The coincidence multiplicity distribution in the peak is not distinguishable from that of
the background. The narrowness in arrival time of the Cherenkov
photons is the same in the peak as in the background.

The signals observed are sufficiently improbable to support
the claim that the Haleakala telescope has observed very high energy
gamma rays from Her X-1. More importantly, if the sharp peak in the
light curve is indeed characteristic of some signals, this suggest
different strategies for searching for such signals.

This research was supported in part by U. S. Department of
Energy contract DEAC0276ER01428.

References

(1) Weekes, T. C.; Whipple Observatory Preprint No. 2312 (1986)

(2) Gorham, P., et al.; U. of Hawaii Preprint (1986)

VHE GAMMA RAYS FROM THE X-RAY PULSAR 4U0115+63

L. Resvanis
Physics Laboratory, The University of Athens
10680 Athens, Greece

J. Learned, V. Stenger, and D. Weeks
Department of Physics, The University of Hawaii
Honolulu, HI 96822, USA

J. Gaidos, F. Loeffler, J. Olson, T. Palfrey, G. Sembroski[*],
and C. Wilson
Department of Physics, Purdue University
West Lafayette, IN 47907, USA

U. Camerini, J. Finley, W. Fry, M. Jaworski, J. Jennings,
A. Kenter, R. Koepsel, M. Lomperski, R. Loveless, R. March,
J. Matthews, R. Morse, D. Reeder, P. Sandler, P. Slane,
and A. Szentgyorgyi
Department of Physics, The University of Wisconsin
Madison, WI 53706, USA

ABSTRACT The x-ray binary pulsar, UHURU designation 4U0115+63, was observed for 39 hours from August through December 1985 at the Haleakala Gamma Observatory. Three intervals of ≤ 1000 seconds exhibited significant (chance probability $\leq 10^{-3}$) pulsation at the characteristic 3.6 second period of the neutron star member of the binary system. The average peak flux above 0.2 TeV for these observations was $(2.0 \pm 0.4) \times 10^{-9}$ cm^{-2} sec^{-1}. These observations are consistent with previous reports of the sporadic nature of this source.

1. INTRODUCTION.

The first observations of the transient x-ray binary 4U0115+63 as a source of pulsed TeV gamma rays were made by the University of Durham group in 1984(1) This source was chosen as a likely emitter of TeV gamma rays due to its close resemblance to Hercules X-1 in luminosity, spin-up rate, and pulsar period. The Durham Group found significant pulsation at the characteristic 3.6 second period of the pulsar at or above an energy of 1 TeV. The emission, as in the Hercules X-system, was found to be sporadic in nature. August through December 1985 af-

---

[*] Presenter of this paper.

K. E. Turver (ed.), Very High Energy Gamma Ray Astronomy, 135–138.
© 1987 by D. Reidel Publishing Company.

forded the Haleakala collaboration its first opportunity to confirm these observations, and we here report on 39 hours of data taken during this period. A description of the apparatus can be found elsewhere in these proceedings(2).

2. ANALYSIS AND RESULTS.

In this analysis a valid event was defined as ≥ 7 of 18 phototubes firing within a 7nsec window, defined in analysis software. This requirement was found to eliminate random triggers due to ambient light(2). Trigger times, accurate to $\pm 2\mu$sec of UTC, were corrected to TDB and propogated to the solar system barycenter. Source barycentering made use of the spin-up, period, and orbital characteristics of the 4U0115+63 system as reported by the Ariel 6 group(3), listed in Table I. Data were scanned over 1000 second intervals and tested for periodicity by forming the Rayleigh vector. Approximately ten independent trial periods were used, ranging from 3.5 to 3.7 seconds.

Those intervals which were incompatible with chance probability at or near the characteristic 3.6 second period were flagged for later analysis, in which they were fourier transformed and the power calculated for frequencies up to and including 5 Hz. This range of frequencies covers 18 harmonics of the characteristic pulsar frequency of the system. Data from a reference aperture separated from the source, taken concurrently as detailed in reference 2, were analyzed in the same manner.

Of the 129 1000 second intervals scanned, three showed statistically significant periodicity with periods near 3.6 seconds. Study of the duration of these emissions lead us to a lower limit of 1.7% for the duty factor of this source. Figure 1 shows the fourier spectrum for these three emissions. The light curves shown in Figure 2 were obtained by folding the data modulo the measured period. Assuming that the low point in each light curve represents the hadronic shower background, we estimate the gamma-ray fluxes from the source. Table II gives the characteristics of each of the emissions. The probabilities quoted take into account the number of independent periods tried.

3. DISCUSSION AND CONCLUSIONS.

From our observations in 1985 of 4U0115+63 we measure an average peak flux during emission of $(2.0 \pm 0.4) \times 10^{-9}$ cm^{-2}sec^{-1}. The observed period is consistent with x-ray determinations to within the uncertainty due to the finite duration of the emission. The broad light curves (see figure 2) are consistent with the observation of Chadwick *et.al.*(1) and lead one to conclude that the emission is entirely at the first harmonic of the 3.6 second period. Indeed, inspection of the fourier spectrum revealed no significant power contained in any harmonic, up to the 18th, other than the first. The lack of any significant signal during any other time but those reported here confirm the sporadic nature of this source. Future observations will be necessary to begin to address the mechanisms responsible for these high energy emissions.

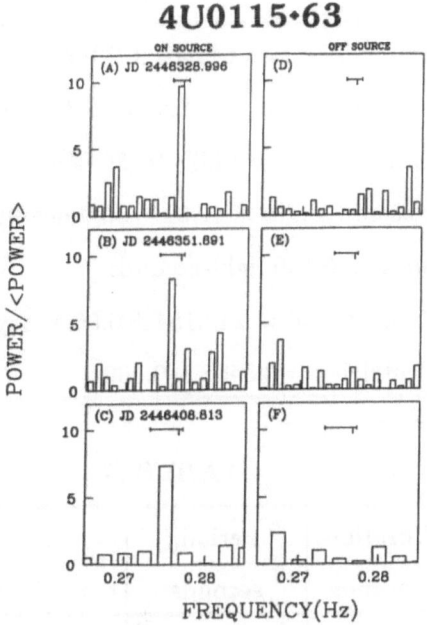

Figure 1 — Fourier spectra of signals in the on-source and off-source apertures for the three episodes reported in this paper.

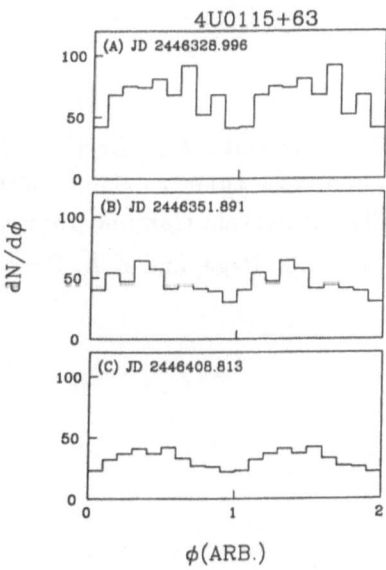

Figure 2 — Light curves for 4U0115+63 during the three episodes reported in this paper.

TABLE I

Pulsar Period $= 3.614664 \pm 0.0000011$ sec
Period Derivative $= -0.000272 \pm 0.000007$ yr^{-1}
Epoch of Orbit $=$ MJD 44586.008
Periastron angle $= 47.08 \pm 0.20$ degrees
$a\sin i = 140.130$ light-seconds
Orbital Period $= 24.3154 \pm 0.0004$ days
Eccentricity $= 0.3402 \pm 0.0004$

TABLE II

MJD of start of emission	Duration, seconds	Period, seconds	Flux, cm^{-2}s^{-1} $\times 10^{-9}$	Probability of randomness
6328.996	1000	3.619 ± 0.013	2.8 ± 0.6	2.2×10^{-4}
6351.891	1000	3.629 ± 0.013	1.4 ± 0.3	1.0×10^{-3}
6408.813	400	3.635 ± 0.032	1.7 ± 0.4	2.0×10^{-3}

REFERENCES

1. P.M. Chadwick, J.C. Dowthwaite, A.B. Harrison, I.W. Kirkman, T.J.L. Mc-Comb, K.J. Orford, and K.E. Turver, Astron. Astrophys. **151**, L1 (1985)

2. L. Resvanis *et.al.*, "The Haleakala Gamma Observatory," in this volume.

3. M.J. Ricketts, R. Paul, C.G. Page, and K.A. Pounds, Space Sci. Rev. **30**, 399 (1981).

PULSED TEV GAMMA RAYS FROM 4U0115+63

R. C. Lamb[1], M. F. Cawley[1], D. J. Fegan[2], K. G. Gibbs[3],
A. G. Gregory[4], P. W. Gorham[5], D. A. Lewis[1], N. A. Porter[2],
V. J. Stenger[5], and T. C. Weeks[3]

[1]Iowa State University, Ames, IA 50011, U.S.A.
[2]University College, Dublin, IRELAND
[3]Harvard-Smithsonian CfA, Amado, AZ, 85645-0097, U.S.A.
[4]University of Adelaide, Adelaide, AUSTRALIA
[5]University of Hawaii, Honolulu, HI, 96822, U.S.A.

ABSTRACT. The transient X-ray binary pulsar, 4U0115+63, has been
observed for 37 hours by the Whipple Observatory Cherenkov imaging
camera between September 1985 and January 1986. These observations
were grouped into nine intervals of three days or less. One interval
shows significant pulsation at the 3.6 second spin period of the
neutron star member of the binary system at a flux level of
$(1.5\pm0.4)\times10^{-10}$ photons above 0.6 TeV $cm^{-2}s^{-1}$. Thus these observations confirm previous detections, with an indication that the TeV
emission may be sporadic.

1. INTRODUCTION

Pulsed TeV photons from the transient X-ray binary pulsar, 4U0115+63,
were first observed by the University of Durham group in 1984
(Chadwick et al. 1985). The Crimean Astrophysical Observatory
(Stepanian et al. 1972, Stepanian et al. 1975) probably observed the
same source at similar energies in 1971 and 1972 (Lamb and Weekes
1986). In this note the Whipple Observatory collaboration reports
its observations conducted 1985 September - 1986 January. These
observations were grouped into nine intervals of three days or less.
One interval shows significant pulsation at the 3.6 second spin period
of the neutron star member of the binary system. Thus these observations confirm previous detections, with an indication that the TeV
emission may be sporadic like the X-ray emission.

2. OBSERVATIONS

The gamma-ray telescope used was the Whipple Observatory 10m
reflector acting as a 37-element camera (Cawley et al. 1985). The
trigger requirement that any two of the inner 19 tubes have a light

139

K. E. Turver (ed.), Very High Energy Gamma Ray Astronomy, 139–142.

signal exceeding a preset threshold corresponds to an effective
energy threshold of approximately 0.6 TeV, a collection area of
6×10^8 cm^2, and an event rate (dominantly cosmic-ray induced) of 150
events/minute at elevation angles near 50 degrees. The time associated with each Cherenkov shower image was recorded to an accuracy
of 0.5ms.

Observations were conducted for 38.4 hours beginning 1985
September 12 through 1986 January 13. The position of the neutron
star in its 24.3-day orbit corresponding to the times of observation
is shown in figure 1.

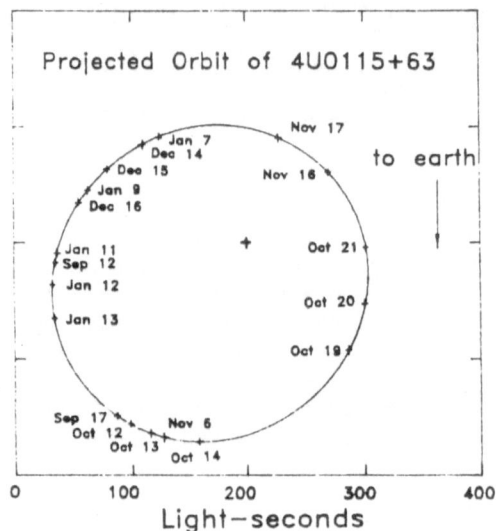

Figure 1. The position of the neutron star in its projected orbit
corresponding to the times of observation. The inclination angle of
the orbit is unknown.

3.   ANALYSIS AND RESULTS

One of the first questions that must be addressed in conducting a
search for periodicity is the range of the search. The three X-ray
determinations (Ricketts et al. 1981, Kelley et al. 1981, Rappaport
et al. 1978) and the TeV detection (Chadwick et al. 1985) show little
apparent variation from a value near 3.6146 seconds. However, significant spin-up rates have been observed; in particular, during the
14-day Ariel 6 observation (Ricketts et al. 1981), a spin-up rate
with a time scale of 4000 years was observed and times of spin-down
must be present in order for the period to remain nearly constant.

Time scales for other accreting X-ray pulsars as short as 100 years have been reported (Joss and Rappaport 1984). To set the limits of the period search we have used a time scale of 500 years, i.e. $\dot{P}/P$ = 2×10^{-3} per year. An extrapolation from the most recent detection by Chadwick et al. in September 1984 gives a search range from 3.607 to 3.622 seconds. For a typical two-hour observation eight independent trial periods are required to cover the search range.

From the 19 nights of observation, nine intervals of no more than three days duration were formed. The data from these intervals were then subjected to periodic analysis in which the only information used was the arrival time of the shower. (In a subsequent report we will describe the effect of selecting showers on the basis of image parameters, followed by periodic analysis.) The times were first corrected to the arrival time at the barycenter of the solar system and then corrected to the barycenter of the binary orbit using the parameters given in Table 1 of Chadwick et al., with the epoch of periastron passage 2,444,586.508 JD.

In the absence of a genuine periodic signal the intensity at different periods will be exponentially distributed and the probability that a given intensity level, I, will be exceeded is: $\exp(-I)$. For two of the nine intervals probability levels of 10^{-3} or smaller were attained. These intervals, October 19-21 and December 14-16, were subjected to further analysis in which a constant spin period derivative term was introduced. We have verified empirically that this is roughly equivalent to the procedure adopted by Chadwick et al. of searching on the orbital period, and in view of the apparent orbital stability and spin instability (cf. the large day-to-day excursions in the timing residuals apparent in the Ricketts et al. observations) a search on the spin period derivative seems appropriate rather than on some orbital parameter.

For the observations of October 19-21 the probability reached a minimum value of 4×10^{-5} for a spin-down rate of 5×10^{-3} per year. The periodogram for this data is shown in figure 2. In view of the number of intervals tested, the number of independent periods, and the degree of freedom due to the use of a period derivative term, the overall probability of the detection is significant at approximately the 99% level. For the December data the peak in the periodogram remained at the 10^{-3} level of probability, which, in view of the additional degrees of freedom, is not judged to be signficant.

Off-source data was subjected to similar analysis with no statistically improbable periods observed in the periodograms over the range of the search.

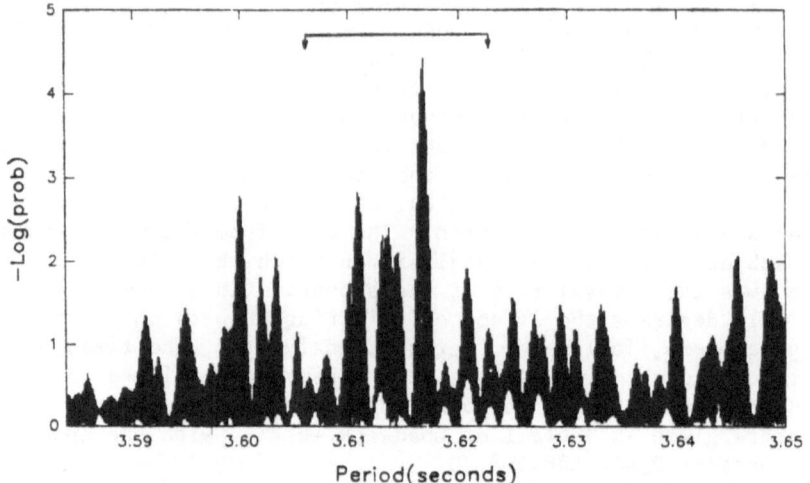

Figure 2.   Periodogram for the observations of 1985 October 19-21.
The range of search is indicated.

4.   DISCUSSION AND CONCLUSIONS

The flux of photons with energies above 0.6 TeV corresponding to the
October detection is $(1.5\pm0.4)\times10^{-10}$ photons $cm^{-2}s^{-1}$, with an additional estimated systematic uncertainty of approximately a factor of
two. At a distance of 2.5 kpc (Rappaport et al. 1978) the luminosity
above 0.6 TeV from 4U0115+63 is 1.8×10^{35} ergs/s assuming an integral
spectrum falling as $E^{-1.6}$. The absence of clear signals at other
times indicates the sporadic nature of the source.
      The period determined from this observation is 3.6166±0.0002
seconds, implying an average spin-down rate of $\dot{P}/P = 5\times10^{-4}$ per year
between September 1984 (Chadwick et al.) and this observation. This
is in contrast to the usual behavior of the X-ray pulsars which generally show spin-ups on timescales from 10^2 to 10^6 years (Joss and
Rappaport 1984). These spin-ups are powered by accretion torques.
The observed slowing down of 4U0115+63 probably indicates that
1984-85 has been a period of diminished accretion.

This work is supported in part by the U.S. Department of Energy and
the N.B.S.T. of Ireland

REFERENCES

Cawley, M. F. et al., 1985, 19th ICRC (LaJolla), 3, 453.
Chadwick, P. M. et al., 1985, Astron. Ap. 151, L1.
Joss, P. C. and Rappaport, S. A. 1984, Ann. Rev. Astr. Ap. 22, 537.
Kelley, R. L. et al., 1981, Ap. J. 251, 630.
Lamb, R. C. and Weekes, T. C., 1986, Astrophysical Letters 25, 67.
Rappaport, S. et al., 1978, Ap. J. Lett. 224, 1.
Ricketts, M. J. et al., 1981, Space Sci. Rev. 30, 399.
Stepanian, A. A. et al., 1972, Nature Phys. Sci. 239, 40.

VERY HIGH ENERGY GAMMA RAYS FROM THE VELA PULSAR

P.N. Bhat, S.K. Gupta, P.V. Ramanamurthy,
B.V. Sreekantan, S.C. Tonwar and P.R. Vishwanath

Tata Institute of Fundamental Research,
Homi Bhabha Road,
Bombay 400 005,
India.

ABSTRACT

Four independent data sets with slightly different gamma ray energy thresholds are separately analysed and the phasograms are derived using the pulsar elements derived from radio measurements. Each phasogram shows a weak double pulse structure at the same phase bins separated by 0.45 of a phase, where the first pulse coincides with the expected position of the optical first pulse. However the statistical significance of the signal improves when lower energy showers only are chosen, based on the pulse height information or the source hour-angle. When the first three 3 phasograms after applying this cut are summed, the resultant phasogram shows 2 peaks at 4.0σ and 1.5σ level. If one multiplies the significances calculated for individual phasograms and after allowing for various degrees for freedom, one obtains a value of 8.0×10^{-4}, for the probability that all the excesses are statistical fluctuations. From the excesses observed at different gamma ray energy thresholds an integral energy spectrum of VHE gamma rays from Vela is derived. The estimated slope of this power law energy spectrum is $-(1.8\pm0.2)$

INTRODUCTION

In the past, the VHE gamma rays from the Vela pulsar had been seen on three occasions : (a) a single pulse in 1975(Grindlay et al, 1975), (b) a double pulse structure by the Ooty group in 1978-79 (Bhat et al, 1980) and (c) a single pulse by the Ooty group in 1980-1981 (Gupta et al, 1982). The latter two observations did not have absolute phase information but the two pulses seen in 1978-79 had a separation of 0.45 in phase. In this paper, we use all the

143

K. E. Turver (ed.), Very High Energy Gamma Ray Astronomy, 143–146.
© *1987 by D. Reidel Publishing Company.*

data from the later Ooty observations which had absolute
phase information.

        Vela pulsar was tracked for 3943 mins and 2147 mins
during 1979-80 and 1982-83 with the full Ooty array (Bhat et
al.,1980). In 1984-85 we split the array into two (Bhat et
al.,1986) and the source was observed from two stations
separated by nearly by 11 km. We observed Vela for 1747 mins
from one site (GRO) and 531 mins from the other(viz: CRL).
We had 398 mins of simultaneous observations at the two
sites. Eight large mirrors were used in the GRO site where
the pulse height information for individual showers was also
avilable. Ten small mirrors were used at the CRL site .

ANALYSIS AND RESULTS

        The data taken during 1979-80, 1982-83 and those
collected at the two sites in 1984-85 form 4 independent data
sets. These 4 data sets were analysed separately. The

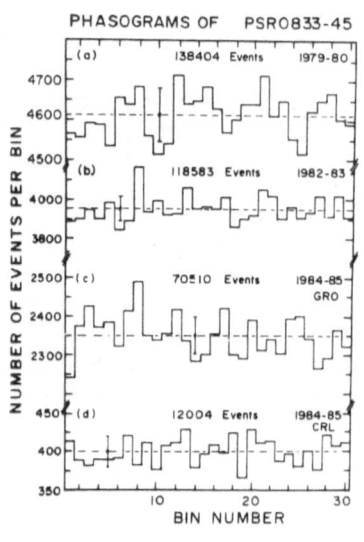

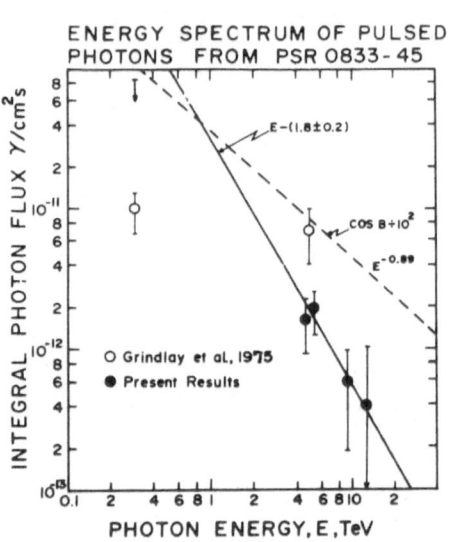

Figure 1.   Figure 2.

arrival time of each event, after suitably correcting for time
calibration, is converted into the corresponding time at the
solar system barycentre. Then the pulsar phase at this time
was deduced using the pulsar elements supplied by R.N.
Manchester (1979-80), R.W.Royle (1982-83) and A.R. Klekociuk

(1984-85) using radio observtions. Thus a phasogram of the events is derived for each data set .

Though there was no significant excess in any of the four phasograms in any phase bin, a small excess was found in channel number 8 and 21,22 in all the 4 data sets. Even though these excesses are not significant by themselves, their repetition over the years in the same phase bin makes them significant. When we summed all the 4 phasograms with respect to the radio phase, we found peaks of 2.4 and 1.3 respectively at channels 8 and 21,22. These phase bins correspond to pulsar phases of 0.25 and 0.7 respectively. The separation of 0.45 is close to the characteristic separation of the pulse and the interpulse seen in the medium energy gamma ray light curve (Bennett K. et al.,1977). However the phase at which the first pulse seems to appear concides with that of optical first pulse (0.24, Buccheri R. et al 1978) but not with that of 1st pulse at GeV energies.

Among the 4 phasograms, the one at the lowest energy threshold (4.9 TeV) showed more prominently (3.1σ and 2.0σ respectively) the existence of a double pulse structure probably suggesting that the lower energy events are richer in the signal. From this clue, we tried to select lower energy events based on the pulse height whenever available and/or the source hour-angle. We chose events collected during which the source hour-angle $|HA| \leqslant 20$. This was applied to all the 4 data sets. However the ADC information on the pulse height was available only for data sets (b) and (c). In these data 50%of the lower energy events only are selected and their phasograms derived. The results are shown in Figure 1. The modified energy thresholds for these subsets are 9.3 TeV, 4.7 TeV, 5.4 TeV and 11.8 Tev respectively.

The phasogram of the sum of the first three data sets after applying the aforesaid cuts shows a 4.0σ peak at the position of main pulse and a 1.5σ peak at the inter pulse position. The probability that this could be a statistical fluctuation is 8.0×10^{-4} assuming that the main pulse could be in any of the 30 bins and the interpulse could be at 2 positions on either side of the main pulse with a separation of ~ 0.45 in phase.

From the observed excess events in both the main and the interpulse positions the integral photon flux recieved from this source can be calculated. The results are plotted in figure 2. Also plotted in the figure are previous measurements of Grindlay et al. The dotted line shows the extrapolation from the lower energy measurements of COSB collaboration (Bennet K. et al.,1977). The results from our previous measurements are not plotted in the figure as the

absolute phase information was not available during these
observations. However our present flux estimates are far too
low compared to the extrapolation from lower energy
measurements. From a fit to the points plotted in figure 2
we estimate the slope of the integral gamma ray spectrum to
be $- (1.8\pm0.2)$, consistent with the above lower limit and
also with our earlier estimate of 2.0 ± 0.2 (Gupta et al).

DISCUSSION

        The position of the main pulse at the pulsar phase of
0.25 is not totally unexpected. There is a hint in the lower
energy measurements of a persistent excess events at the
optical main pulse position (Kanbach G. et al.,1980). An
earlier evidence of having detected gamma rays of energy
>300 GeV by Grindlay et al.(1975) also showed a single pulse
at the position of the first optical pulse. In the present
case too the first pulse appears at the optical phase and in
addition a second pulse also exists with a characteristic phase
difference between the two, which is same as that observed at
medium energy range. This probably indicates that the
production mechanism of medium energy gamma rays and VHE
gamma rays are different. The gamma ray energy spectrum in
figure 2 shows that the energy spectrum steepens discretely
or continuously in the intervening photon energy range of 10
GeV to 5 TeV.

ACKNOWLEGEMENTS

        It is a pleasure to thank Mr.A.R.Apte for setting up
the array. We thank Messrs. N.V.Gopalakrishnan,
R.Mahalingam, S.Ramani, S. Swaminathan and M.Venkateshwarlu
for able maintenance and help in running the experiment. We
thank Drs. R.N. Manchester, R.W. Royle and A.R. Klekociuk
for supplying us the pulsar parameters from their radio
measurements.

REFERENCES

Bennett K. et al, 1977,Astron & Astrophys,**61**,279.
Bhat P.N. et al. 1980,Astron & Astrophys,**81**,L3.
Bhat P.N. et al, 1986, Nature,**319**,127.
Buccheri R. et al, 1978,Astron & Astrophys,**69**,141.
Grindlay J.E. et al , 1975, Astrophys. J,**201**,82.
Gupta S.K. et al., 1982,Proc. Workshop on VHE gamma ray
        astronomy, Ootacamund, 282.
Kanbach G. et al, 1980,Astron & Astrophys,**90**, 163.

VERY HIGH ENERGY GAMMA RAYS FROM THE CRAB PULSAR

P.R. Vishwanath

Tata Institute for Fundamental Research
Bombay 400 005
India

ABSTRACT

The subdivision of the Ooty data on VHE ($>$ 600 GeV) gamma rays into miniruns of one minute duration reveals interesting aspects. Minutes with high gamma ray activity and / or a downward statistical fluctuation of Cosmic Ray background were isolated with a χ^2 analysis. The summed phasogram of such minutes shows two strong peaks coinciding with the Radio Main pulse and the Inter pulse respectively. The probability that these are due to chance is small.

ANALYSIS AND RESULTS

The Atmospheric Cerenkov array at Ooty has been described elsewhere in detail (Gupta et al., 1985 and Bhat et al., 1986). The four data sets used here (for all of which we have absolute phase) -1979-1980, 1982-1983, 1984-1985 (C) and 1984-1985 (G) -had energy threhsolds of 3000 , 600, 2200 and 800 GeV with exposures of 70.6,45.8,31.8 and 26.6 hours respectively. The phasograms for all the four data sets were obtained in the standard manner as described in earlier papers of the Ooty group. When these data were summed in phase, the resulting phasogram did not show time averaged emission from the Crab pulsar .

In one of our earlier papers (Vishwanath, 1982), we had shown the possibility of VHE gamma ray emission in occasional minutely intervals. The new analysis is based on the assumption that the signal, if present, may show up in minutely phasograms with moderate to high χ^2. Such a χ^2 can also be due to upward or downward statistical fluctuation of one or more bins. The total fluctutation in the signal region may be due to both statistics and presence of a

147

K. E. Turver (ed.), Very High Energy Gamma Ray Astronomy, 147–150.
© 1987 by D. Reidel Publishing Company.

signal, whereas in the background region it will be due only to statistics.

The present data were also divided into miniruns of one minute duration . The total number of miniruns was 10489 and the average rate per minirun was 289. When we first computed the χ^2_{19} and classified the miniruns as to their χ^2_{19}, the only interesting phasogram was for the classification $1.4 \lesssim \chi^2_{19} < 1.8$ in which both the Main and the Inter pulses were quite prominent, whereas most the bins in phase region 0.45-0.95 had lower numer of events. When we divided the phasogram into two regions, region A (phase 0.95 - 0.45) and region B (phase 0.45 - 0.95), we found the probability that the the events in the two regions stem from the same population is lower than 1%. Therefore, it is possible that in VHE gamma ray studies region B is populated by background events due to Cosmic Rays only ,whereas the rest of the phase plot has events due to gamma rays also.

For exploring further the differences between regions A and B, events in each minirun with $1.4 \lesssim \chi^2_{19} < 1.8$ were placed either in region A or in region B depending on their phase. For the two bin phasogram thus obtained, χ^2_1 (1 degree of freedom) was computed. Fig.1 displays the $\chi^2_1 > 2$ cut. The phasogram shows considerable excess at both the Main and the Inter pulse phases. The average rate per minirun in region A and B in Fig.1 are 155.03 ± 0.80 and 147.90 ± 0.80 respectively. The average normal background rate, derived by ignoring the bins in the phase interval 0.95 - 0.10 and 0.40-0.45, was 149.9 ± 0.4 for half of the phase plot. Another measure of the normal background rate obtained by comparing the rates just before and just after the miniruns selected is 148.9 ± 0.6 .

The reasons for high rate in region A are interesting. At first, there are more miniruns with excess in region A. Next, when we compared the rate in each selected minirun with normal background average rate of 4 minutes which encompass it, the percentage of miniruns where region A(B) has fluctuated upward by $> 2\sigma$ is 17 (12) and downward by $< 2\sigma$ is 8 (18) whereas one expects it to be 5 from the observed behaviour of the normal Cosmic Ray background (this differs slightly from Poisson flucutation because atmospheric transparency and extraneous lights play an important role in atmospheric Cerenkov methods). Thus, the selection process has preferentially picked out miniruns where regions A and B have shown significant upward and downward fluctuation respectively.

We did a Monte Carlo simulation to see the probability of getting the distribution shown in Fig.1. Only one out of 1000 trials showed a single peak of statistical significance similar to those in Fig.1. Therefore, to get at least two

peaks of similar significance seperated by 0.43 in phase the
probability is 1 x 10^{-7} . Monte Carlo simulations also gave
a rate of 149.9 per region per minirun.

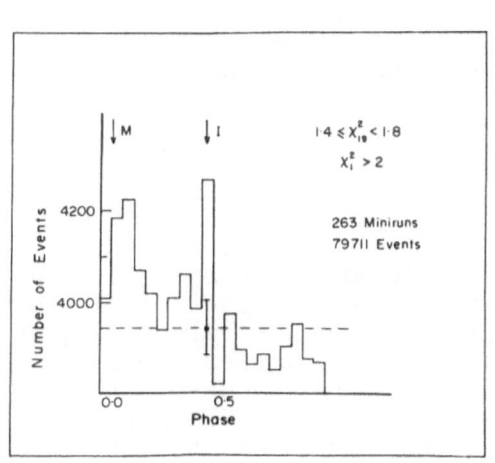

Figure 1.

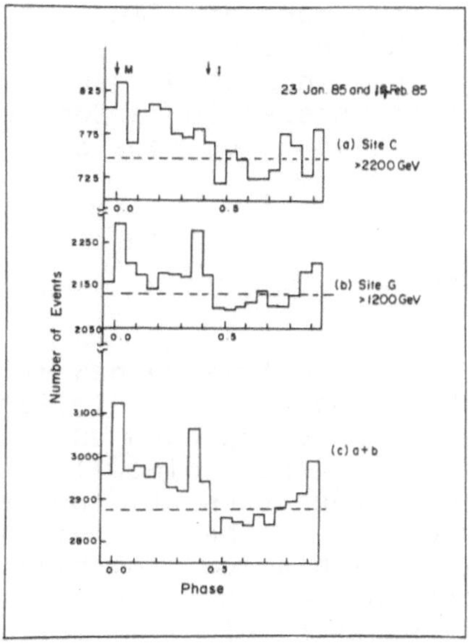

Figure 2

With this background rate, the Main and Inter pulses are
5.9 σ and 5.2 σ above the background. Considering the
number of classifications tried out, the Combined
(Poissonian) probability that the Main and Inter pulses are
due to chance is 1 x 10^{-14} . However, the probability
derived from the Monte Carlo simulations is higher. The Main
pulse at low and high energy thresholds is at significance of
5.7 σ and 1.9 σ respectively. The Inter pulse at the low and
high energy thresholds is at significance of 4.5 σ and 3.0 σ
respectively. We note that the phasogram in region A is very
similar to the light curve seen at lower energies. Also the
present analysis is the only one to have shown the Inter
pulse clearly at VHE gamma ray energies.

During 84-85 observations, when we had simultaneous
observations at two sites there were excess events in region
A in the total phasogram from both the sites on the same two
nights. The combined phasogram from the two nights at sites
C and G and the summed phasogram from these sites are shown
in Figs.2a, 2b and 2c respectively. Considering that there
were 7 nights of simultaneous observations, the probability
of seeing a peak at the Main pulse due to chance at both the
sites is 2 x 10^{-6} . We also note that the 15 minute

burst seen by Bhat et al. occured on one of these nights.
Both Figs.1 and 2c show the Main pulse and a rather bimodal
distribution. Fig.1 is a result of the minirun analysis and
all the data sets contribute to it. Fig.2c represents the
complete data of two nights, and it is highly unlikely that
the background has fluctuated downward on both the nights and
at both the sites. Therefore, Fig.2c can be accounted for
only by increased gamma ray activity on those two nights.
Another interesting aspect of these two phasograms is that
there seems to be some emission from the middle phase regions
also.

SUMMARY

We have tried a new method for locating transient and
weak signals by exploiting the downward flucuation of the
Cosmic Ray background due to statistics. We are also helped
by the rather bimodal behaviour of the pulsar, which has
already been seen at low energies. This results in the
detection of the. Main and the Inter pulses from the Crab
pulsar with a low probability that they are due to chance.
These pulses have been seen at both low and high energy
thresholds and in all four data sets.

ACKNOWLEDGEMENTS

It is a pleasure to thank all the members of the Ooty
group for help and guidance at various stages. Drs
P.T.Wallace and A.J.Lyne are thanked for providing the pulsar
elements .

REFERENCES

Bhat.P.N., Ramanamurthy.P.V., Sreekantan.B.V. and
Vishwanath.P.R., 1986,Nature,319,127

Gupta. S.K., Ramanamurthy.P.V., Tonwar. S.C. and
Vishwanath.P.R., 1985, Astrophysics and Space Science, 115,
163

Vishwanath.P.R., 1982, Proc.of Intl.Workshop on Very
High Energy Gamma Ray Astronomy ed. P.V. Ramanamurthy and
T.C.Weekes (Bombay : TIFR), 21

THE VHE GAMMA RAY LIGHT CURVE OF PSR0531

P.M.Chadwick, N.A.Dipper, I.W.Kirkman,
T.J.L.McComb, K.J.Orford, K.E.Turver.

Physics Department,
University of Durham,
Durham, DH1 3LE,
United Kingdom

ABSTRACT. Measurements of the light curve of the Crab pulsar at TeV energies indicate a very narrow feature coincident with the radio main pulse with little evidence for an interpulse. This result is compared with the predictions of the pulsar model of Cheng, Ho and Ruderman.

The observations of the Crab pulsar using the University of Durham TeV gamma ray telescopes at Dugway have been reported previously (Dowthwaite et al. (1984), Chadwick et al. 1985). A recent detailed model of the gamma ray emission from pulsars (Cheng, Ho and Ruderman 1986(a) - CHRI), and the Crab and Vela pulsars in particular (Cheng, Ho and Ruderman 1986(b) - CHRII), has prompted a further examination of the data. The predictions which derive from the model and which are of interest to TeV observations of the Crab pulsar are:

1. TeV emission could be produced from the outer magnetosphere, just within the velocity of light cylinder.

2. The main pulse of lower energy emission should be coincident in time with this TeV emission.

3. The main pulse of TeV radiation should be narrow.

4. There should be no interpulse at TeV energies.

The energy spectrum of the pulsed emission above optical energies is shown in Figure 1. The experimental points are from various sources and were compiled by Cheng, Ho and Ruderman, 1986(b). The model has been derived to fit these points in a natural way. The emission below about 3 GeV is taken to be due to synchrotron emission by e^+e^- pairs created in outer magnetospheric gaps. Above this energy,

K. E. Turver (ed.), Very High Energy Gamma Ray Astronomy, 151–154.

the emission is due to inverse Compton scattering of soft photons by the pairs, with a cutoff at the maximum energy of the pairs, ~ 5x10^{12} eV. The shape of this part of the spectrum should mirror the synchrotron emission region, but shifted up by about 1000 in energy.

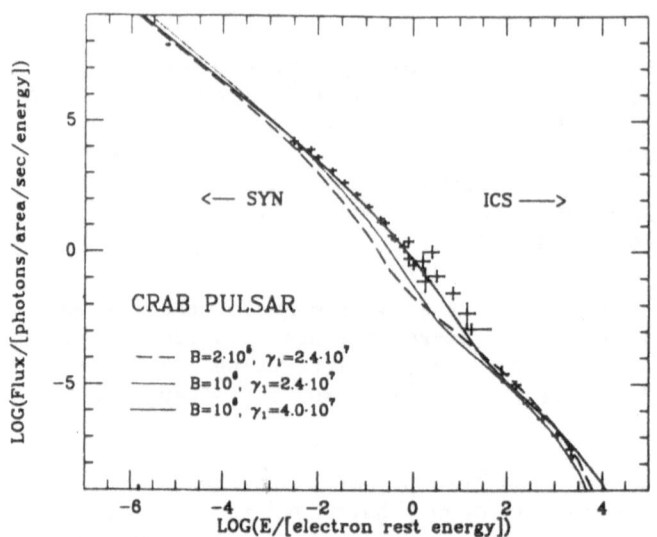

Figure 1. The differential energy spectrum of the Crab pulsar PSR0531 from optical to VHE gamma ray energies. The points are a compilation due to CHRII, the Durham point at 10^{12} eV shows the range of spectral slopes consistent with the data, the solid line is the prediction of CHRII.

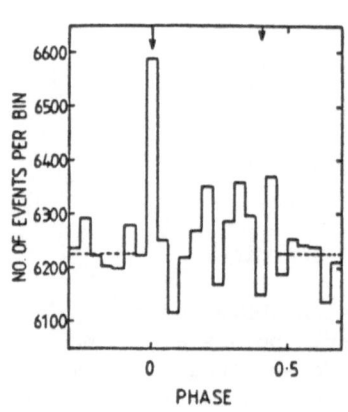

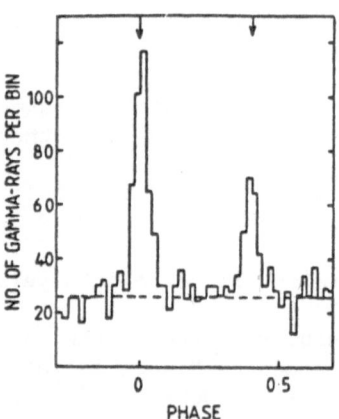

Figure 2(a). The TeV gamma ray light curve of PSR0531 (Chadwick et al. 1985).

Figure 2(b). The low energy gamma ray light curve of PSR0531 (Wills et al. 1982).

The Durham integral energy flux at 1 TeV is shown in Figure 1, together with limits on the integral spectral index derived from an analysis of the gamma ray energy required to generate various multiple telescope responses in the array of four Dugway telescopes (Kirkman,1985).

The time of arrival of each Cerenkov flash was recorded with a resolution of 1 microsecond, and with an absolute uncertainty of < 0.5 ms. This time was converted to the solar system barycentre and further converted to a Crab pulsar phase using a Crab pulsar radio ephemeris (Lyne 1985). The phase distribution of the recorded pulses is shown in Figure 2(a), compared with the position of the radio main pulse from the ephemeris. The low energy gamma ray phase distribution from COS-B (Wills et al. 1982) is shown in Figure 2(b). The coincidence between the emission at all energies up to TeV is strong support for the CHRII model.

The range of rotational longitudes during which emission may be seen should be less at TeV energies than at lower energies. The TeV peak shown in Figure 2(a) is in a histogram bin whose width and position were chosen a priori, on the basis of the binning of the COS B data. Further investigation has shown that all of the TeV emission is confined to about one third of the bin width, or about 33/75 ms (Chadwick et al. 1985). Since the residual uncertainty of the absolute time of arrival of any TeV pulse is of the same order, the pulse width of 0.4 ms must be regarded as an upper limit. It seems likely therefore that the trend to narrower pulses with increasing energy, first observed by McBreen et al. (1973) is confirmed. There are differences in pulse shapes with energy, making it difficult to use a simple width measure in a quantitative way. The full width at half maximum is used here only as an indication of the possible narrowing, from 20-30 degrees of longitude at optical and X-ray energies, through ˜10 degrees at MeV to GeV gamma rays, to less than 5 degrees at TeV energies.

The final point to note about the TeV light curve in Figure 2(a) is the lack of a pronounced interpulse. If the CHRII model is valid, this lack of an interpulse is easily explained. Emission from the Crab pulsar outer magnetsopheric gaps is expected to be observable twice per revolution. One beam is caused by emission from one sign of electron streaming outwards, and another by the opposite charge streaming inwards in the same gap. Photons caused by the inward stream must cross most of the magnetosphere before escaping to be detected at the earth. The magnetic fields to be crossed will absorb the highest energy photons by pair creation, naturally explaining the absence at this energy only of a strong interpulse.

The features noted here which agree well with the CHRII model are capable of greater refinement. A measurement of the exact pulse width, an accurate energy spectrum above a few hundred GeV, a precise value for cutoff energy and a stronger limit on interpulse emission will enable the models to incorporate the TeV spectral range without the present free parameters and enable the components of and conditions in the outer gaps to be probed.

References

1. Chadwick,P.M. et al., Proc, 19th Int. Conf. Cosmic Rays La Jolla,
   OG2.3-9 (1985).
2. Cheng, K.C., Ho, C., Ruderman, M., Ap.J. 300, 500-521 (1986).
3. Cheng, K.C., Ho, C., Ruderman, M., Ap.J. 300, 522-539 (1986).
4. Dowthwaite,J.C. et al., Ap.J. 286, L35-38 (1984).
5. Kirkman, I.W., Ph D Thesis, University of Durham (1985)
6. Lyne,A.G., private communication (1985).
7. McBreen,B. et al., Ap.J. 173, 571 (1973).
8. Wills,R.D. et al., Nature 296, 723 (1982).

VHE GAMMA RAYS FROM THE CRAB PULSAR

L. Resvanis
Physics Laboratory, The University of Athens
10680 Athens, Greece

J. Learned, V. Stenger, and D. Weeks
Department of Physics, The University of Hawaii
Honolulu, HI 96822, USA

J. Gaidos, F. Loeffler, J. Olson, T. Palfrey, G. Sembroski,
and C. Wilson
Department of Physics, Purdue University
West Lafayette, IN 47907, USA

U. Camerini, J. Finley, W. Fry, M. Jaworski, J. Jennings,
A. Kenter, R. Koepsel, M. Lomperski, R. Loveless, R. March,
J. Matthews, R. Morse, D. Reeder, P. Sandler, P. Slane,[*]
and A. Szentgyorgyi
Department of Physics, The University of Wisconsin
Madison, WI 53706, USA

ABSTRACT.   The Haleakala $10m^2$ Cherenkov light telescope, which has an estimated energy threshold for gamma-induced atmospheric cascades of 2×10^{11}eV, has observed the Crab pulsar, PSR0531, for a total of more than 211 hours. The results of analysis of 10% of this data are reported. Using standard methods of periodic analysis, pulsar periodicity consistent with radio ephemerides has been observed on at least two occasions. The light curve obtained exhibits both a primary and a secondary peak. Like other known astrophysical VHE gamma sources, this pulsar appears episodic in its emissions, with episodes lasting less than an hour.

1. INTRODUCTION.

The Haleakala Gamma Observatory[1] is an atmospheric Cherenkov device with $10m^2$ of mirror area situated on Mount Haleakala in Hawaii, at an altitude of 2950m. It is provided with two apertures of angular diameter 0.75° each, separated by 3.6° of declination, which permit it to simultaneously monitor both a candidate source and an adjacent area of the sky. Light from each aperture is collected on 18 photomultiplier tubes (PMTs), each of which views a separate mirror segment. As the telescope tracks the source in right ascension, the roles of the two apertures

---

* Presenter of this paper.

K. E. Turver (ed.), Very High Energy Gamma Ray Astronomy, 155–158.
© *1987 by D. Reidel Publishing Company.*

are periodically interchanged ("wobbled"). This mode of operation helps to eliminate the effects of any systematic differences between the apertures.

The PMTs feed a fast logic system that accepts signals generated
from a single photoelectron. Data digitization and readout are triggered whenever
7 PMTs have a signal within roughly 10ns. The time of each pulse is digitized to
an accuracy of 0.1ns, and pulse heights are digitized as well. A register slaved to a
signal from a Cesium beam atomic clock records the trigger time to a least count
of 1μs. This clock is maintained to within a few μs of UTC.

Data is recorded on magnetic tape by an LSI 11/73 computer. In off-
line analysis, further tests of pulse times eliminate nearly all random coincidences
of ambient light, as verified by tests on samples of events with a lower hardware
trigger multiplicity. What signals remain are primarily from atmospheric showers
of hadronic origin, at a rate of .5 to .7 Hz. We estimate that the threshold of the
instrument is in the vicinity of 200 GeV for gamma-induced cascades.

The signature of the Crab pulsar is its periodicity, which is carefully
and frequently monitored in the radio band. With the aid of a concurrent radio
ephemeris, it is possible to interpolate the period to the time of observation to an
accuracy of better than one part in 10^{11}. This analysis employed ephemerides from
the Jodrell Bank Radio Observatory,[2] which are updated on a monthly basis. It
is particularly important to have a concurrent radio ephemeris, as this source is
subject to frequent sudden small changes of period.

To compare with the ephemeris, signal arrival times were propogated
to the solar system barycenter and converted to the TDB scale, using the JPL numerical integration of the Earth's motion employed in the *Astronomical Almanac*.
Jodrell Bank utilizes the earlier MIT ephemeris, and this discrepancy may give rise
to differences in absolute phase on the order of 10μs.

2. ANALYSIS.

We searched for periodicity in the vicinity of the radio period by means of the
Rayleigh test. The analysis program stepped through a range of trial periods,
forming for each of the N events in the data sample a unit vector at the phase angle determined by the barycentered trigger time modulo the trial period. These
vectors were then summed and the magnitude squared of this vector sum, R^2, was
determined. NR^2 should then be the negative of the natural logarithm of the probability that the apparent periodicity is due to chance.

The analysis was carried out by breaking up the data, which consists
of runs of several hours duration, into intervals of 2000 seconds. The criterion for
an acceptable event demanded at least 6 and no more than 14 tubes be hit within
a span of 10ns. This somewhat arbitrary "weak" cut was chosen because it was
suspected that the Crab pulsar might have a relatively soft gamma ray spectrum.
In this case, a cut that admitted showers of the lowest possible energy would give a
higher signal to noise ratio than a tighter criterion.

Each data sample was tested for periodicity over a band of one hundred trial periods centered about the nominal radio period for that date. The trial period was stepped in intervals of 10ns, insuring that over the duration of the data sample the phases derived from adjacent trial periods would differ by no more than 0.2. As a result, the effective number of statistically independent trials is substantially less than the total number.

Out of 38 2000-second intervals, three gave a probability less than 10^{-4} that the signal was purely random. These are listed in Table I. Two are compatible with the period interpolated from the radio ephemeris, while the third differs from it by one part in 10^4. It is difficult to reliably estimate the errors of our period measurements, since it is possible that the source was not active for the entire interval, but this large a discrepancy is probably statistically significant, and this observation may be spurious. All three came from an aperture that was on the source, and the probability distribution of the off-source data was consistent with random fluctuations.

The three intervals came from data runs designated 424, 463 and 470, and began on January 7, 1986 at 25840s UTC, January 15, 1986 at 31533s UTC, and Feruary 2, 1986 at 24105s UTC respectively. Of the three runs perhaps the most convincing was 470, where the signal continued through a change of apertures, although the improbability of randomness was more significant for 463. The Rayleigh plot for run 470 is shown in Figure 1.

The best estimate of the total flux utilizes the light curve. The curve obtained in run 470 is shown in Figure 2 as an example. Assuming that the lowest bin of the plot represents the hadronic background, for the three runs 461, 333, and 330 events can be attributed to the source. Assuming an effective shower altitude of 11km leads to the flux estimates in Table I. A double-peaked distribution such as that in Figure 2 tends to give a partial cancellation of the Rayleigh vector, so it is likely that the quoted probabilities of randomness are pessimistic. Given the uncertainty in the actual durations of the episodes, we quote no errors on the flux estimates.

TABLE I

run #	episode start (JD-2440000)	probability of randomness	obs. period, $msec$	radio period, $msec$	flux, $s^{-1}cm^{-2}$
424	6436.79908	1.6×10^{-5}	33.32306	33.31944	2.7×10^{-9}
463	6445.86497	5.1×10^{-6}	33.31984	33.31973	1.9×10^{-9}
470	6463.77900	1.0×10^{-4}	33.32045	33.32039	1.9×10^{-9}

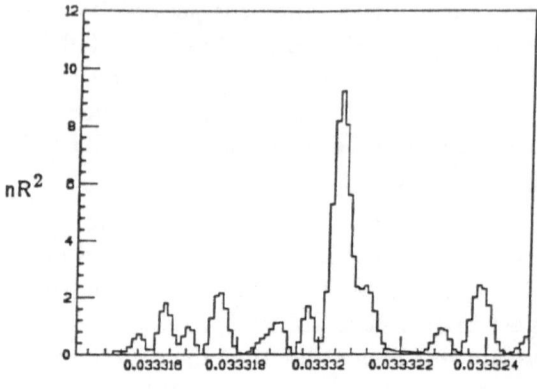

PERIOD, SECONDS

Figure 1 — Rayleigh vector magnitude vs. period for run 470.

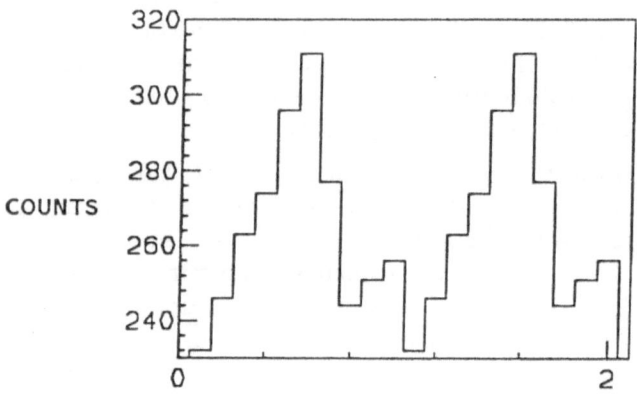

PHASE

Figure 2 — Light curve obtained from run 470.

References

1. A more complete description of this device may be found in U. Camerini *et al.*, *The Haleakala Gamma Observatory*, in this volume.

2. A. Lyne, private communication.

1000 GeV GAMMA RAYS FROM THE 1.5 ms PULSAR

P. M. Chadwick, N. A. Dipper, I. W. Kirkman,
T. J. L. McComb, K. J. Orford, K. E. Turver
and S. E. Turver

Department of Physics
University of Durham
Durham DH1 3LE
United Kingdom

ABSTRACT. The 1.5 ms pulsar PSR1937+21 has been observed using the Dugway 1000 GeV gamma ray telescopes and the observations show evidence for pulsed VHE gamma ray emission, significant at better than the 3.5 sd level. The signal comprises two pulses separated by half the period which is similar to the radio emission. The total energy flux in VHE gamma rays is 2x10 erg s^{-1} and agrees with the prediction of a current model that the peak emission of electromagnetic energy should be near 100 GeV.

Following the discovery of PSR1937+21, the 1.5 ms pulsar, by Backer et al. (1) we began a series of observations using the 1000 GeV (VHE) gamma ray telescope array at Dugway, Utah - see Gibson et al. (2). Although in genesis, and its value of magnetic field, PSR1937 is very unlike the Crab pulsar (which is a well established emitter of VHE gamma rays) it has been suggested by Usov (3) and Cheng et al (4,5) that the 1.5 ms pulsar may have a significant part of its electromagnetic energy emission in the form of VHE gamma rays around 100 - 1000 GeV. The 1.5 ms period of this pulsar has created a number of technical problems for our experiment, the timing system of which was designed primarily for observing pulsars no faster than that in the Crab. Moreover, since on the basis of other observations of weak VHE gamma ray emitters we would expect that data taken over a considerable time would need to be summed in order to detect a significant VHE gamma ray signal, the pulsar ephemeris must possess very small errors. The accuracy should be such that the error in predicted pulsar phase over several months will be considerably less than one pulsar cycle. Such an ephemeris of unprecedented accuracy is available with a residual phase uncertainty over two years of 1-2 microsec (6).

Our first VHE gamma ray observations of PSR1937+21 were made in 1983, but the size of the dataset and the system timing uncertainties limit their usefulness. Further substantial amounts of data were taken with an improved timing system in 1984. During 1984 July-September a total of 200840 atmospheric Cerenkov light events were recorded with

159

K. E. Turver (ed.), Very High Energy Gamma Ray Astronomy, 159–162.
© *1987 by D. Reidel Publishing Company.*

PSR1937+21 in the field of view of the telescopes in a total of 85 hours of observation. The threshold gamma ray energy of the telescopes was dependent on zenith angle, but for most of the data was ˜1000 GeV. The time of each event was recorded with a resolution of 1 microsec using an oven-controlled crystal as a timebase (accurate after correction to 1 part in 10^9). The clock was regularly synchronised using an off-air signal from the close-by radio station WWV at Fort Collins, Colorado. The time (UT) of each event was determined, limited by the accuracy of clock rate and its off-air synchronisation, to 0.3 ms. (The validity of this estimate is confirmed by our successful VHE gamma ray measurements with the same system of the absolute phase of the Crab pulsar emission - see Dowthwaite et al (7)). Each event time in UT was corrected to TAI and then to the solar system barycentre using the MIT Earth ephemeris (8).

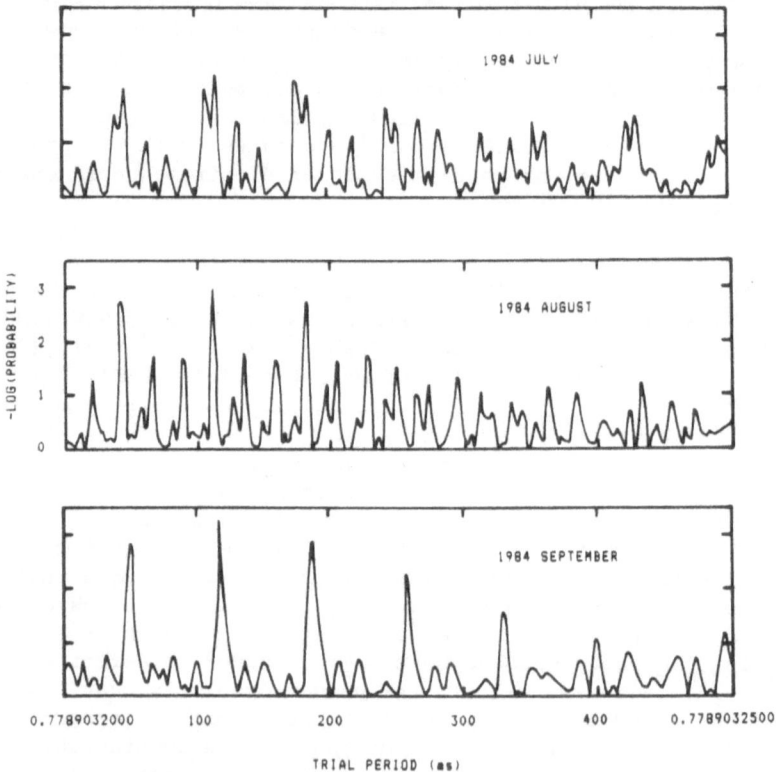

Figure 1. The probability of chance origin of the periodicity in three independent datasets as a function of the trial period. The side bands arise from the 24 hour seperation between observations each of typically 3-4 hours duration. The 1984 September data were all taken on the same clock synchronisation, while the clock was resynchronised several times during each of the 1984 July and August observations. The additional structure in these two datasets is due to the timing uncertainties associated with these synchronisations.

   We have considered two possible approaches to the analysis of our
data. First, we have considered the epoch folding technique in which
the barycentre time (TDB) of each event was converted to a pulsar
phase using the accurate pulsar radio ephemeris. The danger here is
that small errors in the clock rate or synchronization will smear the
light curve.

   The second approach to the data analysis which we have employed
is to take the data in 3 sets each corresponding to observations
spread over 7 or 8 days in the months of 1984 July, August and
September and test for uniformity in phase over the appropriate range
of periods allowing for the possible effects on periodicity of the
timing uncertainties. We estimate that these could cause a true
periodicity at the predicted value of 1.55780645 ms to appear anywhere
between about 1.55780643 and 1.55780647 ms in our data. It is also
desirable to allow for the presence of significant power in the second
harmonic since the radio light curve shows two symmetric peaks of
equal strength. The result of this procedure shows independent
evidence for non-uniformity in phase at levels of significance of 5×10^{-3}
, 1×10^{-3} and 5×10^{-4} for the data in July, August and September at a
period of 1.55780642 $\pm$ 0.00000003 ms (the value expected from the
radio ephemeris is 1.557806453 ms). The indications are that the
emission is equally strong in each observation and that there is
significant power in the second harmonic. The results of the period
sweeps (at half the pulsar period) for the three independent datasets
are shown in Figure 1. These three have been combined (after allowing
for the effects of the small period derivative) and the result is
shown in Figure 2. The combined probability of these non-uniformities
arising by chance is 1×10^{-4}. (Additional and independent data of less
statistical significance and with less reliable timing taken earlier
show similar effects significant at the 1×10^{-2} level).

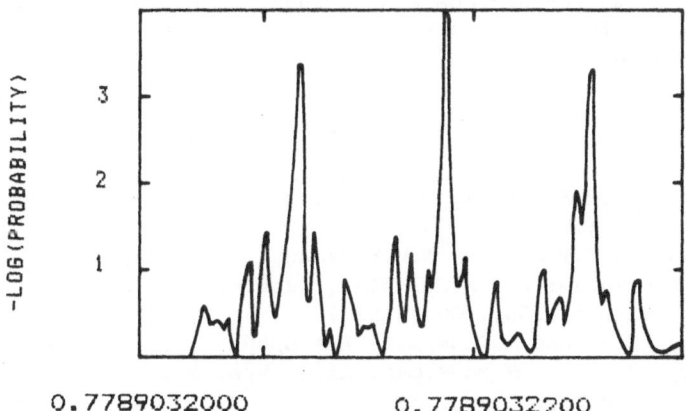

TRIAL PERIOD (ms)

Figure 2. The combined probability of chance origin of the periodicity
as a function of the trial period. The three independent datasets
shown in Figure 1 have been combined after allowing for the effects of
the period derivative.

Thus although our experiment is unable because of limitations in the timing accuracy to investigate the detail of the emission, we have evidence for significant and reproducible VHE gamma ray emission from PSR 1937 with the general double pulse feature seen at radio wavelengths. The recent model of Cheng et al (4,5) is successful in explaining the separation of the pulse and interpulse and their variation with energy for the Crab and Vela pulsars. The observation of an interpulse at TeV energies is according to this model an indication that the gamma rays are produced in the outer magnetic gap, emitted backwards and cross the whole magnetosphere. This will only be possible if the magnetic field traversed is, as in the case of PSR1937+21, small enough to avoid e^+e^- production.

The time averaged gamma ray flux may be determined from the ratio of the gamma ray count rate to the (known) rate of cosmic ray protons in the telescopes' fields of view. The observed fraction is 1 ± 0.5 % of the cosmic ray flux. For a distance to PSR1937+21 of 5 kpc (9) and assuming a differential spectral index of 3, a flux of 1 % corresponds to a luminosity above 1000 GeV of $3 \pm 0.6 \times 10^{35} \mathrm{erg\ s}^{-1}$. This is about an order of magnitude less than that observed from the Crab pulsar, in agreement with the model of Usov (3). If the source distance is 5 kpc, the energy flux of $2 \times 10^{35} \mathrm{erg\ s}^{-1}$ above 1000 GeV is about 10% of the total energy flux from the pulsar deduced from the period derivative (assuming $I = 2 \times 10^{45} \mathrm{g\ cm}^{-2}$). If the value of the differential spectral index assumed to estimate the energy flux were as steep as 6, the value of the flux would only be reduced by a factor of 2. It is therefore necessary that the energy spectrum reaches a peak somewhere near our energy threshold in order to limit the VHE energy flux. This suggests that a significant fraction of the EM energy flux is in the 100-1000 GeV range, in agreement with the emission model of Usov (3).

We are grateful to the Science and Engineering Research Council for the provision of funds and to the Commander and Staff of Dugway Proving Ground for their help.

REFERENCES

(1) D.C.Backer, S.R.Kulkarni, C.Heiles, M M.Davis and W.M.Goss, Nature, 300, 615, (1983).
(2) I.A.Gibson, A.B.Harrison, I.W.Kirkman, A.P.Lotts, J.H.Macrae, K.J.Orford, K.E.Turver and M.Walmsley, Nature, 296, 833, (1982).
(3) V.V.Usov, Nature, 305, 409, (1983).
(4) K.S.Cheng, C.Ho and M.Ruderman, Astrophys. J., 300, 500, (1985).
(5) K.S.Cheng, C.Ho and M.Ruderman, Astrophys. J., 300, 522, (1985).
(6) M.M.Davis, J.H.Taylor, J.M.Weisberg and D.C.Backer, Nature, 315, 547, (1985).
(7) J.C.Dowthwaite, A.B.Harrison, I.W.Kirkman, J.H.Macrae, T.J.L.McComb, K.J.Orford, K.E.Turver and M.Walmsley, Astrophys. J., 286, L35, (1984).
(8) M.E.Ash, I.I.Shapiro and W.B.Smith, Astron. J., 72, 338, (1967).
(9) C.Heiles, S.R.Kulkarni, M.A.Stevens and D.C.Backer, Astrophys. J., 273, L75, (1983).

PULSED TeV GAMMA RAYS FROM VELA X-1

A.R.North, O.C.de Jager, B.C.Raubenheimer, A.J.van Tonder
and G.van Urk
PU-CSIR Cosmic Ray Research Unit, Potchefstroom University,
2520 Potchefstroom, South Africa.

Evidence is presented that the 283 s pulsar in the X-ray binary
system Vela X-1 radiates TeV γ-rays. The observations taken over
one month indicate a time-averaged flux of 2.2×10^{-11} γ cm^{-2} s^{-1} with
a probability that it is due to chance of 6×10^{-5}. On 4 May 1986 a
periodic outburst occured with a maximum flux of
9.3×10^{-11} γ cm^{-2} s^{-1}. The 283 s modulated outburst was directly
observable in the counting rate and the probability for occurence
due to chance is 2.5×10^{-5}. In both these cases a broad, unsymmetric
light curve was observed, consistent with X-ray information on this
source and results on the two other binaries emitting TeV γ-rays.

The binary X-ray source Vela X-1 was discovered by the Uhuru and
SAS-3 satellites[1] during 1975 and has been continuously monitored since
then by a variety of satellites[2,3]. The optical counterpart was identified[4]
as the B supergiant HD 77581 at a distance of (1.9 ± 0.2) kpc from Earth.
The orbital period is 8.964 days[2] whilst the pulse period of the neutron
star is 282.92 s. Large secular variations in the pulse period over time
scales of days and years have been observed[2] but from 1980 up to the
newest measurements[3] during 1984 it has remained nearly constant although a small spin-up is evident. The X-ray emission shows rapid and
complicated variations as may be expected from a wind driven, accreting
binary[1,3] and the radiation is not totally extinguished during the eclipse,
lasting ~20% of the orbit[2,5,6]. The pulsed light curve is highly variable
in the X-ray region changing from a very broad structure with five
discernable peaks in the 5 keV range[5] to a broad double peak in the
(15-30) keV range[2,6]. No evidence of radiation from Vela X-1 was reported from the γ-ray telescopes on the SAS-2 and COS-B satellites, as
was the case with all other X-ray binary systems. Vela X-1 also radiates
in the PeV energy range[7,8].
Observations of Vela X-1 were made on eleven nights during the
period from 2 April to 10 May 1986 with the VHE γ-ray telescope at
Potchefstroom[9]. The threshold energy of this telescope is ~1 TeV at the
zenith and the field of view 2.2° On all occasions the source was tracked

163

K. E. Turver (ed.), Very High Energy Gamma Ray Astronomy, 163–167.
© *1987 by D. Reidel Publishing Company.*

for a period of between 95 and 190 minutes at zenith angles between 17°
and 36°. A total number of 54450 Cerenkov events were registered during
the observation time of 26.2 hours. The arrival time of each event was
registered with a resolution of 0.1 μs. Due to the long pulse period involved in this source and considering the duration of each of our observations it was not necessary to correct the arrival times to either the
Solar System barycentre or to the binary focus. The TeV-observations
on the binary systems Her X-1 and 4U0115+63 indicates that the light
curve of the pulsed emission as seen in the X-ray region is duplicated
in the TeV γ-ray region[10,11]. As Vela X-1 shows at least a double peak
structure it was decided to use the Z_2-test[12] (which combines the Fourier

power in the fundamental and first harmonic) in analyzing the data. The
individual tracks were tested over a period range equal to one independent period around the known pulsar period of 282.92 s. Six of the
eleven tracks showed deviations from uniformity at a confidence level of
90% or better (three of these were at a confidence level larger than 99%).
The binomial probability that this deviation from uniformity is due to

chance is 1.78×10^{-4}. Combining[13] the chance probabilities of all the
tracks, Fig. 1 results. A peak at (281.8 ± 4.3) s with a chance probability

of 5.9×10^{-5} can be seen. The time-averaged pulsed γ-ray component of
the 11 tracks is (3.5 ± 2.3)%. The error in the signal strength is significantly different from that expected from a source with a constant radiation pattern, indicating that the radiation is time variable. Since the
effective collection area of the telescope[9] is 9×10^{8} cm^2 the resulting
time-averaged   pulsed   flux   during   our   observations   is

$(2.2 \pm 1.4) \times 10^{-11}$ γ cm^{-2} s^{-1}.

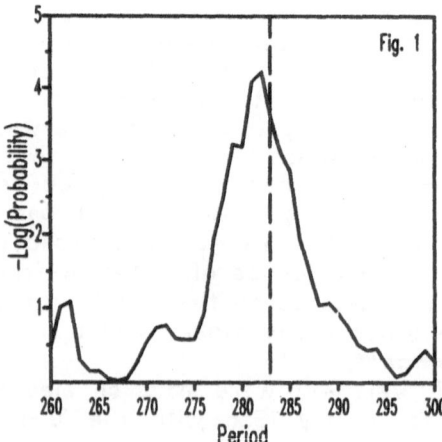

Fig. 1: The combined Z_2-probabilities for all 11 tracks on Vela X-1 as a
function of the test period over four independent periods. The dashed
line are multiples of the pulsar period measured in the X-ray region.

   With this assurance that pulsed emission was present the data was searched for any DC enhancement of the counting rate. Such an event was identified in the eigth track, on 4 May 1986, where the counting rate of all three telescope units in operation at the time increased for ~1.0 minute by ~100% over the mean value of 44 counts min.$^{-1}$. This is equal to a 5.7σ deviation from the mean rate. The chance probability for such an occurence in our whole data set is 8.8×10^{-4}. Moreover, as may be seen from Fig. 2, it seems as if this enhancement is modulated with the pulsar period and that the enhancement persists over at least four pulsar rotations. This result, we believe, is the first direct illustration of periodic emission of TeV γ-rays from an X-ray binary. This excess of (42±5) photons over a period of one minute, is equal to a DC flux of $(9.3 \pm 1.1) \times 10^{-11}$ γ cm^{-2} s^{-1}. This enhancement of the flux is of the same order as that seen by various observers[3,5,6] in the X-ray region. The whole track containing the outburst was then analysed for periodic emission. The data was split into 14 overlapping sections of length five times the pulsar period. The result indicated activity only for the 24 minutes after the beginning of the outburst. During this time the probability that this emission at 283 s was due to chance is 2.0×10^{-3}, or 0.028 when all the degrees of freedom were taken into account. The pulsed γ-ray content of this event is (9±3)% resulting in a pulsed flux of

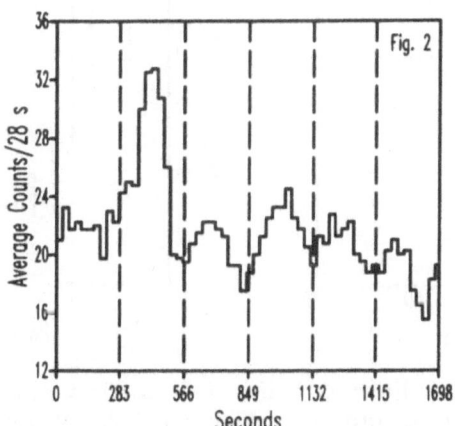

Fig. 2: The average number of counts (calculated over 112 s) as a function of time for the outburst on 4 May 1986. Zero time corresponds to 17:14 UT. The dashed lines indicate the pulsar period.

$(5.4 \pm 1.8) \times 10^{-11}$ γ cm^{-2} s^{-1}. Since the DC and periodic tests on the outburst are independent, they may be combined. We conclude that the probability that this pulsed outburst is due to chance is 2.5×10^{-5}.

In Fig. 3 the pulsed light curves of the outburst and the track with the strongest signal are illustrated. In both cases a broad unsymmetric structure can be seen which is not inconsistent with the X-ray light curves, especially at the lower energies[4,5]. These light curves show a resemblance with the other two TeV γ-ray binaries[10,11] Her X-1 and 4U0115+63. Assuming a mean threshold energy of 1.5 TeV for our telescope during these observations the γ-ray luminosity of Vela X-1 varies from $(2.6 \pm 0.1) \times 10^{34}$ erg s^{-1} averaged over all our observations to a maximum of $(9.3 \pm 1.1) \times 10^{34}$ erg s^{-1} during the outburst. This luminosity is about a factor of 10 lower than those measured so far for the abovementioned X-ray binaries in the TeV range [14]. It must, however, be kept in mind that Vela X-1 is not only closer to Earth but, due to it's longer orbital as well as pulsar period, it may be expected that it's TeV γ-ray luminosity could be lower than those for the other two systems. Analysing the activity of Vela X-1 modulo the orbital phase, no definite structure was discernable although indications point to a higher activity during the phase-interval 0.3-0.8. More observations are, however, needed to establish this.

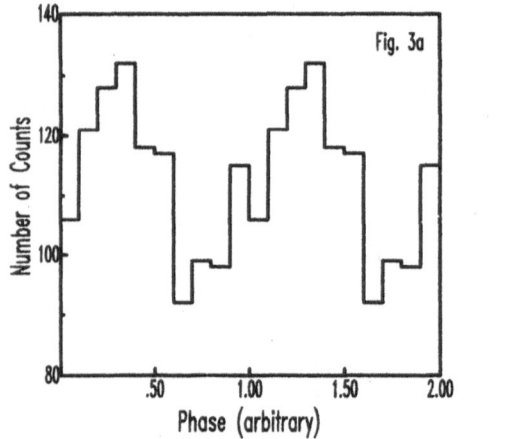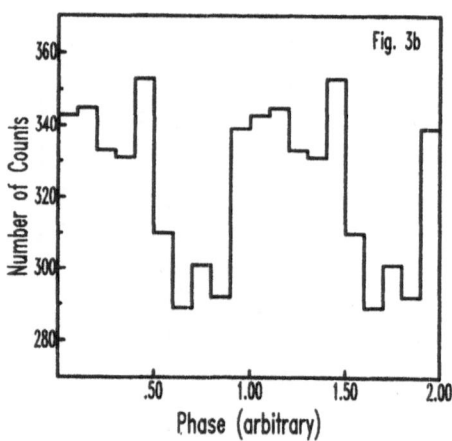

Fig. 3: The 283 s γ-ray light curves, (a) During the 24 min. outburst on 4 May 1986 and (b) during the track on 3 April 1986.

We are extremely grateful to all our colleagues who offered their valuable time in realizing these measurements.

References

1. Joss,P.C. and Rappoport,S.A., Ann. Rev. Astron. Astrophys., 22, 537-592 (1984)
2. Nagase,F. et al, Astrophys. J. (Lett), 280, 259-268 (1984)
3. Nagase,F., IAU Symp. No. 125, preprint, Nanjing, China (1987)
4. Sadakane,K. et al, Astrophys. J., 288, 284-291 (1985)
5. Van der Klis,M. and Bonnet-Bidaud,J.M., Adv. Space Res., 3, 39-42 (1984)
6. Bautz,M. et al, Astrophys. J., 266, 794-805 (1983)
7. Protheroe,R.J., Clay,R.W. and Gerhardy,P.R., Astrophys. J. (Lett), 280, L47-L50 (1984)
8. Suga,K. et al, Proc. Workshop on Techniques in UHE Gamma-Ray Astronomy, La Jolla, 48-53 (1985)
9. De Jager,H.I. et al, South African J. Phys., 9, in press (1986)
10. Dowthwaite,J.C. et al, Nature, 309 691-693 (1984)
11. Chadwick,P.M. et al, Astron. Astrophys., 151, L1-L5 (1985)
12. Buccheri,R., Proc. Workshop on Techniques in UHE Gamma-Ray Astronomy, La Jolla, 98-103 (1985)
13. Eadie,A.T. et al, Statistical Methods in Experimental Physics, (Amsterdam:North Holland) (1972)
14. Weekes,T.C., Proc. of the VIth. Astrophysics Meeting, preprint, Les Arcs, France (1986)

Search for TeV Gamma rays from the Galactic Plane near Cygnus X-3.

P.T.Reynolds[1], M.F.Cawley[2], D.J.Fegan[1], K.G.Gibbs[3],
R.C.Lamb[3], D.A.Lewis[2], N.A.Porter[1],T.C.Weekes[3].

1   Physics Department, University College, Dublin, Ireland
2   Department of Physics, Iowa State University, Ames, Iowa
3   Whipple Obsevatory, Harvard-Smithsonian Center for
        Astrophysics, Amado, Arizona

ABSTRACT.   The null results of a search for gamma-ray emission at TeV
energies from the galactic plane near Cygnus X-3 are reported.

INTRODUCTION

The Crimean Astrophysical Observatory group reported evidence for a
0.5% deficit in the distribution of cosmic ray air showers coming from
within ± 1.5° latitude of the galactic plane (Fomin et al. 1977). This
result was based on their own observations taken between 1971 and 1975
and other published results. Their interpretation assumed that there
was a uniform source of gamma rays within ±5° of the galactic plane
which originated from the inverse Compton scattering of cosmic
electrons on the interstellar sea of optical photons from stars; in the
plane itself this sea of photons was strongly attenuated by dust and
hence few gamma rays resulted. No evidence was presented for this broad
source of photons which should dominate the TeV gamma-ray sky as the
Milky Way does at lower energies.
        The Whipple Observatory Sky Survey, which was primarily designed
to search for discrete sources in the Northern Hemisphere, scanned a
large portion of the galactic plane (Weekes et al. 1979). Two results
were apparent from the galactic plane scans: a broad excess in the
galactic longitude range 170° to 210° and a narrow deficit between
longitude 40° and 80°. However the detectors used were sensitive to
changes in the night-sky light background due to different star fields;
this is a second order effect in the search for candidate discrete
sources but in broad regions of extra brightness such as the Milky Way
it is serious enough to cast doubt on the reality of any apparent
anisotropy found in the cosmic ray air shower distribution.
        The University of Durham group, in a single long scan across the
galactic plane in the Cygnus region, saw a broad excess within ±5° of
the galactic plane with a dip back to the off-plane average close to
the plane itself (Dowthwaite et al. 1985). The excess was (8±3)% of the
cosmic ray background, considerably larger than the previously reported

169

K. E. Turver (ed.), Very High Energy Gamma Ray Astronomy, 169–172.

effects. It corresponded to a flux of 3×10^{-7} photons cm^{-2} s^{-1} $ster^{-1}$ at an energy threshold of 1 TeV.

An analysis of underground muon data from the Soudan Mine Experiment showed evidence for the same kind of distribution in the galactic plane region near Cygnus (Marshak 1986).

On the basis of their observation Dowthwaite et al.(1985) have suggested that Cygnus X-3 lies in a hole in the galactic plane so that observations which compare the intensity of cosmic ray air showers coming from this region with nearby regions will underestimate the Cygnus X-3 flux.

The unambiguous detection of TeV gamma rays from the galactic plane would be one of the most important results in VHE gamma-ray astronomy and would open up the possibility of detailed mapping of the TeV cosmic ray distribution. As the first stage in such a verification we have attempted to duplicate the reported distribution around galactic longitude 80°.

OBSERVATIONS

Six nights of observation were taken under clear skies between June 6 and June 13, 1986. The gamma-ray telescope used was the Whipple Observatory 10m Reflector acting as a 37 element camera (Cawley et al. 1985b). The trigger requirement was that any two of the inner 19 tubes have a light signal that exceeded a preset threshold. Because of the relative brightness of the Milky Way this threshold was 30% higher than that normally used in the Whipple Observatory Gamma-ray Program.

Each night the reflector was positioned to scan the galactic plane at longitude 80° (declination(1950) $40^{\circ}47'$) approximately 30 minutes after the beginning of the observation which lasted for 2 to 3 hours (fig.1). A fixed elevation of 64° was used so that the average counting rate was 100/minute throughout. Padding lamps were used in the inner 19 tubes and were set to compensate for variations in the

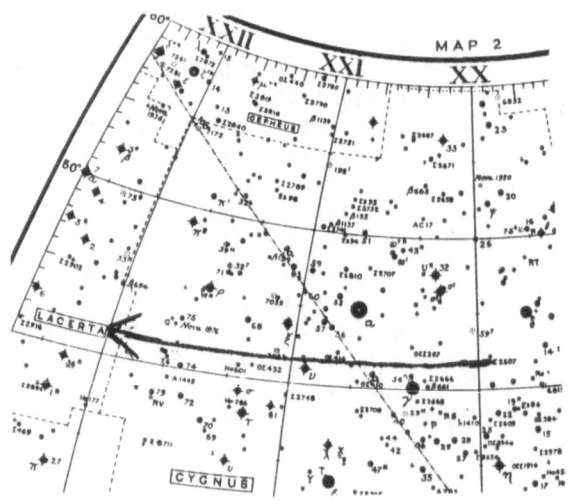

Fig.1. The region of sky scanned during these observations.

background current due to stars in the Milky Way. The 2.3 magnitude
star, Gamma Cygni, was the only star which exceeded the padding lamp
settings; no effect in the total counting rate was found to correlate
with the presence of this star in one of the trigger tubes.

Images of each event were recorded with ten bit resolution; also the
time was recorded with an accuracy of 0.5 ms. For this analysis no
cuts were made on the shower image size, magnitude or orientation. The
one minute total rates for each scan were checked for consistency as
were the four and eight minute totals. The observed variations were all
found to be statistical. The scans were also checked for linear drifts
in counting rate (sometimes found in long scans due to atmospheric
changes or instrumental drifts) but none were found.

The one minute totals for the six nights were summed in minutes of
right ascension with the results shown in figure 2. There are no
significant deviations from a random distribution on any time scale. In
particular we find no evidence to support the reported deficit at the
galactic equator at this longitude nor is there any evidence for an
excess within $\pm 5^{\circ}$ about the plane. Defining the ON region as the 80
minutes of scan time centered on the galactic plane and the OFF region
as the following 80 minutes, we get ON total = 48,264, OFF total =
48,201, difference = +63. The 3 sigma upper limit is 4.0×10^{-11} photon
$cm^{-2}\ s^{-1}$ above an energy of 0.9 TeV corresponds to less than 1.5% of
the cosmic ray background.

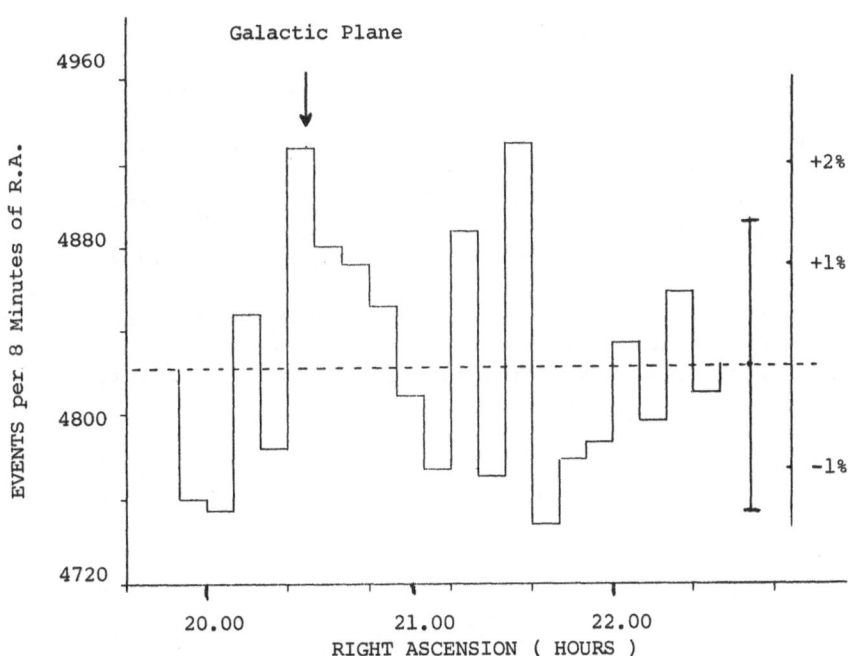

Figure 2. The shower totals per eight minutes of Right Ascension
across galactic longitude 81°.

DISCUSSION

This result applies only at this particular longitude but since
this is the region that includes the important source, Cygnus
X-3, and its comparison OFF regions, it is important to establish
if non-uniformities exist. We note that the Crimean results
(Fomin et al. 1977) show less than a one sigma effect at this
longitude. Also this is just at the edge of the 40°-80° range of
longitude where an effect was reported by the Whipple Observatory
(Weekes et al. 1979). In fact at this longtitude the Sky Survey
does not show any significant effect. The Cygnus X-3 galactic
plane detection was quoted as evidence to support the marginal
detection of M31 (Dowthwaite et al. 1984); however this
observation has not been confirmed (Cawley et al. 1985a).
       It is important that the question of TeV gamma-ray emission
from the galactic plane at any longitude be clarified; a program
to observe the galactic plane at all longitudes will be a major
undertaking but should be a priority at all observatories with
suitable telescopes. Not only is the galactic plane emission
important in its own right but a non-uniform background may
seriously limit the detection of discrete sources close to the
plane.

   This work is supported by the U.S.D.O.E. and the N.B.S.T. of
Ireland.

                        References.

Cawley,M.F. et al.;1985a, Proc.Int. Cosmic Ray Conf.(La
        Jolla).3, 453.
Cawley,M.F. et al.;1985b, Proc.Int. Cosmic Ray Conf.(La
        Jolla).1, 264.
Dowthwaite,J.C. et al., 1984, Astron. Astrophys.,136, L14.
Dowthwaite,J.C. et al.; 1985, Astron. Astrophys.,142, 55.
Fomin,V.P. et al.; 1977, Proc.Int.Cosmic Ray Conf.(Plovdiv)
        1, 12.
Marshak,M.; 1986, Proceedings of XXIth Recontre Moriond,Les
        Arces,France. (in press).
Weekes,T.C. et al.; 1979, Proc. Int. Cosmic Ray Conf.(Kyoto) 1,
        133.

OBSERVATION OF CYG X-3 FROM OCTOBER 1985 TO MAY 1986 AT AKENO

T. Kifune, M. Mori*, T. Hara & K. Nishijima,
Institute for Cosmic Ray Research,
University of Tokyo,
3-2-1 Midori, Tanashi, Tokyo, Japan.
*Department of Physics,
Kyoto University, Kyoto, Japan.

ABSTRACT. The EAS Array of Akeno is described about its present
status on observation of ultra high energy gamma rays. The high rate
observation by a new recording system has started in December 1985. A
preliminary analysis shows that the PeV gamma ray flux from Cyg X-3
may have been in an enhancement of a factor of five since the radio
outburst in October 1985.

1. INTRODUCTION

It is not an easy task to improve the poor statistics, the main cause
of the present somewhat ambiguous status in PeV gamma ray astronomy.
However, it is also true that the extensive air shower (EAS) arrays do
not usually record all the small showers hitting the array. In the
case of Akeno array (1), the detection area for a few PeV EAS was made
much smaller than the available in order to have an 'equi-partition'
of a counting rate over a wide energy range. With a capability of
recording a large number of events at high rate, the observation
deducted for gamma ray astronomy has been in operation at Akeno array
since December 1985.

On Cyg X-3, the best studied point source in PeV energy region,
the observations by different groups (2),(3),(4) are not in a full
agreement with each other about the time variability of the flux, the
phase of 4.8 hrs. period at the peak flux and the amount of muon
content in the EAS cascade.

Recently, it is reported by Tien Shien group(5) that the radio
outburst at Cyg X-3 in October 1985 was associated with the flux
increase in ultra high energy gamma rays. This encouraging
indication, if the observations by many groups are available, should
be fully utilized for establishing a common view on Cyg X-3.

K. E. Turver (ed.), Very High Energy Gamma Ray Astronomy, 173–178.
© *1987 by D. Reidel Publishing Company.*

2. NEW OBSERVATION SYSTEM AT AKENO FOR GAMMA RAY

Fig. 1. shows the detectors which are linked with the new recording system, controlled by a micro computer PDP 11/73. All these detectors are the ones constructed for multi-purpose EAS measurements of Akeno array (hereafter called in this paper as 'common array' or 'common recording'). In the central area, the scintillation counters are re-arranged to distribute in the grid of 20m separation. Most of them are with facility of measuring the arrival direction by fast timing. However, the time resolution is to be improved, because they were constructed for common array but not for gamma ray research. It is now under way to add new fast timing detectors with a better resolution. The total area of 225m^2 for muon detection (shown by squares in Fig. 1) will soon be linked. Only 25m^2 area (a black square) is already done by the present time of analysis. A clock with a precision as good as 0.1msec will also join the recording soon.

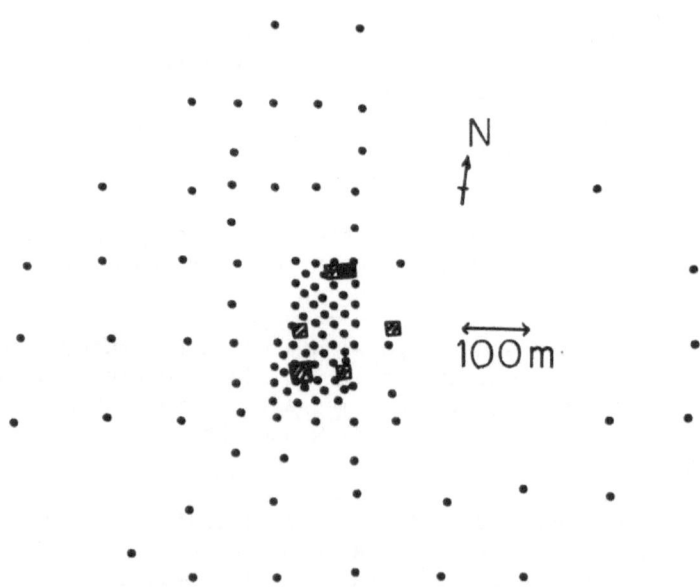

Figure 1. The array for new recording system.

Although a substantial part of the new system was not yet
completed, the data recording has started in December 1985. The
data-taking is continuously done except a short break once a week when
magnetic tape for recording data is replaced. The trigger rate is
about 250 per hour. This rate is 10 times higher than the small EAS
rate in common recording. Owing to the increase of detection area and
to the somewhat lower detection threshold. The threshold energy is
estimated as a little lower than 1 PeV.

3. RESULTS

The data in common array were analysed for the period of 1981-1984,
giving 18 muon-poor showers correlated with 4.8 hours X-ray variation.
 The data-taking continues still at the present time in common
recording and the full analysis will be done with the more data
accumulation.

In this paper the data in common recording during the period of
radio outburst are picked up for analysis because of its special
interest, together with a part of the data by the new recording system
which appears outstanding beyond the observational errors we have so
far checked.

3.1 Data during radio outburst

In 1985, a radio outburst occurred in Cyg X-3 reaching the maximum on
October 12 lasting for about 10 days(6). Common array was in
operation during this period. The conditions on data-taking are the
same as the ones during 1981-1984.

Analysed are the data of one month live time from the beginning
of October to the middle of November. Our previous result predicts
about 0.4 muon poor shower from Cyg X-3 for this time span. We have
observed four muon poor showers as shown in Fig. 2. The muon content
of two showers from the direction of Cyg X-3 (out of 5 showers from
all the directions in the sky) is smaller than 1/20 of the normal
showers and the other two has 1/10 of muon content of the normal.
These showers are surrounded by a dotted circle in Fig. 2. All the
showers from $\pm$ 10° window around Cyg X-3 are plotted as a function
of muon to electron ratio in the vertical axis and shower age in the
horizontal axis.

The figure supports our previous result that the events from Cyg
X-3 are explained by gamma-initiated EAS of conventional particle
interactions. The most of the events are consistent with the general
trend of nuclear active particles, i.e. older ages for high muon to
electron ratios.

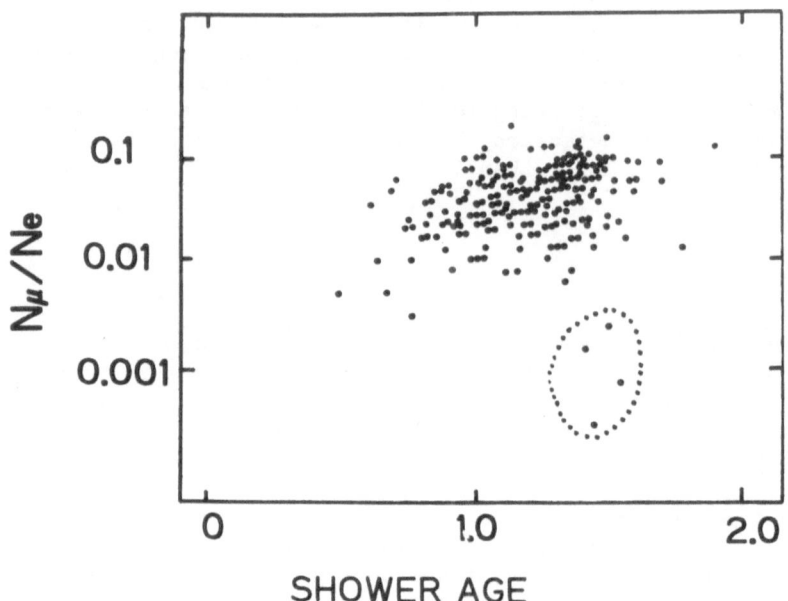

Figure 2. Scatter plot of muon to electron ratio vs shower age.

     The phase of 4.8 hrs. period (7) is 0 for one shower and about
0.6 for three showers.

3.2 Data for new recording system

About a million of EAS have been accumulated during six months
operation of December 1985 to May 1986. About thirteen thousand
showers are fallen within $\pm$ 10° window around the direction of Cyg
X-3. Since the available area $25m^2$ of muon detectors at present is
too small, the muon content is not analysed yet. The phase of 4.8
hrs. period is calculated for all the showers within the direction
window. The result, as presented in Fig. 3, shows an excess of 130
events at the phase of about 0.6. The excess gives 3.4σ effect (2.7σ
in 0.55 - 0.60 bin and 2.1σ in 0.60 - 0.65 bin).

     In the analysis of 4.8 hrs. period, spurious effects are brought
in when; (a) the observation time per day is not equal to 4.8 hrs.,
and (b) the time span of the whole observation is not much larger than
half a year, the period that the uniform distribution of zenith angle
is achieved in all the phase bins. The analysis is done by requesting
a window of observation time per day exactly to be the same value with

4.8 hrs. period to avoid the effect (a). The observation period of about a half year of ours is considered to minimize the effect (b).

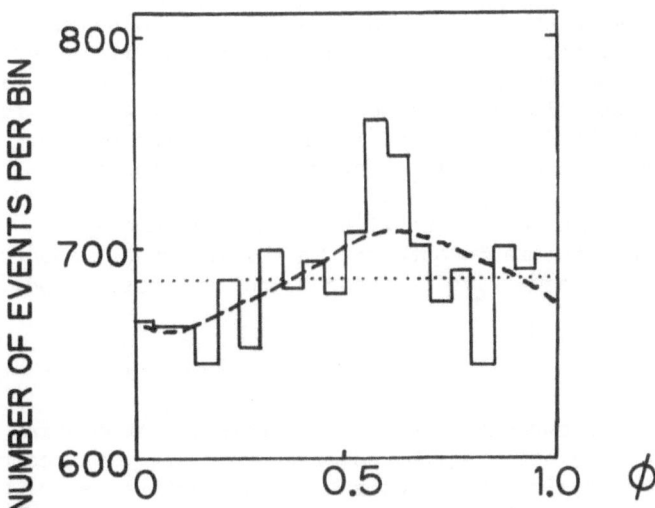

Figure 3.   Events by new recording system.

However, the obtained result is still influenced by the zenith angle effect, noticed in the figure as a sinusoidal change of counting rate vs phase. The variation of counting rate due to zenith angle change is observationally obtained by folding the whole data by the period of 1/5 of the sidereal day. By using this distribution and by assuming that one tenth (15 days) of the whole data are peaked at phase 0.6, the maximal effect of non-uniform zenith angle contribution was estimated. The resulting effect is shown in Fig. 3. by dashed curve. The observed peak is seen higher and narrower than this effect.

The peak is not, however, sharp enough to produce the peak precisely and exclusively at the X-ray period, because the short time span of observation can produce in any test period near the X-ray period.

4. DISCUSSIONS

Our previous result(4) in common array predicts 0.4 muon poor showers
per month, of which muon to electron ratio less than 0.001. If we
compare this rate with two events in Figure 2, the flux enhancement is
obtained as $2/0.4 = 5.0 \pm 3.5$. The four muon poor events encourage
our previous conclusion that the gamma ray showers exhibit the
ordinary expected feature, i.e. muon poor and old aged EAS in contrast
with the normal showers of charged particle.

The results in the new recording system should be compared with
the 'temporal trigger' in our previous report(4). If the excess
events of about one hundred in the new recording system is taken, the
enhancement in the flux is obtained as a factor of 4 - 5. It is
consistent with the flux increase in radio outburst period. The phase
of 4.8 hrs variation period is about 0.6 in both cases.

Our results are consistent with a view that Cyg X-3 is now in a
high state, since the radio outburst in the last year, and emitting
ultra high energy gamma rays at phase 0.6 of 4.8 hrs period.

ACKNOWLEDGEMENT

The authors are indebted to the other members of Akeno array for their
co-operations in maintaining the observation. The data reductions
were done by FACOM M380 at the Computer Room, Institute for Nuclear
Study, University of Tokyo.

REFERENCES

(1) T. Hara et al. 1979, Proc. 16th International Cosmic Ray
     Conference (Kyoto) 8, 135.
(2) M. Samorski and W. Stamm 1983, Ap.J. (Letters) 268, L17.
(3) J. Lloyd-Evans et al. 1984, Nature 305, 784.
(4) T. Kifune et al. 1986, Ap.J. 302, 230.
(5) A.E. Chudakov 1986, Private communication.
(6) L.A. Molnar 1985, Private communication.
(7) M. van der Klis and J.M. Bonnet-Bidauld 1981, Astr. Ap. 96 L5.

Observations of Cygnus X-3 near 10^{15}eV made at Haverah Park : 1979-1985

P J V Eames, A Lambert, J C Perrett, R J O Reid,
N J T Smith, A A Watson and A A West

Department of Physics, University of Leeds, Leeds LS2 9JT UK

Abstract

Observations of PeV γ-rays from Cygnus X-3 made at Haverah Park between 1979 and 1985 are described. Evidence for a change in the light curve over this period is summarized.

1 Introduction

During the construction of a purpose built γ-ray detector at Haverah Park (Brooke et al 1985), the series of observations first described by Lloyd-Evans et al (1983) has been continued. In addition to the 50 m water-Cerenkov array used in the earlier work, and from which data are now available from January 1979 to December 1985, a second array was operated during July to October 1984 and again from June 1985. The original array comprises 4 x 13.5 m^2 water-Cerenkov detectors and lies 2 km from the centre of the giant shower array. The new array of 4 x 6.7 m^2 detectors is of similar construction and is at the centre of Haverah Park, as is the purpose built detector.

Data collected since 1982 provide some support for the original claims (Samorski and Stamm 1983, Lloyd-Evans et al 1983) for a 4.8 hr pulsed signal from Cygnus X-3 at PeV energies. However, the statistical significance of the new results is low and the tentative interpretations discussed here must await confirmation (or otherwise) from the new observations with improved equipment. In most of the discussion below we have departed from previous practice by using a revised ephemeris for Cygnus X-3 (K O Mason, private communication) in place of the van der Klis-Bonnet-Bidaud (VKB) ephemeris adopted previously. This ephemeris was derived by applying a consistent method of analysis to the more recent X-ray satellite data. The ephemeris parameters (in the usual notation) are t_0 = 2442946.739 JD, p_0 = 0.1996851 days and $\dot{p}_0$ = 7.84 x 10^{-10}.

K. E. Turver (ed.), Very High Energy Gamma Ray Astronomy, 179–184.

2 Comparison of Haverah Park (1979-83) and Kiel PeV light curves

The Kiel results, which first drew attention to Cygnus X-3 as a PeV γ-ray source (Samorski and Stamm 1983), were derived from showers recorded between 18 March 1976 and 7 January 1980. Modifications to the original analysis have been described (Samorski and Stamm 1985). The Kiel phaseogram, recomputed with the Mason ephemeris, is shown in Figure 1a. There is a peak of 10 events in the phase bin at $\emptyset$ = 0.25 compared with an expected number of 1.44 derived from the off-source background. Using Hearn's (1969) relative likelihood method the confidence level, C, of this result is 99.98% allowing for the fact that there was no 'a priori' expectation of one bin being favoured over another.

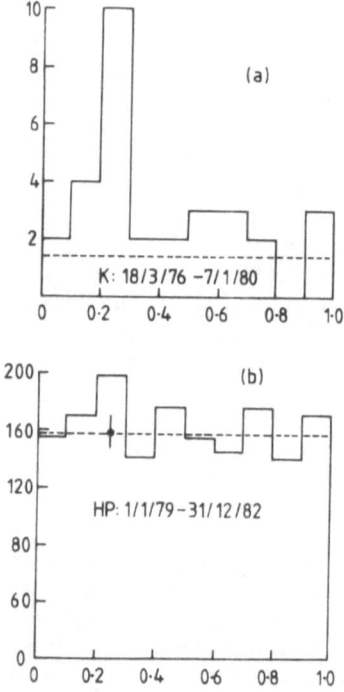

Figure 1: Comparison of Kiel and Haverah Park data
in 10 phase bins (with K O Mason's ephemeris)

The Haverah Park data for 4 years (1 January 1979 - 31 December 1982) were first discussed by Lloyd-Evans et al (1983) where they were binned in 40 intervals using the VKB-ephemeris. These data, re-analysed with the Mason ephemeris, are shown, in 10 phase bins, in Figure 1b. The peak of 198 events over an 'off-source' background of 158 is in the same 0.25 phase interval as the Kiel peak. The confidence level, assuming the bin to have been predicted by the Kiel result, is 99%. The 40 bin phaseogram formed using the Mason ephemeris is compared with the original Haverah Park result in Figure 2. The very striking sharp peak in phase bin 10 of the original analysis is now less so (61 counts as against 73) and the maximum count has shifted from bin 10 to 11. However, the general agreement of the PeV light curve with the Kiel data, for what is the most nearly contemporaneous PeV data set, is encouraging.

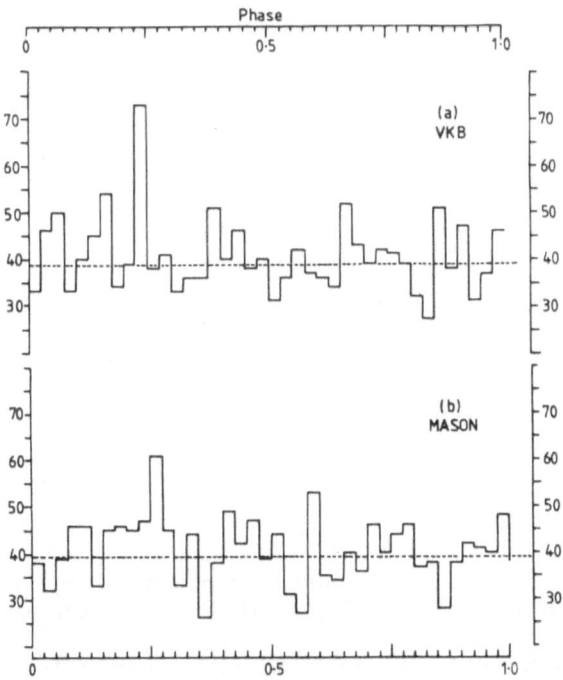

Figure 2: Comparison of analysis of Haverah Park data (1979 - 1982) in 40 bins using van der Klis-Bonnet-Bidaud ephemeris and the Mason ephemeris

3 Haverah Park measurements : 1983 - 1985

Measurements with the array used by Lloyd-Evans et al (1983) are continuing. The numbers of events in phase bin 11 (of figure 2) and in phase bins 9 to 12 (of figure 1) are plotted, together with the time-average flux values, in Figure 3 as a function of year. The upper limits shown are 95% limits (Hearn 1969). It is clear that the intensity in phase bin 11 post 1980 has fallen below that observed in 1979 and 1980. However, the earlier intensity is consistent with the average value of the intensity above 10¹⁵eV (3.2 (± 0.4) x 10⁻¹⁴ cm⁻² s⁻¹) deduced from the DC excess of 76 events seen with this array over the 7 years in question. This result leads to the conclusion that the shape of the light-curve has altered in the post 1980 period.

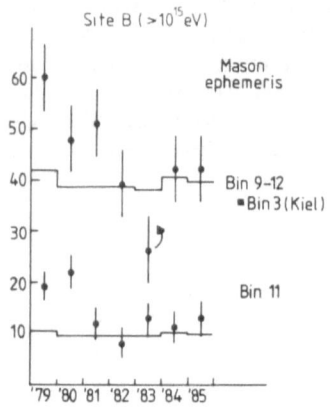

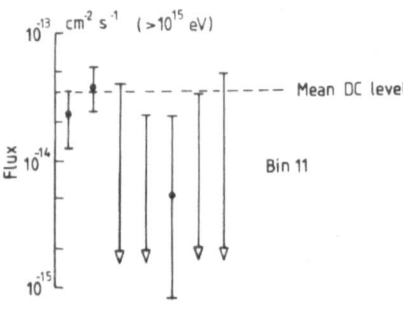

Figure 3:   Variation of flux and observed counts with time.   The solid
            lines indicate the background level of each year

    For part of 1984 and 1985 an additional array, with a slightly
lower energy threshold, was brought into operation.   The summed
histograms, found using the Mason ephemeris, are shown in figure 4
for 1984 and 1985 respectively.   The angular resolution is about the
same ($\sigma(\theta)\sim3^\circ$) for both arrays.   Marked on each histogram are phase
windows ($0.18 < \emptyset < 0.34$ and $0.58 < \emptyset\ 0.75$) defined by measurements
made at energies $< 10^{14}$eV (see Watson 1985 for details) after
correction for the Mason ephemeris.   These windows indicate the
spread in phase about 0.25 and 0.65 in which significant peaks have
been seen in other experiments.   In total the excess number of events
is 55 above the 'off-source' background.   Although neither histogram
shows any very striking peaks, it is of interest that there are
adjacent bins in 1984 and 1985 which are above the background
within the two windows.   All of the net excess over the background
expectation can be found in those windows which are near $\emptyset = 0.65$ in
1984 and $\emptyset = 0.25$ in 1985.   We have noted this change in the light
curve previously (Lambert et al 1985) in the 1984 data.   Redistribution
of the target material in which the parent π^0's are created may be the
reason.   Such redistributions may be associated with radio outbursts
at the source and since September 1980 there have been 13 flares
above 1 Jy at 11 cm, of which 3 have been above 10 Jy, a level
comparable to that reached during the famous 1972 flare.

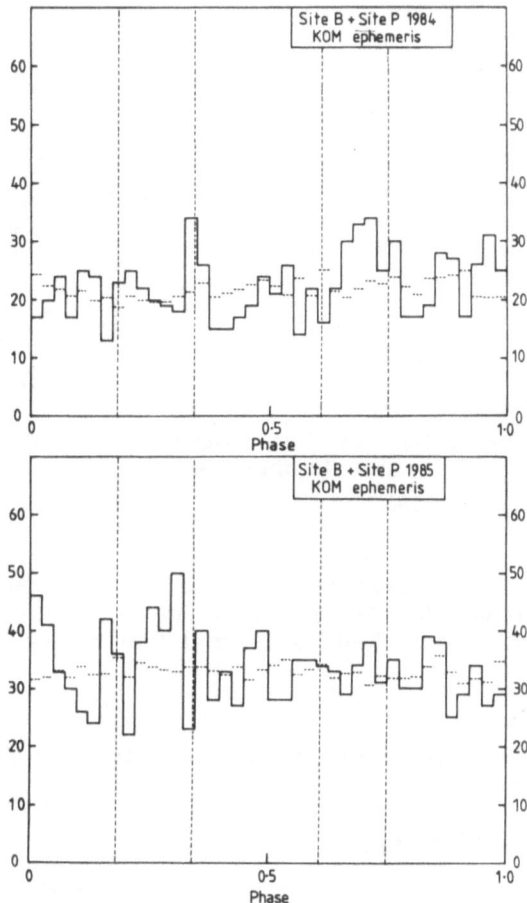

Figure 4:   Phaseograms for data from the two water–Cerenkov arrays
(B and P) during 1984 and 1985

Flares occurred on 30 September 1984 (11 Jy) and on 3 and 7
October 1985 (5 and 17 Jy). The array with the lower energy
threshold was operational during October 1984 and October 1985:
phaseograms for these periods are shown in Figure 5. Enhancements
are visible in both histograms within the phase windows defined
above. The time averaged flux levels above 5 x 10^{14}eV are about 1 x
10^{-12} cm^{-2} s^{-1}. Extrapolating this intensity (with an E^{-1} spectrum)
to 10^{15}eV gives a flux level of about 5 x 10^{13} cm^{-2} s^{-1} which is
somewhat greater than that found, on average, between 1979 – 1985.
These flux levels are the highest found in 10 months operation of the
low threshold array. We expect to operate the new array during the
next predicted radioflare period.

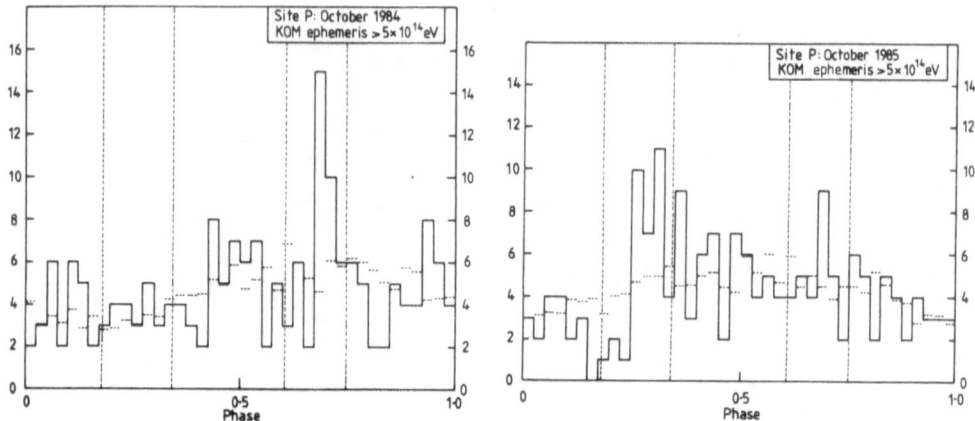

Figure 5:   Phaseograms for data from site P for October 1984
and October 1985

Acknowledgements

        The continuing support of SERC to the Haverah Park project.
Efficient operation of the arrays have greatly benefited from the
expertise of Mansukh Patel, Paul Ogden and Don Pearce.

References

Brooke G et al, 1985, Proc 19th Int Cosmic Ray Conf (La Jolla) **3** 426
1985

Lloyd–Evans J et al, 1983, <u>Nature</u> **305** 609 1983

Samorski M and W Stamm, 1983, <u>Ap J Lett</u> **268** L17 1983

Samorski M and W Stamm, 1985, Proc Workshop on 'Techniques in Ultra
High Energy Gamma Ray Astrophysics', La Jolla 1985 p85

Hearn D, 1969, <u>Nucl Inst Meth</u> **70** 200 1969

Watson A A, 1985, Rapporteur Paper, Proc 19th Int Cosmic Ray Conf (La
Jolla), in press

Lambert A et al, 1985, Proc 19th Int Cosmic Ray Conf (La Jolla) **1** 71
1985

van der Klis M and Bonnet–Bidaud J M, <u>Astron Ap Lett</u> **95** L5 1981

FURTHER STUDIES OF UNDERGROUND MUONS AT SOUDAN

K. Ruddick
School of Physics and Astronomy
University of Minnesota
Minneapolis, MN, USA 55455.
     For the Soudan I collaboration(*)

ABSTRACT. The Soudan I experiment has obtained confirmatory evidence for the existence of underground muons associated with Cygnus X-3. A significant signal was observed during a recent period of very high radio emission. Additional possible astronomical sources of the underground muons are being studied.

The Soudan I and NUSEX nucleon decay experiments have reported the observation of underground muons originating from the direction of Cygnus X-3, the well-known X-ray binary system (1,2). These muons show the characteristic 4.79 hour periodicity of Cygnus X-3, presumed to be the orbital period of a neutron star about a large companion star. These observations, if confirmed, would have great significance in the fields of particle physics, cosmic ray physics, and astrophysics. No conventional explanation is possible within our present understanding of particle physics. In order to maintain the requisite phase coherence of the signal between Cygnus and earth, the signal carrier must have a low ratio of mass to energy (3). For the observed energies of at least tens of GeV, the carrier mass must be less than a few MeV. The carrier must be electrically neutral to have avoided deflection in the galactic magnetic field. The only possible conventional candidates for the carrier, photons or neutrinos, must have impossibly high interaction cross-sections to explain the phenomenon.

The original Soudan data were obtained in 1981-83, the NUSEX data in 1982-84. More recently, the Kamioka and Frejus underground experiments have reported null results for the effect, with fluxes significantly below those reported (4,5). Those data were obtained during 1984-85.

The Soudan I detector resumed operation at the beginning of 1985. During most of 1985, this detector also showed no appreciable signal, possibly indicating that the Cygnus X-3 system was in a period of low emission. However, in September-October, 1985, an unusually large radio outburst was reported from Cygnus X-3. We have searched for underground muons associated with this outburst.

Figure 1 shows the phases of muons arriving within 5 degrees of the Cygnus X-3 direction, for the period 24 September to 7 October, 1985,

K. E. Turver (ed.), Very High Energy Gamma Ray Astronomy, 185–188.

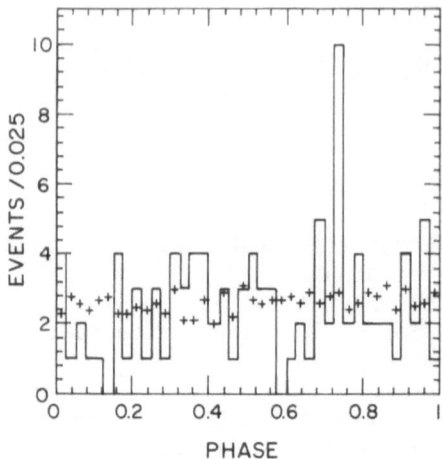

Figure 1. The phase of muons arriving from within 5 degrees of the direction of Cygnus X-3 during the period 24 Sept. to 7 Oct., 1985.

using the Van der Klis, Bonnet-Bidaud ephemeris (6). The plot shows a significant excess in the phase region 0.725 to 0.75. (In the earlier Soudan data, the muons arrived over a wider phase interval, 0.6 to 0.9). In Figure 2, we show the muon flux from this narrow phase interval in 1 week intervals for the whole of 1985. There is a strong correlation between the muon signal and the radio signal, with the muons preceding the radio emission by about one week. We believe that this is confirmatory evidence for the presence of muons associated with Cygnus X-3. Conservatively, we estimate the probability of the signal being a random fluctuation as 0.7% (7).

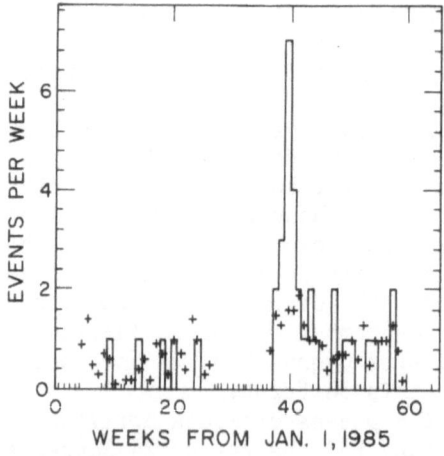

Figure 2. Muon arrival rate from Cugnus X-3 during 1985. (Within the phase interval 0.725 to 0.75)

The Frejus collaboration observed no excess during this active period of Cygnus X-3. From the Soudan data, we estimate that they should have observed a signal of 6+3 events within phase 0.7 to 0.8, above their background of approximately 2 events per phase interval. They do observe a fluctuation as large as this, but at phase 0.9 to 1.0.

In a continuing effort to study the possible muon flux from astrophysical sources, we have installed a 1 millisecond accuracy clock at Soudan. This enables us to investigate a much wider range of short time period systems. One such system is the X-ray binary 1E2259+586, which is located at the centre of an observable supernova remnant. This object is particularly favourable for study by Soudan I, since it has a favourable pulsar period of 7 seconds, but particularly because the plane of its orbit is aligned almost perpendicular to the line of sight. This makes Doppler corrections to the observed pulsar period very small (<0.19 sec), and thus minimizes possible systematic errors due to uncertainties in the orbit parameters. (Since the Soudan measurements are made over periods of typically one year duration, the precise orbit parameters are much more significant than in the case of the TeV gamma ray observations, which make observations over minutes, generally corresponding to only a fraction of the full orbit).

Figure 3 shows a plot of the Rayleigh power for pulsar 1E2259+586 searching over periods consistent with the x-ray ephemeris obtained by Fahlman and Gregory: $p=6.978632$ sec, $\dot{p}=(-7$ or $+2)\mathrm{x}10^{-14}$ sec/sec (8). In determining the power, we have used the sum of the first and second harmonics, since the known phase plot indicates that significant power should appear in both harmonics. An apparently significant peak appears at period 6.978627 sec. The probability that this peak corresponds to a statistical fluctuation has been estimated by monte-carlo calculation, and is approximately 2% over the appropriate search range. The analysis is in a very preliminary stage, but the signal is tantalizing enough to warrant considerable further investigation.

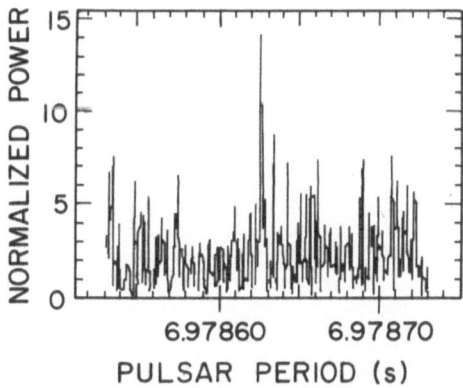

Figure 3. Rayleigh power versus search period for pulsar 1E2259+586.

Our future studies of underground muons will be considerably enhanced by the commissioning of Soudan II, a new 1 kiloton nucleon decay detector, which will have ten times the effective area of Soudan I, and which will begin initial operation by early 1987. The effects described here will be verified (or negated) by the end of 1987.

ACKNOWLEDGEMENTS. The Soudan experiment is supported by the U.S. Department of Energy and the Graduate School of the University of Minnesota. We have enjoyed the cooperation of the Minnesota Department of Natural Resources, and particularly the staff of the Tower-Soudan State Park.

REFERENCES.
* The Soudan collaboration consists of H.Courant, K.Heller, S.Heppelmann, K.Johns, T.Joyce, M.Marshak, E.Peterson, K.Ruddick, M.Shupe, of the University of Minnesota, and D.Ayres, J.Dawson, E.May, L.Price, of Argonne National Laboratory.

1. M.L.Marshak et al., Phys. Rev. Lett. 54,2079(1985) and 55,1965(1985)
2. G.Battistoni et al.,Phys. Lett. 155B,465(1985)
3. K.Ruddick, Phys. Rev. Lett. 57,531(1986)
4. K.Oyama et al., Phys. Rev. Lett. 56,991(1986)
5. Ch.Berger et al., Phys. Lett. 174,118(1986)
6. M.van der Klis and J.M.Bonnet-Bidaud, Astron.Astrophys.95,L5(1981)
7. D.Ayres, Proc. 2nd Int.Conf. on Intersections between Nuclear and Particle Physics, Lake Louise, Canada, May 1986.
8. G.G.Fahlman and P.C.Gregory, Supernova Remnants and their X-Ray Emission, J.Danziger and P.Gorenstein, eds., IAU Symposium No. 101.

SEARCH FOR MUONS FROM THE DIRECTION OF CYGNUS X-3

FREJUS Collaboration
presented by G.Chardin, DPhPE-Saclay,
91191 Gif-sur-Yvette, France

Abstract : We have analyzed muons and multimuons detected in the Fréjus
underground nucleon decay detector between February 1984 and June 1986. No
excess events are observed in the direction of Cygnus X-3, which yields a
90 % confidence level upper flux limit of $0.7 \ 10^{-12}$ cm^{-2} s^{-1}, for an average
rock overburden of 5000 hg cm^{-2} corresponding to energies > 3 TeV. Using the
4.79 hour periodicity of Cygnus X-3, no signal is found in any phase
interval.

1. INTRODUCTION

Two nucleon decay dedicated experiments have reported evidence for
underground muons from the direction of Cygnus X-3 with a characteristic
4.79 h period[1-2]. The status of the Fréjus experiment concerning the
search for muons from the direction of Cygnus X-3 is reported.

2. THE FREJUS DETECTOR

The Fréjus nucleon decay detector is installed in an underground laboratory
located in the middle of the Fréjus alpine road tunnel near Modane, France.
The coordinates of the laboratory are 6.7° E and 45.1° N. The detector
weighs 900 tons, is 6 meters high and presents a horizontal surface of (6 x
12.3) m^2. It is a very fine grain tracking calorimeter with ≈ 1 million
(5mm x 5mm) cells and a calorimetric sampling of 3 mm of iron. The number
of detected muons is ≈ 20 /hour and corresponds to a flux of $4.9 \ 10^{-9}$ cm^{-2}
s^{-1}.

3. DATA SAMPLE AND ANALYSIS

Events were recorded between February 19, 1984 and June 1, 1986. Two thirds
of these data were taken after June 85, with the full size detector. Before
that date, the detector was gradually increased from 240 to 900 tons. The
detector was active 78% of the time. Among the muons coming from the path
of Cygnus X-3, a small fraction (5%) are almost parallel to the vertical
detector planes and are lost because these muons do not satisfy the trigger

189

K. E. Turver (ed.), Very High Energy Gamma Ray Astronomy, 189–192.

requirements. In addition, short tracks or tracks with less than 8 flash
chamber planes per view as well as tracks which undergo large multiple
scattering have been removed. These cuts reduce the data sample by 11% and
ensure that the angular accuracy of the muon angle is better than the mean
scattering angle in the mountain rock. The data sample contains 219 273
muons and 5 636 multimuons. The orientation of the detector is known with
an accuracy of 0.1°. The 1-σ overall angular resolution of the muon
direction was determined to be 1.2°.

To look for a possible signal from Cygnus X-3 (α=307.65°;
δ=40.78° in A.D. 1950), we calculate the direction of each muon in
equatorial celestial coordinates, α and δ. The angular resolution and the
accumulated statistics make it possible to consider a cone of half angle 2°
centered on Cygnus X-3. We also present data in a 5° cone for direct
comparison with Ref.2.

We have compared the distributions of rock thickness crossed by
muons coming from the direction of Cygnus X-3 and by muons coming from the
entire path followed by Cygnus X-3. We found good agreement. The average
thickness, 1835 m or ≈5000 hg/cm^2, corresponds to muon energies > 3 TeV.
It is larger than the average amount of rock (1780 m) crossed by the entire
sample of muons from all directions.

The phase of each event is calculated from ref.3. The phase
histograms are shown in Fig.1 for 2° and 5° windows centered on Cygnus X-3.
The determination of the background estimate can be found in ref.4.

4. RESULTS

4.1 Full data sample

The phase distribution in the 2° window (Fig.1a) is compatible with the
background expectation. The number of observed muons is 88 and the expected
background 82.2. Using a Monte Carlo simulation of the likelihood function,
we obtain a confidence level of 45 % for the background hypothesis.

In the 5° window (Fig.1b), the number of detected muons and
multimuons is 542 whereas the expected background is 526. The chi-squared
for the background interpretation is 11.5 for 10 degrees of freedom (32 %
confidence level). The 90% C.L. flux limits corresponding to the 2° and
5° windows are given in Table I for 10 phase intervals. We have looked at
the time distribution of the events and found it to be compatible with a
uniform distribution. We have applied the Rayleigh test to our data,
looking for a unimodal contribution. Applying this test at every point on
the sky map defines a field of vectors which allows to find the origin of a
possible contribution from a point source (Fig.2). No such contribution is
found.

4.2 1984 data

The NUSEX experiment has recently provided the phase distribution in 82,
83, 84 and 85 (Ref.2b). The maximum effect is observed in 1984. For the
same period of observation, at the same depth and with a similar detector,
the Fréjus data is in direct contradiction with the NUSEX data. The 90%
C.L. limit (0.4 10^{-12} cm^{-2}s^{-1}) is a factor 10 lower than the NUSEX flux
during the same period.

In conclusion, analysis of the Fréjus data yields new upper limits on the excess muon flux from the direction of Cygnus X-3. The positive signals reported in Refs. 1 and 2 are not confirmed. Moreover, the observation reported in ref.2 is in direct contradiction with the Fréjus data.

REFERENCES

[1a] M.L. Marshak et al., Phys. Rev. Letters, **54**, 2079 (1985)
[1b] M.L. Marshak et al., Phys. Rev. Letters, **55**, 1965 (1985)
[2a] G. Battistoni et al., Phys. Letters, **155B**, 465 (1985)
[2b] B. d'Ettorre-Piazzoli, Summer Study on the Physics of the SSC, Snowmass,   CO., June 23-July 11, (1986)
[3] M. van der Klis, J.M. Bonnet-Bidaud, Astron. Astrophys., **95**, L5 (1981)
[4] Ch. Berger et al., Phys. Lett.B, **174**, 1, 118, (1986)

Table 1.   90% confidence level upper flux limits for the 2° and 5° windows in 10 phase bins.
The limits are calculated by considering that Cygnus X-3 is not detectable between 0.36 and 0.72 of sidereal day. This arbitrary convention is very similar to the one used in Ref. 2.

90% C.L. flux limits in units of 10^{-12} cm^{-2} s^{-1}		
	2° window	5° window
0.0 - 0.1	0.45	0.8
0.1 - 0.2	0.2	0.4
0.2 - 0.3	0.2	0.7
0.3 - 0.4	0.1	0.3
0.4 - 0.5	0.25	0.25
0.5 - 0.6	0.2	0.3
0.6 - 0.7	0.2	0.85
0.7 - 0.8	0.1	0.22
0.8 - 0.9	0.05	0.07
0.9 - 1.0	0.33	0.75
All phases	0.7	1.7

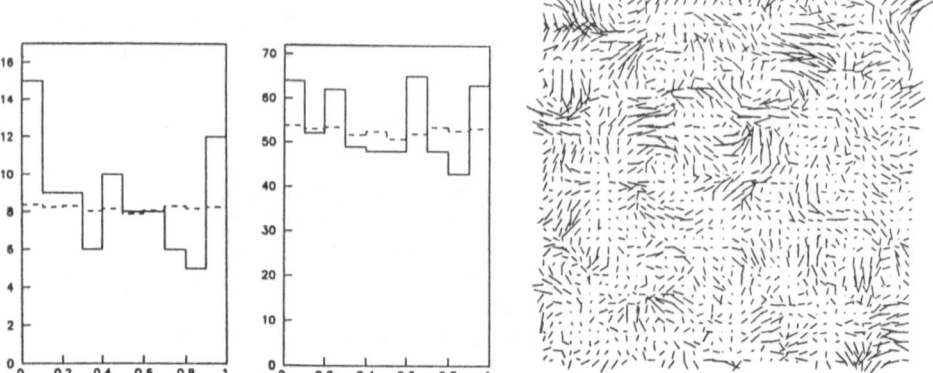

Fig. 1 Phase distribution of muons and multimuons in cones of half angle 2° (a) and 5° (b) centered on Cygnus X-3. Solid line: detected muons and multimuons, dashed line: expected number of atmospheric muons and multimuons.

Fig. 2 Rayleigh field of vectors in a 20°x20° window centered on Cygnus X-3. Each point is associated with a 2° radius window.

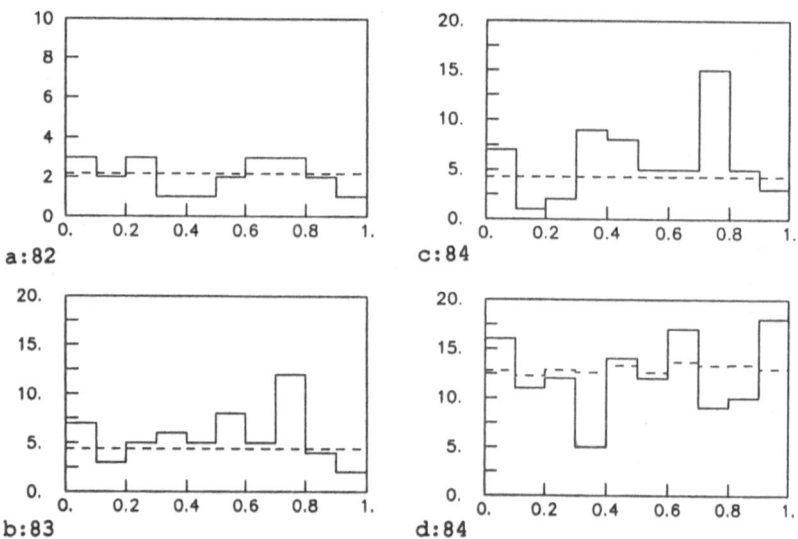

Fig. 3 Phase distribution for the NUSEX experiment: a:82, b:83, c:84, and for the Fréjus experiment: d:84.

UNDERGROUND SEARCH FOR PARTICLES FROM CYGNUS X-3

G.Czapek, B.Hahn, W.Krebs, J.Lauber, H.Scheidiger and
P.Schlatter

Laboratory for High Energy Physics, University of Berne,
Switzerland

ABSTRACT. Particle tracks observed in an underground experiment between August 1985 and July 1986 have been analysed with respect to the direction and phase of Cygnus X-3. The detector, having an area of (4×2) m^2, consists of proportional tubes and plastic scintillation counters. No excess flux was observed from Cyg X-3 in any phase interval which results in a 90 % C.L. upper flux limit of 1.7×10^{-7} $m^{-2}s^{-1}$ for an average rock overburden of 3.4×10^5 g/cm^2.

1. INTRODUCTION

Cyg X-3 ($\alpha = 307.6°$, $\delta = 40.8°$) is considered to be a binary system consisting of a compact object and a companion star with an orbital period of 4.8 h. Several experiments [1] have observed γ-rays from Cyg X-3 at energies up to 10^{16} eV with the characteristic 4.8 h periodicity established in the X-ray energy region. The recent observation of a muon signal correlated with direction and time modulation of Cygnus X-3, reported by the two underground experiments Soudan I [2] and NUSEX [3], and the negative results of similar experiments [4-6] have lead us to search for a signal in the Gotthard underground laboratory.

2. DETECTOR

Our detector is located in the Gotthard alpine highway tunnel (46.6° N, 8.6° E). The minimum rock overburden along the path of Cyg X-3 is 3.1×10^5 g/cm^2 of standard rock, which corresponds to a muon threshold energy of 1.5 TeV.

The detector consists of three horizontal planes of (2×4) m^2 area separated by 2×0.33 m. Each detector plane is subdivided into a layer of plastic scintillation counters and two crossed layers of proportional tubes. In such a plane the 2 cm thick scintillation counters are viewed by 20 photomultipliers. Their signals are used to trigger the experiment as well as to measure the energy loss of particles penetrating the

K. E. Turver (ed.), Very High Energy Gamma Ray Astronomy, 193–196.
© *1987 by D. Reidel Publishing Company.*

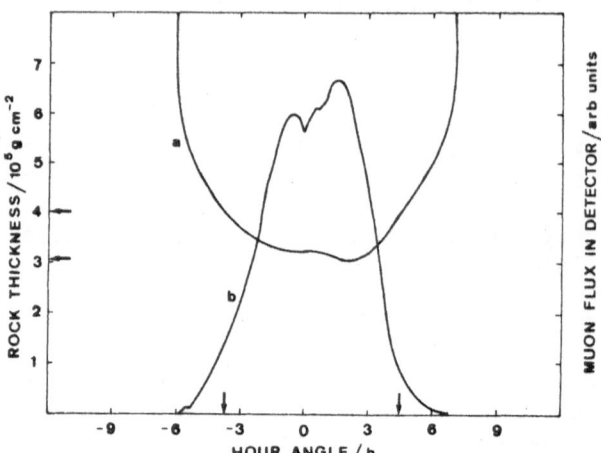

<u>Figure 1</u>.   Rock thickness crossed by particles with the same
declination as Cyg X-3 (a), corresponding muon flux (b).
The arrows indicate the cut applied to the event selection.

detector. The proportional tube wires are separated by 1.20 cm. They
provide two orthogonal views of a particle track with an angular resolution of about 1°. There is a total of 1920 proportional tubes in the
detector.

       We use a CAMAC readout system to monitor the setup and to acquire
data. An event is defined by a 2 out of 3 majority coincidence between
the three scintillator planes. Events with at least 4 proportional tube
hits are accepted and recorded on magnetic tape. An event record contains the following information : 60 scintillator pulse heights, the
addresses of the hit wires and the date and time with an accuracy of
about one second.

3. DATA ANALYSIS AND RESULTS

Our data sample represents events taken between August 1985 and July
1986 with a running efficiency of 90 %. During this period a total
number of 44724 muon tracks was reconstructed. For the present analysis
we considered events within the hour angle interval shown in Figure 1.
This interval corresponds to an average rock thickness of 3.4×10^5 g/cm^2
for tracks pointing to Cyg X-3. Using the quadratic ephemeris [7]

$$T_0 = \text{JD } 2440949.8986$$
$$P_0 = 0.1996830 \text{ d}$$
$$\dot{P} = 1.18 \times 10^{-9}$$

the phase of each event was calculated from the arrival time which was
reduced to the position of the sun. Figure 2 shows the phase distribution for two subsets of events, lying within 3° and 5° of the direction
of Cyg X-3.

The background contribution in the phase plots is expected to be flat as the measuring time is close to twice the beat period of 170 days between the sidereal day and the Cyg X-3 period. In fact the measured background including all events in the declination band $\delta\pm3°$ and $\delta\pm5°$ did not show any significant fluctuation.

In the phase plots there are no deviations from the background exceeding 2.5 σ. For each phase bin we derive an upper limit for the particle flux from Cyg X-3 using Poisson statistics (see Figure 3). Without considering the phase the upper flux limits are 2.6×10^{-7} $m^{-2}s^{-1}$ (3° cone) and 1.7×10^{-7} $m^{-2}s^{-1}$ (5° cone) at a

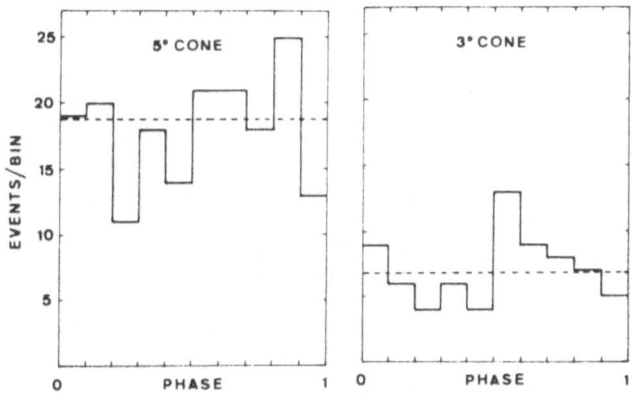

Figure 2. Phase distribution of particles in cones around Cyg X-3 (half angles 5° and 3°). The dashed line shows the average measured background.

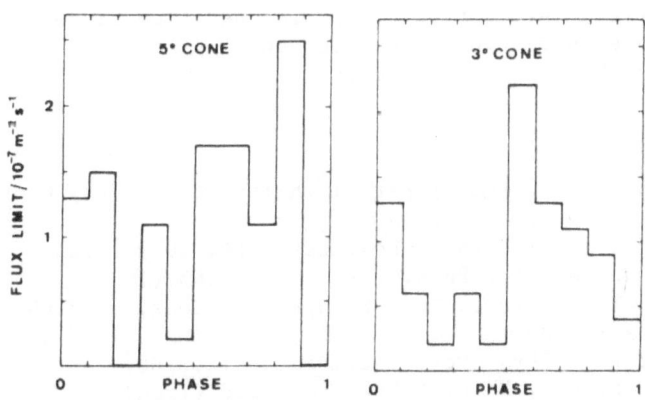

Figure 3. Time averaged flux limits (90 % C.L.) calculated for each phase bin.

confidence level of 90 %. Figure 4 compares our result with the flux measurements of other experiments.

The amount of energy lost in the scintillators by events pointing to Cyg X-3 is compatible with minimum ionising particles.

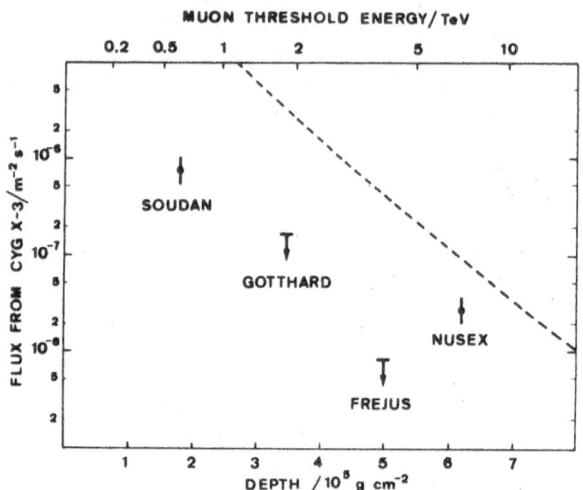

Figure 4. Summary of results obtained by ionisation detectors. The muon threshold energy was calculated using dE/dx values taken from ref. 8. The dashed line shows the background muon flux from a 5° cone.

4. CONCLUSION AND OUTLOOK

So far we do not observe a similar effect in our data as reported by the Soudan I and NUSEX collaborations. In order to improve our event statistics in a reasonable time we built a second detector of the same size and similar construction. It is in operation since July 1986. The total muon rate in both detectors is now 15 h^{-1} and we expect to double our statistics by the end of 1986.

REFERENCES

1. A.A. Watson, rapporteur paper, 19th Int. Cosmic Ray Conf., La Jolla 1985, and references therein
2. M.L. Marshak et al., Phys. Rev. Lett. 54, 2079 (1985)
3. G. Battistoni et al., Phys. Lett. 155B, 465 (1985)
4. E. Aprile et al., Proc. Int. Europhysics Conf. on High Energy Physics, Bari 1985
5. Y.Oyama et al., Phys. Rev. Lett. 56, 991 (1986)
6. Ch. Berger et al., Phys. Lett. 174B, 118 (1986)
7. M. van der Klis and J.M. Bonnet-Bidaud, Astron. Astrophys. 95, L5 - L7 (1981)
8. W.Lohmann et al., CERN 85-03 (1985)

TOO MANY MUONS FROM COSMIC ACCELERATORS?

R. MORSE
Department of Physics, University of Wisconsin
Madison, WI 53706, USA

The evidence for both surface and underground muons produced by radiation from Cyg X3 is reviewed. The number of surface muons if real require impossibly large muon to electron ratios if they are produced by gammas from Cygnus. The underground muons, because of their number, angular spread, and implied energy from depth form such a contradictory set of circumstances that rule out atmospheric production by gammas or any other particle! The facts suggest the muons have low energy and are produced close to the detectors. Light particles called *cygnets* with electron-like masses and cross sections $\sigma = 10 - 100\mu b$ are conjectured as responsibile for both the surface and underground muons. Models, using associated production, heavy parents, or endothermic mechanism as *escapes* are presented to reconcile cygnets with beam dump experiments. The difficulties associated with new particle interpretations of Cygnus muons are substantial!

I. Introduction

Underground muons excesses presumably produced by radiation from Cygnus X3 and other cosmic sources have been reported by several underground detector groups–most notably Soudan, NUSEX, and recently IMB[1]. These signals, if authentic, present a serious challenge to the standard model of quarks and leptons. One, however, does not have to go *underground* to get into trouble, for the Kiel-Haverah Park result[2]–or suspected *muon-rich* γ-showers from Cygnus already argues for QED violations and a possible *composite* model of leptons.

Other underground experiments such as Kamioka and HPW[3] however report no such muon excesses from Cygnus X3 and presumably other sources while the experiments at Frejus and Baksan[4] have given tantilizing yet inconclusive results. It is possible that Cygnus X3 is a capricious source as suggested by VHE gamma ray results from Cerenkov devices[5], and that the null results are due to either: i) not looking at the correct time or ii) diluting an already weak signal with lots of noise when there was no signal present at all! Since significance goes as $\sqrt{f}$ where f is the duty factor, a 10% duty factor can hide a 3 σ effect, even when a phase analysis is done!

Surface EAS (extensive air shower) experiments[6] have established the existence of energetic presumably $> 10^{15}eV$ γ-ray showers from point sources such as Cygnus, Vela, and LMCX-4, and many of those results depend on the expected *muon-poor* nature of γ-initiated showers, e. g. , the Tien-Shan and Akeno results[7]. Thus, there seems to be no doubt that Cygnus radiates VHE gammas

197

K. E. Turver (ed.), Very High Energy Gamma Ray Astronomy, 197–220.
© *1987 by D. Reidel Publishing Company.*

up to $10^{15}eV$. However, the question of surface muons or *Kiel effect* like the underground muons or *Soudan effect* are not established and evidence, either for or against, is one of the most important problems facing astro-particle physics at this time.

Both effects have been claimed in data on Cygnus X-3, an object 10kpc or 10^{1} seconds distant. *Seeing* Cygnus manifests itself in two ways, first by rate or number excess that points back at the source, and second by a phase or time enhancement from periodic pulses or burst from the source. Such absolute phase information is usually tested against an ephemeris that has been established via radio or X-ray observations[8].

II. Present Experimental Situation

What is the present experimental situation that we are confronting? First, Cygnus and other sources seem certain to produce lots of $\geq 10^{12}eV$ gammas as reported by air Cerenkov techniques. Less certain, but probable is that they also produce $\geq 10^{15}eV$ gammas since there are many claimed sightings by EAS arrays. So far nothing is too far afield, though some pulsar models might be strained to produce $E_\gamma \geq 10^{15}eV$[9]. It is the *number* of surface muons, and for the underground muons, their *number*, *apparent energy* because of observed depth, and lack of *close pointing* back to the source, that cause the dilemma[10].

The situation as depicted in Figure 1 for gammas and Figure 2 for the marvelous X^o particle whose properties we will specify as our needs arise!

1. High-Energy Gamma Rays from Cygnus have been detected by Air Cerenkov devices yield $E_\gamma > 200\ GeV$, reports from MHO, HGO, and Dugway-Durham.

2. Detection of gammas from Cygnus by Extended Air Shower (EAS) Arrays yield $E_\gamma > 10^{15}eV$. Signal enhanced by demanding *muon-poor* showers, the *Akeno effect*. Akeno, Tien-Shan have reported this; so far no problem!

3. Detection of *gamma-like* radiation via pointing excess except showers are found to be *muon-rich* and they hold Cygnus orbit phase, the insidious *Kiel effect*. Verification of Cygnus signal is provided by Haverah Park, though they can't really make a definite statement about muon content. Now we have a problem with γ-production of muons.

4. Soudan-IMB effect. Suspected γ-produced high energy muons of energies > 0.6 TeV observed deep underground which roughly point ($\pm 5°$) to Cygnus and hold Cygnus phase. This effect, though energetically possible, is highly improbable. Its like every photon of $E_\gamma > 10^{12}eV$ from Cygnus turned into a muon of the same energy. Further, the $\pm 5°$ scatter is too big to be explained by multiple scattering. Either it is a high P_T muon process, or it is happening very close—that is, underground!

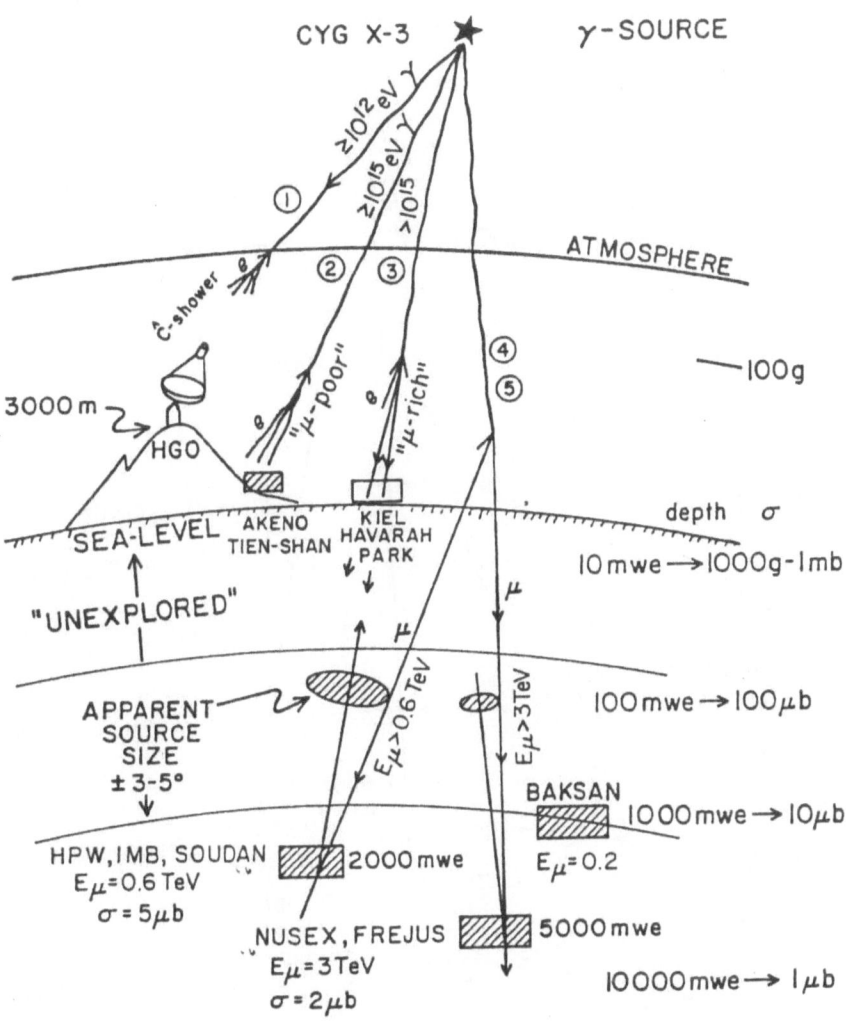

Fig. 1. Schematic Presentation of Signals reported from Cygnus X–3, assuming the messenger is a photon: (1) $10^{12}eV$ Air Cerenkov Detection, HGO, etc; (2) $10^{15}eV$ EAS muon poor detection, Akeno, etc; (3) $10^{15}eV$ EAS muon-rich detection, Kiel, etc; (4) and (5) Deep underground detection via TeV muons, the Soudan/NUSEX effect.

200

R. MORSE

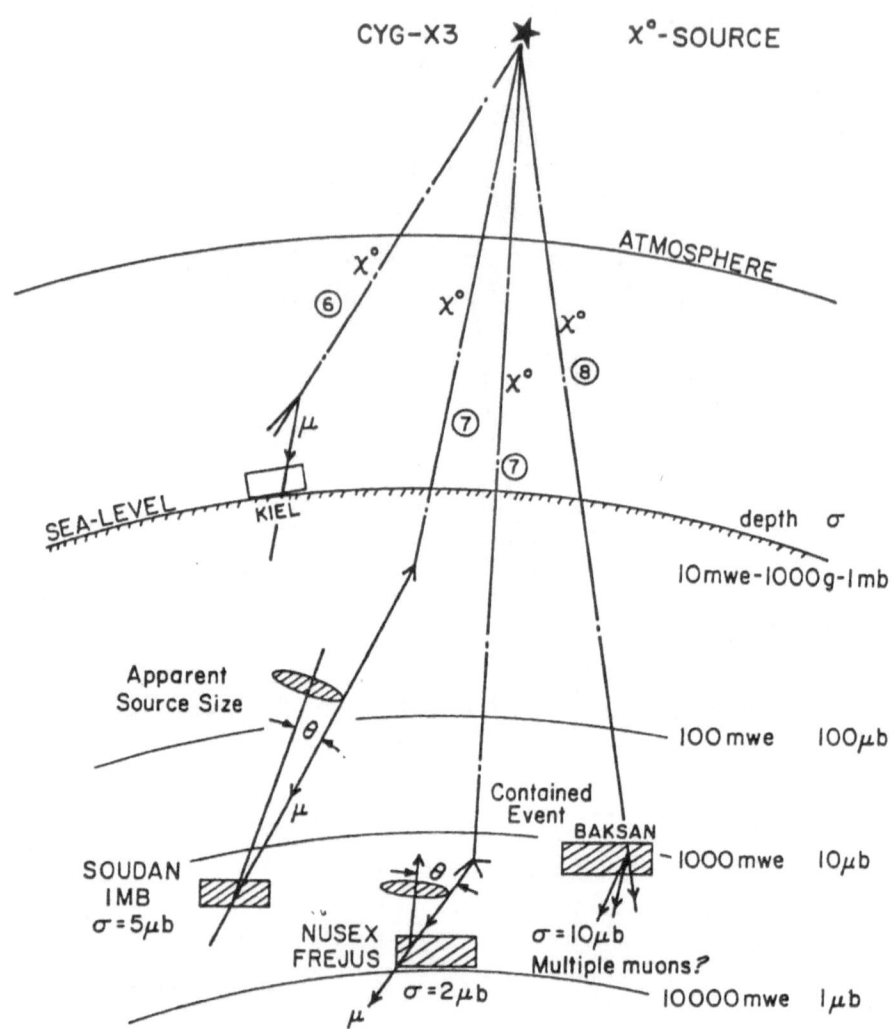

Fig. 2. Schematic Presentation of Signals reported from Cygnus X–3, assuming the messenger is a X° (Cygnet): (6) Surface production of muon's at small angles, Kiel; (7) Deep underground X° production of low energy muons close to detector, Soudan, IMB, Nusex, Frejus (?); (8) Multiple muons via X° production. Are these almost contained events? BAKSAN.

5. NUSEX/Frejus Effect. Underground muons which point back to Cygnus X3. If γ-initiated, these muons must have an energy of $\geq 3TeV$ to get to these depths. If produced by X^o interactions, muons could be produced uniformly in the rock. Such a production mechanism could explain the large apparent footprint of Cygnus.

6. Kiel Effect. In Figure 2 we illustrate where the messenger is the X^o. If the X^o had a $\sigma = 10\mu b$ cross section, the 1% of them would interact in the atmosphere to produce muons, and the rest would penetrate the earth to depths $\sim 2000mwe$ and produce the Soudan effect muons. Recall 1 atm $= 10$ mwe.

7. Soudan Effect. Here the muons detected deep underground are produced via X^o interactions in the rock. The fact that they are produced locally explains the large muon footprint of about $\pm 5^o$.

8. Baksan or multi-muon events. Are these evidence for contained events? Every 3 meters of underground detector is equivalent to an atmosphere so if $\sigma(X^o) \sim 10\mu b$ then the contained event rate should be similar to the Kiel rate,that is every 3 m of detector is 1 atm in rate.

III. Experimental Constraints

What are the options available to establish a Kiel or Soudan effect? First, the messenger particle must be neutral since the effects point back to the source, and thus is not bent by the galaxy's magnetic field. Second, the messenger must be almost stable ($\tau \sim 1$ year) since the Lorentz modified lifetime $t = \gamma t_o$ implies

$$\frac{m}{\tau} < \frac{E}{T} = \frac{10^{15} eV}{10^{12} sec} = 10^3$$

which says that $10^{15} eV$ particles of mass $= 1$ GeV (e. g. , neutrons) must live about 1 year.

Requiring the messenger to almost hold X-ray phase yields even more stringent limits since a particle of mass m and energy $E = \gamma m$ loses phase or time with respect to a photon as

$$\Delta t = d(\frac{1}{v} - \frac{1}{c}) \simeq \frac{t}{2\gamma^2}$$

where t is the transit time (10^{12}sec). This requiring the messen ger to hold Cygnus X3 orbit phase (4.8 hours) to 10% or $\sim 10^3$ seconds means $\gamma \geq 3 \times 10^4$ whereas demanding it to hold 10% pulsar phase (thought to be 12.59 ms) means $\gamma > 10^7$.

Requiring the messenger to hold phase also limits the amount of multiple scattering that it can suffer due to randomly oriented magnetic fields in interstellar space. This then places a limit on the messenger's charge and the accuracy to which it *points back* to Cygnus. Such calculations by Berezinsky et al[10] give $q < 10^{-7}e$ and θ scatt $< \frac{1}{4}^\circ$.

Thus the pointing and phase requirements put severe limitations on the charge, mass, and lifetime of the messenger. Holding orbit phase requires the particle to have about a nucleon mass whereas holding pulsar phase requires the particle to have an electron mass. We summarize all of these features in Table 1.

All the arguments thus point to the messenger particle being a photon, however, *photons do not make muons!* – not very efficiently anyway, and we will find we have problems, not only with muon numbers, but their energy, angular distributions, and apparent source size! Photon-showers develop very high in the atmosphere (about 10 km), and after a few radiation lengths or about 100 grams, a primary photon thru pair production ($\gamma \rightarrow e^+e^-$) and bremsstrahlung ($e \rightarrow e\gamma$) has generated a shower of $e^\pm$ and γ's. The prominent feature that sets γ-showers apart from hadron showers is their low muon content. The number of muons in a γ-shower is typically a few percent of the number in a hadron shower in which muons are abundantly produced by the decays of produced $\pi^\pm$. For γ-showers the $\gamma \rightarrow \mu$ process proceeds through decays of photon-produced hadrons,

$$\gamma \rightarrow \pi^\pm \rightarrow \mu^\pm$$

$$\gamma \rightarrow \rho^o \rightarrow \pi^\pm \rightarrow \mu^\pm$$

which proceed to order $\alpha = 137^{-1}$. Since a sizeable amount of the shower energy is carried by the electron-position component we note that $e \rightarrow \mu$ process really proceeds via $e \rightarrow \gamma \rightarrow \mu$ of order $\alpha^2 = 137^{-2}$. The entire process $\gamma \rightarrow \mu$ proceeds very inefficiently at the 2–10% level so that typically $N_\mu \approx \frac{1}{50}Ne$ to $\frac{1}{10}Ne$. We exhibit the number of muons vs electrons as a function of primary energy and primary type, either γ or hadron in Figure 3.

III.1 Surface Muons

Now the Kiel effect, after detecting a 4σ enhancement of the cosmic ray flux in the direction of Cygnus performed two tests to confirm the signal. They found that *on-source* showers indeed remember the 4.8 hour binary period, but also found the showers are not *muon-poor* and the expected 2% muon content was measured to be 70% of the content of the hadronic background.

About 20 experiments have found evidence for the emission of VHE γ-rays from the direction and with the time structure of Cygnus ranging from $10^{12}eV$ all the way to $10^{17}eV$. The integral flux can be approximated as[20]

$$F(> E) = \frac{4 \times 10^{-11}}{E(TeV)} cm^{-2}sec^{-1}$$

for $E \geq 10^{11}eV$. The data used to obtain this flux is shown in Figure 4. This is

Table 1
Messenger Characteristics

	Orbit Phase ($\Delta t = 10^3\ sec$)	Pulsar Phase ($\Delta t = 1\ sec$)
Mass	$0.1 - 10.0 GeV/c^2$	$0.1 - 10 MeV/c^2$
Charge	$< 10^{-7}e$	$< 10^{-10}e$
Lifetime	$> 10^6 - 10^8\ seconds$	$> 10^3 - 10^5\ seconds$
Pointing	$< 0.25^o$	$< 0.025^o$

Table II
Some TeV Results - Cerenkov Results

	Flux	Energy	Phase
HGO	$1.0x10^{-9}$*	$\geq 0.2\ TeV$	.68
MHO	$4.0x10^{-11}$	$\geq 0.8\ TeV$	.60
Durham-Dugway	$3.0x10^{-10}$	$\geq 1.0\ TeV$	.62
Fly's Eye	$4x10^{-13}$	$\geq 1000\ TeV$	.28

* Instantaneous rate

TABLE III
Some Surface Results - PeV Result

	Flux $(cm^{-2}sec^{-1})$	Energy (eV)	Phase
Fly's Eye	$4x10^{-13}$	$> 10^{15}$	$\sim .28$
Kiel	$7.4x10^{-14}$	$\geq 10^{15}$	$0.1 - 0.$
Haverah Park	$7x10^{-14}$	$\geq 10^{15}$	$.6 - .7$
Akeno	$1.1x10^{-14}*$	$\geq 10^{15}eV$	$.55 - .8$

$+$ phase reanalyzed using Van der Klis ephemeris
*muon poor result.

Table IV
Some Underground Muon Results

	Area	Depth	Energy	Flux	Phase
Soudan	$8m^2$	1800 mwe	$> .6\,TeV$	$7x10^{-11}$	$\sim .7 - .8$
IMB	$400m^2$	1700 mwe	$> .6\,TeV$	$7.2x10^{-10}$	$.5-.65$
NUSEX	$12m^2$	5000 mwe	$> 3\,TeV$	10^{-11}	$.68 - .76$
Frejus	$96m^2$	4800 mwe	$> 3\,TeV$	—	—
Baksan	$250m^2$	1000 mwe	$> .2\,TeV$	$2x10^{-11}$	$.7 - .8$
HPW	$120m^2$	1600 mwe	$> .6\,TeV$	$< 10^{-11}$	—
Kamioka	$130m^2$	3000 mwe	$> 1.5\,TeV$	$< 10^{-11}$	—

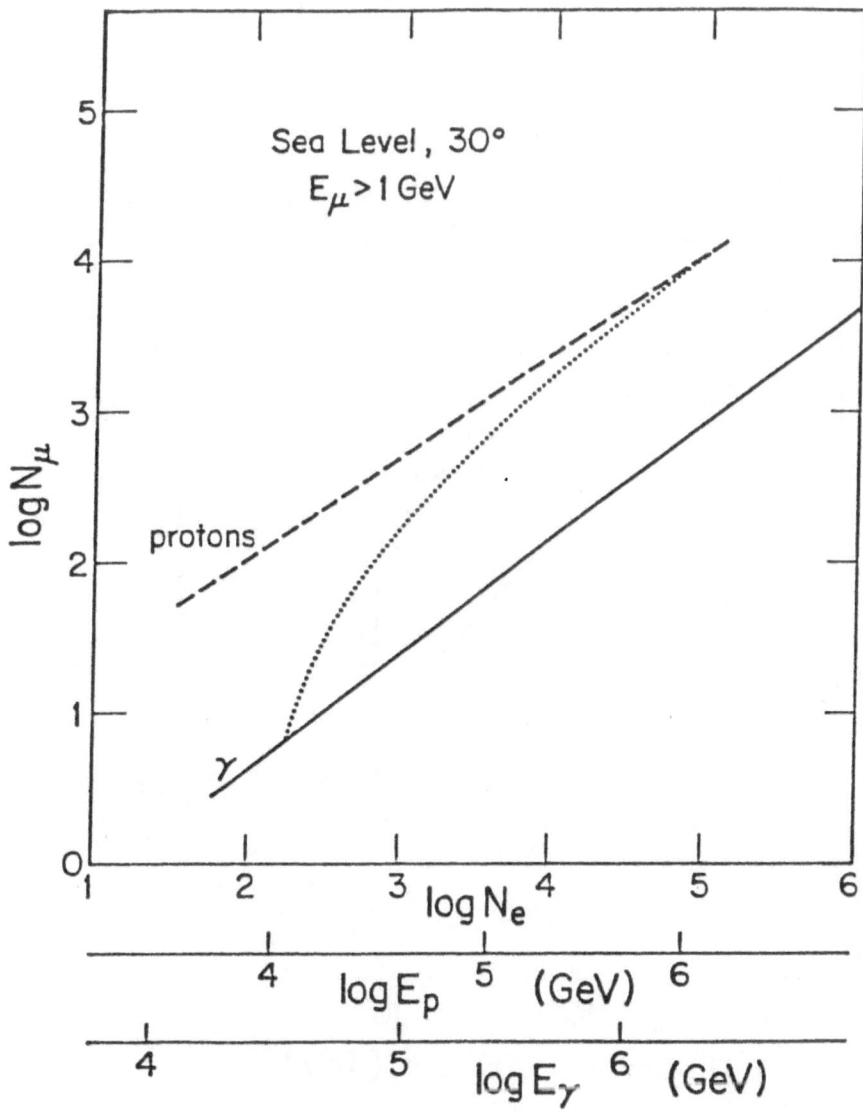

Fig. 3. Number of muons in excess of 1 GeV (N_μ) versus shower size (*i.e.* number of electrons N_e) for γ- (solid line) and proton- (dashed line) initiated showers. The dotted line shows the muon signature in a model where the photon becomes strongly interacting at very high energies. The horizontal energy scales show the conversion of shower size to initial proton or γ-ray energy.

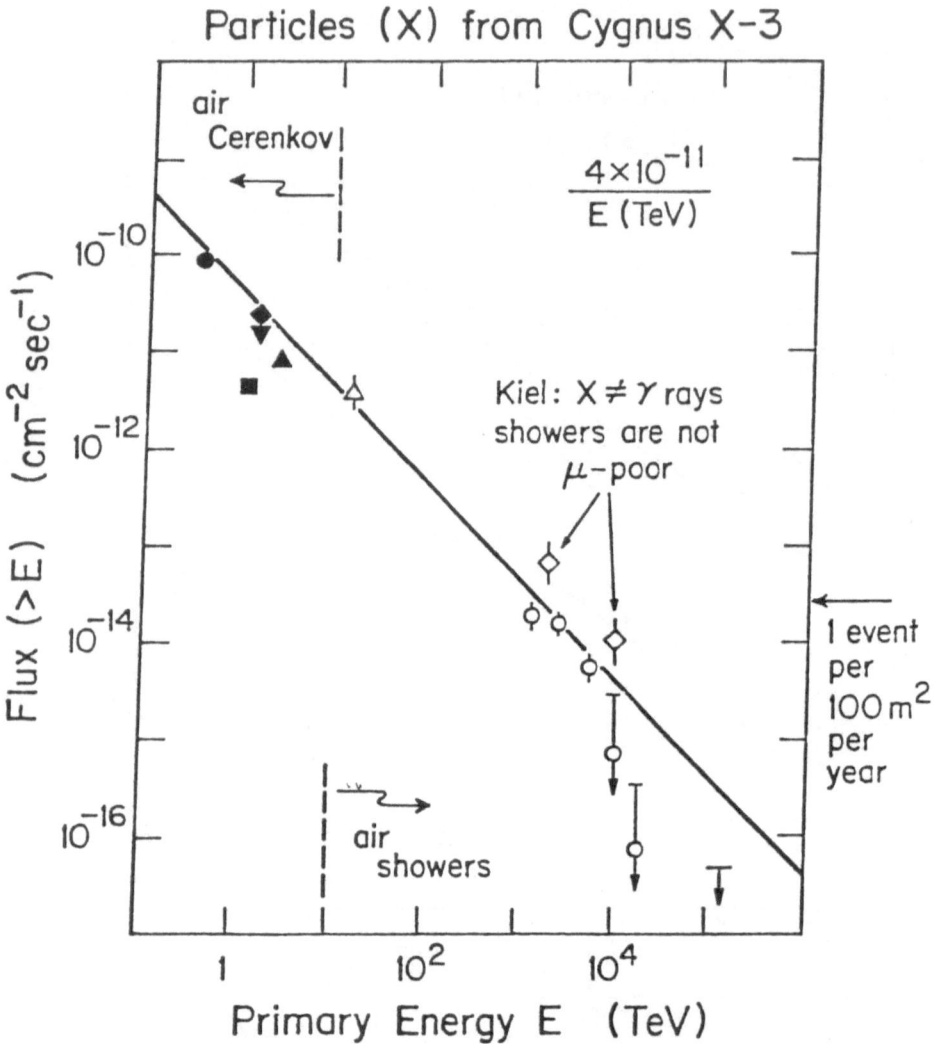

Fig. 4. Integral flux of very high energy particles from the X-ray binary
Cygnus X-3. We will call these particles γ-rays although the Kiel
experiment challenges this identification. Note the flatter E^{-1} energy
dependence compared to the $E^{-1.7}$ fall-off of the cosmic ray flux.

a time average result, and EAS arrays such as Kiel yield for certain epochs fluxes higher by a factor of ten, and some TeV experiments have identified increased activity over periods of minutes[5].

Most experiments except Kiel cannot firmly establish a signal with pointing excess alone, and thus exploit the 4.8 hour modulation of the Cygnus signal on a continuous and typically 10–100 times larger cosmic background. Using an X-ray ephemeris, a light curve is established which is just the accumulated data folded by the suspected orbital period of the pulsar. In Figure 5 we display the emission phases of showers detected by Haverah Park, and a compilation of other measured phase peaks. This H.P. light curve serves as a warning that the structure is complex, and we could have energy dependent phases, as well as a different phase if there were an X^o component.

Just how seriously should we take all of this? If the Kiel effect were due to γ's and e's that *punched thru* the some 30 radiation lengths of shielding and thus faked *muons*, then there would be no surface muon problem. The Haverah Park Nottingham $10m^2$ muon detector data[11] is consistent with the *muon-rich* signal, but both results are inconsistent with Akeno and Tien-Shan who require a $Nu/Ne < .001$ ratio for showers whose core is greater than 50 m from the muon detectors. Further Akeno studies show for core distances < 50 m there is substantial punchthru probability. Akeno also finds $Nu/Ne = 0.03$ for normal hadron showers.

Taking the eighteen Akeno events as a real result and not an upper limit, one deduces a flux of 0.15 of the Haverah Park result. The integral spectrum has been following E^{-1} law from Cerenkov light measurements at $10^{12} eV$ with no evidence for the expected $3^o K$ microwave absorption of gamma rays due to $e^+ e^-$ production at about $10^{15} eV$[12]. The absorption factor should be about a factor of 5 for Cygnus if we assume a 12 kpc distance. This absorption factor is really hard to measure since the spectrum is falling, the flux is barely detectable[13], and the absorption is sharp in energy and only a factor of 5 in rate. But in spite of these hardships, maybe Akeno is seeing γ's and Kiel is seeing γ's and something new which the 50 m Akeno muon cut has removed from their sample. A reaction $X^o + N \rightarrow \mu = \ldots$ would be characterized by forward going muons hitting the detector within a 50 m radius of the core. Kinematics require the muon scattering angle θ_μ in the above inclusive reaction to be[10]

$$\theta_\mu < \sqrt{\frac{2M_N}{E_\mu}} < 1^o \ for \ E_\mu > 1 \, TeV$$

or very sharply peaked in the forward direction, even if the muon is produced by secondary decays.

Thus Kiel could be witnessing two distinct phenomena and it would seem that the highest priority of an EAS array with muon detection would be to see if the

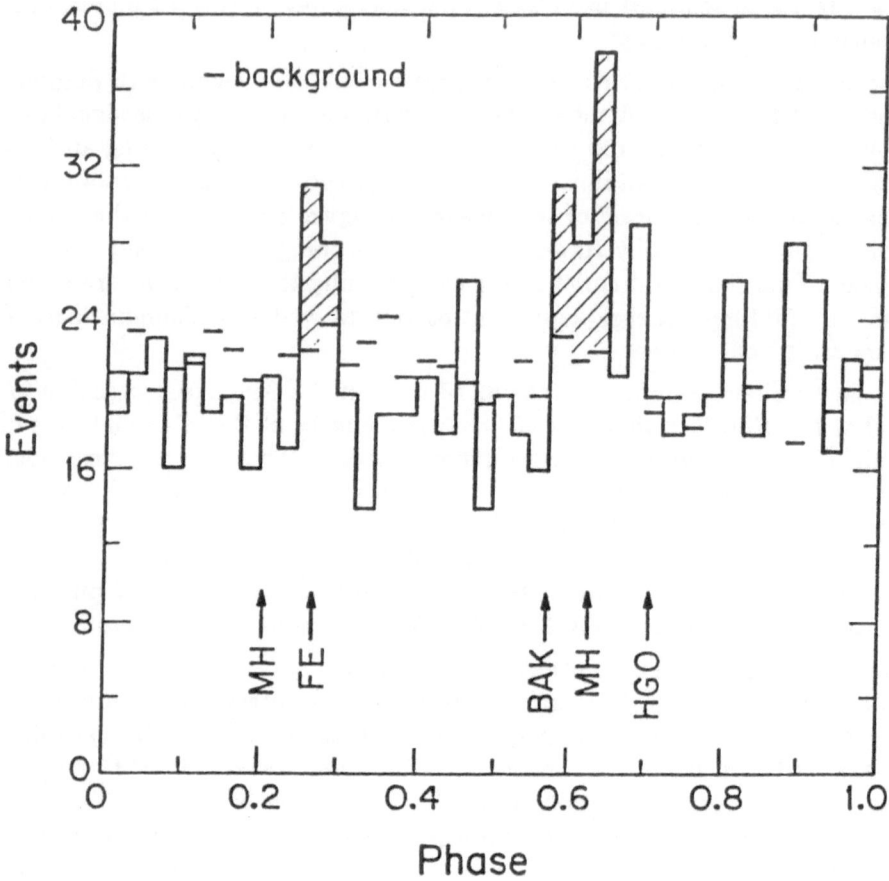

Fig. 5. Phase distribution of the radiation in the direction of Cygnus X-3 as observed by the Haverah park extensive air shower array. Emission is also observed in the general vicinity of $\phi \simeq 0.2$ and $0.6 \sim 0.7$ by Baksan, Fly's Eye, Mount Hopkins and Haleakela (HGO).

muon-rich signal, from a binary system held pulsar phase as well as orbit phase for this would tighten the mass constraints on the messenger even further. At present masses from 0.1 to 10 GeV/c^2 are possible, whereas holding pulsar phase would constrain masses to about 0.1 to 10 MeV/c^2.

In conclusion, the Kiel result however could be flawed by *punchthru*. Even if flawed in this way, the Kiel result is still a remarkable result for it established a sizable flux of presumably photons of $E > 10^{15}eV$ emission from Cygnus. More EAS array work of this nature is needed to establish the very high energy content (or lack of it) for other sources such as Her X1, 4U0115+63, and especially the higher rate pulsars such as PSR1953+ etc. , which are not supposed to radiate at these higher energies.

However, if the Kiel muons are real, then we have some choices: i) QED violations where photons are coupling to lepton-quark like constituents which like heavy quarks give rise to lots of strange particles and muons; or ii) we could be seeing a new X^o-particle, the *Cygnet*, a light neutral which has an interaction cross-section of .01–0.1 mb and would have a mass of $\sim 1 - 10$ GeV to hold Cygnus orbit phase, or 1–10 MeV if it holds pulsar phase.

III.2 Underground Muons

Now if the Kiel effect is a mild annoyance, the Soudan, IMB, NUSEX effect is a real dilemma. Basically, Soudan at a depth of 0.6 TeV is seeing muons from Cygnus that point and hold phase, with instantaneous fluxes of $10^{-9}cm^{-2}sec^{-1}$. These fluxes[14] are as large if not larger than the EAS γ-fluxes measured by Cerenkov Telescopes with $10^{12}eV$ thresholds. It is as if *every photon of energy $E_o \geq 10^{12}eV$ turned into a corresponding muon of the same energy*, a somewhat outrageous notion[15]! The observed underground muon fluxes are shown in Figure 6.

Standard QED production of muons from γ's is shown in Figure 7 from which we see a $E \approx 10^{12}eV$ photon produces an 0.6 TeV muon capable of getting into the Soudan detector with probability of 10^{-5} so that Soudan fluxes should be $10^{-14}cm^2/sec$, or 1 muon every century! Further, the muons have a rather large footprint of $\pm 5^o$, far greater than the 1° one would expect from the multiple scattering effects of the rock, and the detector resolution.

Underground muons should in fact reveal the primary direction through the relation[20]

$$\theta(in\ mrad) = \frac{\pi}{\sqrt{E_\mu(TeV)}}.$$

Although γ-rays are very inefficient at producing muons, the VHE γ-spectrum shown in Figure 4 will generate some Tev muons that will produce a calculable

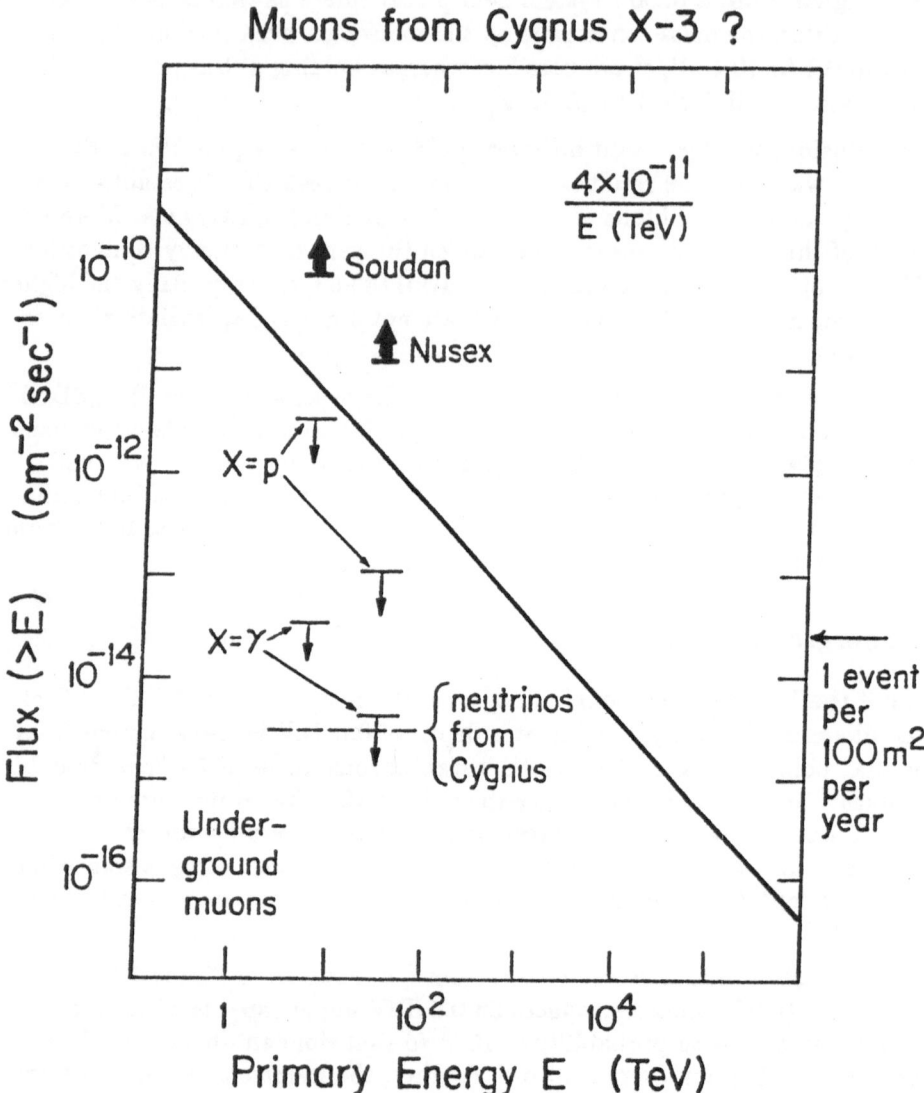

Fig. 6 . Observed and calculated underground muon fluxes assuming that the source of the muons are alternatively γ-rays ($X = \gamma$), hadrons ($X = p$) interacting with the atmosphere or neutrinos produced in the Cygnus beam dump of Fig. 5 (dashed line).

signal in underground detectors. Refering to Figure 7 which graphically presents VHE μ-production probabilities adapted from calculations by Stanev[16] we see that

$$Prob. \sim (E_\nu/E_\mu)^n$$

with $n \sim 1 - 1.7$ for most energies so that with a falling E^{-1} spectrum, the μ-production is fairly energy independent for γ-energies above threshold. For Soudan we expect then a ratio $\mu/\gamma = 10^{-4}$ of the γ-flux at 1 TeV or about $10^{-14}\mu's\ cm^{-2}sec^{-1}$, or *about three orders of magnitudes less than what is seen!* To quote Halzen, underground detectors have rediscovered an old puzzle.

What about these muons? Unlike the Kiel effect, these muons are not in doubt, only their production mechanism. Since they point and hold phase, presumably they are produced by radiation from Cygnus, but is it γ-radiation?

In Figure 8 we exhibit the muon *light curve* for the NUSEX data. The events in the signal at phase $\phi = 0.7 - 0.8$ seem to turn on and off with time. Soudan confirms this sporadic activity as is shown in Figure 9. The *art of detection* consists of limiting the data sample to such *periods of activity* as to successfully pull out the signal. Most of the Frejus data were taken during the off-period and if one limits the Frejus data to the suspected 81–84 on-period as seen in Figure 8b, Frejus could have possibly seen a marginal signal in their initial turnon data. Their data reported at Bari (1985) is shown in Figure 10 and is consistent with a marginal signal, and consistent with an analysis comparing Frejus in Summer 1985 with the expected phase plot from NUSEX data[17].

These muons, if genuine Cygnus muons, defy any conventional explaination and though there is no *smoking gun* evidence, the circumstantial evidence is mounting. The data are based on pointing excesses from the Cygnus direction, which seem to hold the 4.8 hour orbit phase, and periods of increased activity during the various radio outbursts. This problem of source variability or its sporatic nature is at best vexing. It would certainly explain why some groups initially *saw* Cygnus, and then as they accumulate more data the effect get buried in the subsequent noise since the source was mostly *off*.

The usual figure of merit for a signal rate S and background rate B $(B >> S)$ is

$$\frac{N_S}{\sigma} = \sqrt{\frac{S^2T}{B}} \rightarrow \sqrt{f}\frac{N_S}{\sigma}$$

where T is the viewing time. For sporatic effects it is degraded by $\sqrt{f}$ where f is the viewing fraction, so that a *lucky* 3σ, 10% of the data effect can be washed out as more mostly *off* data is accumulated.

A casual glance at Table 1 will convince you that the only candidate for the messenger is a photon. This conclusion is fatally flawed as we list below.

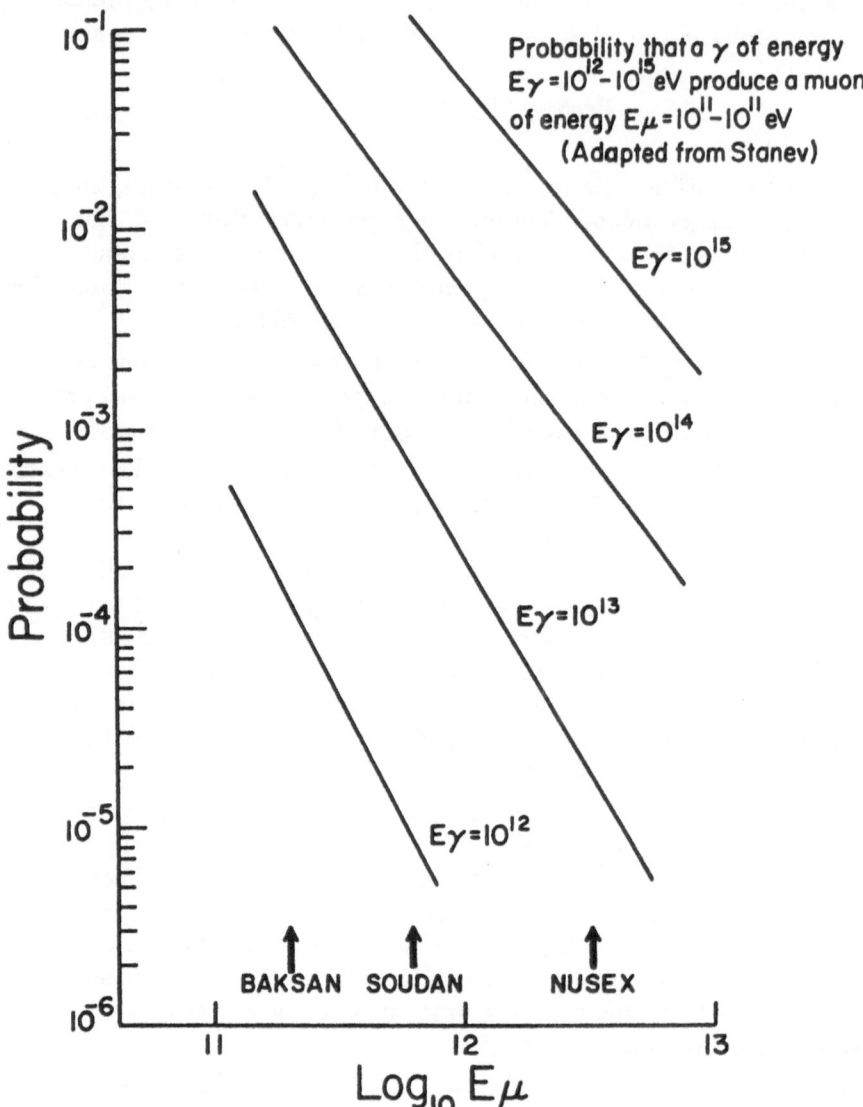

Fig. 7. Probability that an atmospheric γ of energy $E_\gamma = 10^{12} - 10^{15} eV$ will produce a muon of sufficient energy $E_\mu = 10^{11} - 10^{13} eV$ to penetrate to the
deep underground detectors.

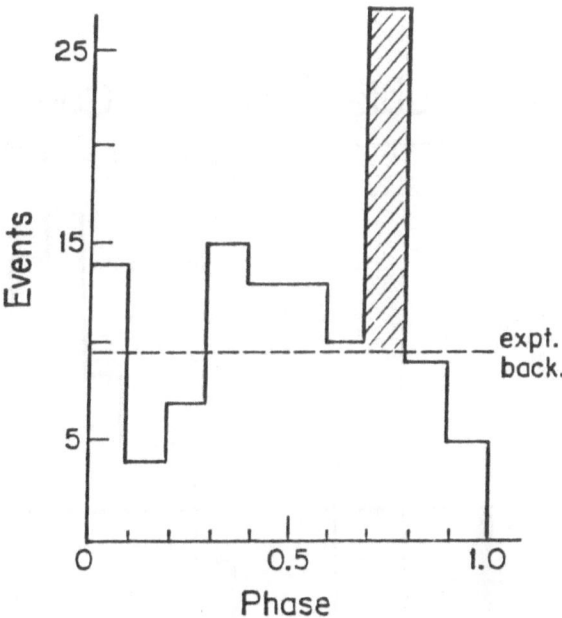

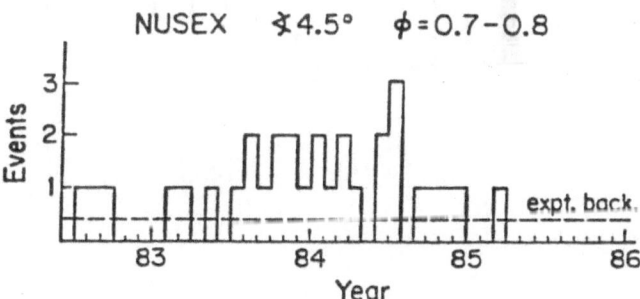

Fig. 8 . (a) Phase plot of the events during the active period 1983–84. On-source data are in a cone of radius 4.5° and phase $\phi = 0.7$–0.8. (b) Muons from the direction of Cygnus X-3 as a function of their time of observation since the start of operation of the Nusex detector.

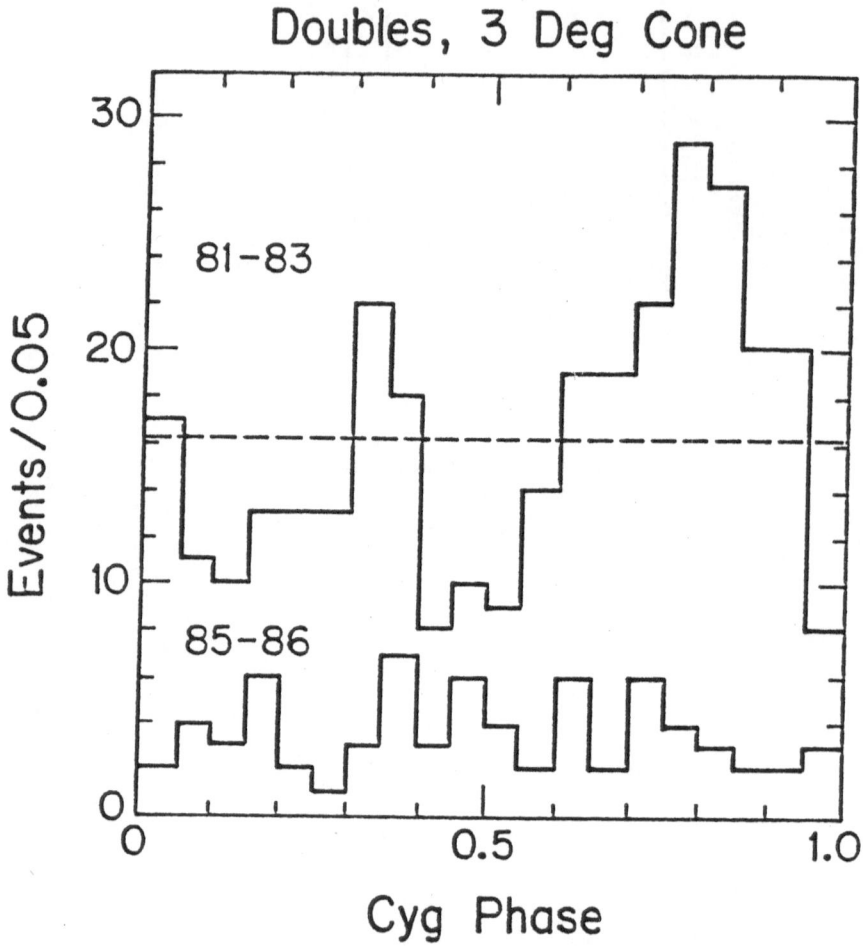

Fig. 9. Confirmation by the Soudan experiment that emission observed during 1981–83 period is not present during 1985–86. Shown is the phase
of dimuons (two muons in a single 4.8 hour period) in a 3° circle
surrounding the source.

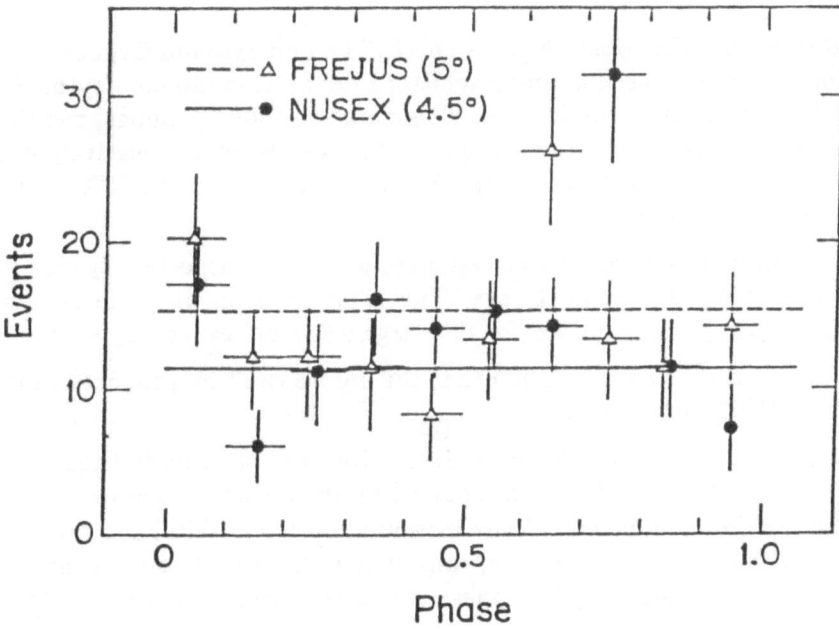

Fig. 10 . Comparison of Frejus and Nusex data preceding the "quiet period" which began in 1985.

1. The *muon-rich* Kiel effect we have already seen.

2. The *absence* of any $3°$ microwave absorption.

3. The *energy/number* problem. The underground muons energy and number are inconsistent with standard production models by 3–4 orders of magnitude.

4. The *minimum depth/zenith angle effect*. The underground Cygnus muons are mostly seen at small zenith angles or at minimum depth. The fact is used to rule out ν-production. There are no stopping muons, but this must be a pointing-reconstruction problem. However, Soudan's relatively flat overburden means at $65°$ from zenith, they are at equivalent NUSEX depth, yet the fluxes don't agree.

5. The *pointing* problem. The arrival direction of the suspected Cygnus muons fills a $5° \times 5°$ *footprint* in the sky. This is far too large an angle to be a multiple scattering effect. For angles this large we would expect $E_\mu \leq 150\,GeV$.

We thus arrive at the conclusion that the muons can't be produced in the atmosphere by VHE gamma rays.

To further confound an already confused situation, the source 1E2259 + 586 (the Soudan nail) has recently been reported by the Soudan group to be another source of underground muons[18]. They measured both its orbital and pulsar period to be $\tau_o = 2310 sec$ and $\tau_p = 6.978 sec$, respectively. These values are consistent with x-ray observations. Holding pulsar phase means the messenger is now constrained to mass values

$$M \leq 2\,MeV$$

since it keeps about a 1 second precision over 3×10^{11} seconds or 3.6 kpc.

Most important however is that they measured the time variations of both periods $\dot{\tau}_o = (-1.2 \pm .4) \times 10^{-8}$ and $\dot{\tau}_p = (-5 \pm 3) \times 10^{-14}$. All this is new information and can be checked by any observation, not just another muon experiment.

IV. New Mechanism

One way out is to postulate particle X^o, a neutral long life ($> 10^6 - 10^8$ seconds) particle and arrange a cross section length to be greater than the thickness of atmosphere, so that only a small amount of production occurs with regular air showers. (Recall 1 atmosphere = 1000 grams = 10 mwe so that $\sigma = 1$ mb guarantees something happens in the atmosphere.) At Soudan depths of 2000 mwe or 200 atm $\sigma = 5 - 10\mu b$ guarantees that they would be contained events. Thus we can *tune* the particles interaction to fit these needs, but we will get in trouble since it should be seen in beam dump experiments[10].

Since there are too many muons at too high an energy to be produced by *gamma*'s in the atmosphere, we investigate *local* in the rock production mechanisms as outlined in Figure 2. Consider the reaction

$$X^o + N \rightarrow \mu + h.$$

Then the scattering angle of the muon θ_μ is bounded, even if produced by a secondary decay by

$$\theta_\mu < \sqrt{\frac{2M_N}{E_\mu}}$$

which gives $\theta_\mu < 2^o$ if $E_\mu = 10^2 eV$ which is much smaller than the 5° usually quoted. This just underscores that no mechanism will allow for the production of the wide angle underground muons in the atmosphere.

To give the 5° angular spread the muons must have $E_\mu < 200 \ GeV$ or about a range of 800 mwe, so that the muons are produced locally. Thus the range of the X^o has to be 4000 mwe or about $\sigma = 2 - 10\mu b$.

V. Beam Dump Arguments

If the reaction $X^o + N \rightarrow \mu + h$ occurs with $\sigma = 2 - 10\mu b$ cross section, then the ratio of $(X^o \rightarrow \mu)/(\nu \rightarrow \mu)$ should be enormous in beam dump experiments. It is not! Below are some of the vague ways out of this dilemma.

1. Associated Production. Since for beam dump experiments $E_{cm} \simeq 30 GeV$ production of the light X^o with a heavy associate ($> 10 \ GeV$) would effectively freeze it out at current beam dump experiments. This must be a variation on Ruddick's scheme[21] where he needs a massive secondary particle, the S with mass $\approx 40 \ GeV$.

2. Massive Parent. Cygnets are not produced directly via $pp \rightarrow X^o+$, but indirectly via $pp \rightarrow \Gamma+$ and the $\Gamma \rightarrow X^o+$. The particle should be semistrongly interacting with $M \leq 10 \ GeV$ so as to provide the correct cygnet flux yet $M \geq 4 \ GeV$ so not to be seen in beam dumps. It needs a long lifetime $> 10^{-10}$ seconds, so that its lifetime is longer than its mean free path for interaction. In fact by tuning on its lifetime we can arrange for the X^o to escape any beam dump experiment, in fact a lifetime $> 10^{-5}$ seconds works fine!

3. *Weak Photon.* A weak photon is a particle whose mass $\rightarrow$ 0 that is stable until it gets around matter because its decay products are more massive than it is, like $\gamma \rightarrow e^+e^-$ pair-production, which requires a heavy nucleus to make it kinematically possible. It might possible have strange semi-weak couplings like $\gamma_W \rightarrow e^+\mu^-$, however, probably it should have been discovered! This may also be a variation of the Ruddick scheme if the electron becomes the S.

VI. Conclusions

The problems raised by the Kiel and Soudan effects are enormous. Kiel, if correct, argues for QED violations at the least, and possible for a new particle, the X^o. The underground results, because of the large angular spreads of the muons and these large fluxes compound with EAS fluxes dictate that the muons have low energies and are produced deep in the earth close to the detector.

DEDICATION

This work is dedicated to the memory of my father, Malcolm McKay Morse, who passed away suddenly and unexpectedly on November 14, 1986 at the age of 71 in Washington, D.C.

ACKNOWLEDGEMENTS

This manuscript has been prepared with the help and warnings of many people. I would like to thank all of my friends and colleagues in the Haleakala group, notably Ugo Camerini, James Matthews, Jack Fry, and John Learned. I would also like to thank C. Goebel and T. Stanev for numerous discussions, and I would like to especially thank Francis Halzen whose active interest, insights, and generosity provided me with lots of material.

REFERENCES

1. Battistoni et al, Phys. Lett. 155B (1985) 465. Marshak et al, Phys. Rev. Lett. 54 (1985) 2079. Bionta et al, preprint, Univ. of Michigan UM PDK 86-5 (1986), submitted to Phys. Rev. D.

2. Samorski et al, Ap. J. Lett. 268 (1983) L17. Lloyd-Evans et al, Nature 305 (1983) 784. Blake et al, preprint, Univ. of Nottingham (1985).

3. Oyama et al, (Kamoikande) University of Tokyo preprint UT-ICEPP-85-03. E. Aprile et al, (HPW) Proceedings of Europhysics Conference on High Energy Physics, Bari (1985).

4. Berger et al, (Frejus) Proceedings of Europhysics Conference on High Energy Physics, Bari (1985). Chudakov, 19th ICRC (La Jolla) 1985, Vol. 9 (441). Francis Halzen, John Learned, private communications.

5. Turver, Highlight paper, 19th ICRC (La Jolla) 1985 Vol. 9. Camerini et al, Haleakala (HGO), NATO Workshop on VHE Gamma Rays, Durham (1986). Cawley et al, 19th ICRC (La Jolla) 1985 OG2.1-11. Fry, Proceedings of the 1986 Aspen Winter Conference, M. Block, editor. Chadwick et al, 19th ICRC (La Jolla) 1985, OG2.1-8. Bhat et al, ibid. OG2.1-1.

6. Protheroe et al, Ap. J. 280 L47(1984). Morello et al, 18th ICRC (Bangalore) 1983, Vol. 1, 91. Baltrusaitus et al, ibid. p.234. Lambert et al, ibid. p.71.

7. Kifune et al, Proc. of 19th ICRC (La Jolla) 1985 Vol. 1, 67. Kirov et al, ibid. p. 135.

8. Van der Klis et al, Astr. Astrophysics. Lett. 97, L5 (1981).

9. Hillas, Highlight talk, 19th ICRC (La Jolla) 1985, Vol. 9. Brecher, NATO Workshop on VHE Gamma Rays, Ddurham (1986). Eichler et al, Nature 307 (1984) 613.

10. For an excellent discussion of all of the problems see Berezinsky et al, Moscow preprint ITEP-127 (1985). De Rujula preprint CERN-TH 4267 (1985).

11. Blake et al, preprint, University of Nottingham (1985). Blake et al, 19th ICRC (La Jolla) 1985, Vol. 1 OG2.1-4.

12. Samorski, preprint, *Chances to Observe VHEGR Sources Beyond 10^{14} eV with EAS Arrays*, Kiel (1984).

13. It is interesting to speculate on the Bhat et al effect, 19th ICRC (La Jolla) 1985 Vol. 1, OG2.1-10 where they report on the decreasing flux of Cygnus with time. It seems to be falling as the area and sensitivity of the reporting detectors becomes greater. Are we destined to be always working with *just above threshold effects?*.

14. M. Marshak, private communication.

15. Stanev, preprint, *Are Underground Detectors Good VHEGR Telescopes?*, Bartol Research BA-86-7 (1986). Halzen et al, University of Wisconsin preprint MAD/DH/260 (1985).

16. E_γ/E_μ ratios derived from work by Stanev, private communication.

17. In deference to the Frejus non-result on Cygnus presented at the NATO workshop on VHEGR Astronomy, Durham (1986) it has recently been argued by Marshak (23rd Int. Conf. on Particle Physics - Berkeley - 86) that the muon signal shows signs of variability. Durham-Dugway and Haleakala have seen clear indications that the gamma ray flux peaks over intervals as short as one mintye. Most of the Frejus data are taken outside the *on* period, so given this *shutdown* Frejus could have small signal in their data taken at the initial stage of operation. See Halzen, University of Wisconsin preprint MAD/PH/243 (1985), unpublished.

18. Marshak, talk at the 23rd Int. Conf. on Particle Physics, Berkeley (1986). Courant et al preprint, Soudan Collaboration, ANL-HEP-PR-86-XX (1986).

19. We used *all nucleon* integral spectrum

$$N_>(E) = 6.8x10^3 \ E^{-1.65}(GeV)m^{-2}sr^{-1}s^{-1}$$

   of Hillas, Proc. Bartol Conf. (1978), p. 373, and Proc. 16th ICRC Kyoto (1979). We calculate the acceptance of cosmic rays into the Cerenkov telescope view by assuming 1° full view aperture of $\Delta\Omega = 2.4x10^{-4}sr$.

20. Halzen, talk at Int. Europhysics Conf., Bari (1985) and preprint, Univ. of Wisconsin (1985) MAD/PH/260.

21. Ruddick, Phys. Rev. Lett. 51 (1986), 531.

*For an excellent overview of the field see: Watson, Rapporteur paper Cosmic γ-Rays above 1 TeV, 19th ICRC (La Jolla) Vol. 9. Turver Highlight Talk, *Ground Based VHEGR Astronomy* ibid. Hillas, Highlight Talk, *Why is Cygnus X3 a Highlight of Ccosmic Rays* ibid.

THE UNIVERSITY OF DURHAM NEW VHE GAMMA RAY TELESCOPES

I. Carstairs, P. M. Chadwick, N. A. Dipper, E. W. Lincoln,
T. J. L. McComb, K. J. Orford and K. E. Turver.

Department of Physics
University of Durham
Durham DH1 3LE
United Kingdom

ABSTRACT. The specification and performance of two new VHE gamma ray
telescopes are described. The larger of the telescopes (MK III) has a
threshold of 240 GeV and is at present en route for a series of
observations in the Southern hemisphere. The MK IV telescope now
nearing completion has a slightly higher threshold (300 GeV) and will
be used in 1987 for Northern hemisphere observations.

The TeV gamma ray facility at Dugway, Utah constructed by the
University of Durham ceased operation in October 1984. The new
telescopes described here will replace the Dugway telescopes and are
conventional in design and incorporate the lessons learned during the
Dugway project. The main development which has been incorporated in
both telescopes is the new lightweight and low cost mirror of
sufficient optical quality for Cerenkov light studies. This is a 60 cm
diameter mirror made from anodised aluminium (reflectivity 80 %)
stretched over a former and bonded to a sandwich honeycomb backing
structure. The focal length of the mirror is 254 cm and the size for
an on-axis stellar image is < 1 cm for an individual facet and < 2 cm
for a composite image from an assembly of 50 mirrors. In both
telescopes we have retained the requirement for a fast threefold
coincidence to define a Cerenkov event so allowing operation at high
gain and noise levels without accidental noise events. The choice of
PMTs (RCA 8575) has been made mainly on the basis of their low
noise-illumination characteristics and stable operation. We have
progessively reduced the geometrical aperture of our telescopes in
recent years to improve their sensitivity. The new telescopes have an
aperture of 0.9 deg FWHM, consistent in our experience with accepting
the Cerenkov light from the gamma rays but rejecting the maximum
number of signals from off-axis protons. Both telescopes have
provision for maintaining constant background light (using servoed
LEDs) and simultaneous background (off axis) measurements in adjacent
areas of sky. A common design of steering the telescopes on their
alt-azimuth mounts is employed which uses DC electric motors and
absolute digital shaft encoders, giving position resolution of < 0.1

K. E. Turver (ed.), Very High Energy Gamma Ray Astronomy, 221–223.
© *1987 by D. Reidel Publishing Company.*

deg. The data record for each event comprises the time of detection (derived from a Rubidium oscillator), the amplitude of each PMT signal, the individual PMT rates and anode currents.

THE MK III TELESCOPE

The specification of the Mk III telescope is given in Table I. The characteristics of the 12 sq m flux collecting dishes apply to each of the three units comprising the telescope. The telescope will be set up at Narrabri, NSW, Australia (30 S, 149 E) during September 1986.

The indication of the sensitivity of the new telescope is given by its response to claimed detections of emission of VHE gamma rays noted during the Dugway project. For example, the time to observe to the 3 sd level of significance the weak but apparently persistent pulsed emission from the Crab pulsar and the significance of the outburst of emission from Cygnus X-3 which led to the claim for a 12 ms pulsar period.

MARK III TELESCOPE

(SOUTHERN HEMISPHERE TELESCOPE)

CHARACTERISTICS

AREA OF EACH COMPOSITE MIRROR:	0.27 x 43 SQ. M. (11.61 SQ. M. TOTAL)
APERTURE:	40 mm.
FOCAL LENGTH:	2.54 M.
FIELD OF VIEW:	0.9 DEG.
EFFECTIVE REFLECTIVITY:	0.8
COUNT RATE:	75 C.P.M.
THRESHOLD:	240 GeV

SENSITIVITY:

CRAB (PULSED) SEEN TO 3 σ	1.1 hr
450 s OF CYG X-3 (12 ms PULSED) SEEN AT PROB.	1.6×10^{-37}

TABLE I: The specification and performance of the Mark III telescope.

FOOTNOTE: The first observations made in Australia with the MK III telescope in September 1986 confirm that the count rate near the zenith is 75 cpm.

THE MK IV TELESCOPE

This telescope differs from the Mk III only in having smaller (5 sq m) flux collecting dishes and simplified recording electronics and its characteristics are summarized in Table II. The telescope is designed to be readily portable and suited to short duration observing campaigns, the first of which is planned for mid-1987 to continue our observations of Cygnus X-3.

MARK IV TELESCOPE

(LIGHTWEIGHT TELESCOPE: CYGNUS X-3 MONITOR?)

CHARACTERISTICS

AREA OF EACH COMPOSITE MIRROR:	19 x 0.27 SQ. M. (5.1 SQ. M. TOTAL)
APERTURE:	40 mm.
FOCAL LENGTH:	2.54 M.
FIELD OF VIEW:	0.9 DEG.
EFFECTIVE REFLECTIVITY:	0.8
COUNT RATE: (PREDICTED)	44 C.P.M.
THRESHOLD: (PREDICTED)	300 GeV

SENSITIVITY:

CRAB (PULSED) SEEN TO 3 σ	3 hr
450 s OF CYG X-3 (12 ms PULSED) SEEN AT PROB.	1.6×10^{-13}

TABLE II: The specification and performance of the Mk IV telescope.

ACKNOWLEDGEMENTS: We are grateful to the Science and Engineering Research Council and the University of Durham for their continuing support.

THE HALEAKALA GAMMA OBSERVATORY

L. Resvanis, S. Tzamarias, and G. Voulgaris
Physics Laboratory, The University of Athens
10680 Athens, Greece

J. Learned, V. Stenger, and D. Weeks
Department of Physics, The University of Hawaii
Honolulu, HI 96822, USA

J. Gaidos, F. Loeffler, J. Olson, T. Palfrey, G. Sembroski,
and C. Wilson
Department of Physics, Purdue University
West Lafayette, IN 47907, USA

U. Camerini, J. Finley, M. Frankowski, W. Fry, M. Jaworski,
J. Jennings, A. Kenter, R. Koepsel, M. Lomperski, R. Loveless,
R. March, J. Matthews, R. Morse,* D. Reeder, P. Sandler,
P. Slane, and A. Szentgyorgyi
Department of Physics, The University of Wisconsin
Madison, WI 53706, USA

ABSTRACT.   The Haleakala Gamma Observatory is a $10m^2$ multi-mirror telescope for observing Cherenkov light from electromagnetic cascades in the atmosphere. It is situated at an altitude of 2950 meters at $20.7°N$, $156°W$ on Mount Haleakala, Maui, Hawaii. It differs from most Cherenkov devices in accepting single photoelectron pulses. It employs two sets of 18 phototubes observing seperate regions of the sky to continuously monitor hadronic background. Hardware coincidence resolution is 10ns, and digital filtering can reduce this substantially, effectively eliminating random signals from ambient light. Events are timed to within $±2\mu s$ of UTC by a Cesium beam atomic clock. Hadronic showers are observed at rates of 0.5 to 0.7 Hz, implying a threshold for gamma-induced showers of about 200 GeV.

1. DESCRIPTION.

The Haleakala Gamma Observatory (HGO) detects atmospheric cascades at energies in the range above $10^{11}eV$ by means of Cherenkov light. Background from ambient light is eliminated by means of fast hardware coincidences and subsequent digital filtering of signals from photomultiplier tubes (PMTs) operating at single photon sensitivity. This design achieves the lowest energy thresholds compatible

---

* Presenter of this paper.

K. E. Turver (ed.), Very High Energy Gamma Ray Astronomy, 225–229.
© *1987 by D. Reidel Publishing Company.*

with a given mirror area. Only 2–3% of the PMTs will be occupied by an ambient light signal over the duration of a typical shower, ~5ns. For a fixed mirror size the occupancy varies as τ/N where τ is the resolving time, and N is the number of PMTs viewing separate segments of the mirrors. The τ/N ratio for the HGO is about 2–5% of the ratio for conventional Cherenkov devices, in which each PMT views all of the mirror area.

The telescope employs two sets of 18 PMTs. Each set is directed at one of two sectors of the sky of angular diameter 0.75°, separated by 3.6° in declination. One set monitors isotropic background while the other is focussed on the source. The mirrors are coplanar on a common equatorial mount to preserve time coherence. Triggering hardware demands a designated minimum number of PMT signals (usually ~7) in a tight time window, in place of the more common practice of requiring a larger signal on one or a few PMTs.

Figures 1 and 2 show the physical layout of the apparatus. The 6 spherical glass mirrors are f/1 devices of 1.5m diameter. The image of a point source at infinity has an angular spread of about 0.5°. Each mirror has a 30cm diameter cylindrical can mounted coaxial to the optic axis, housing 6 PMTs and associated power supplies and amplifiers. Two focal plane apertures, each displaced by 1.8° declination from the optic axis, admit light to a tight grouping of 3 PMTs with 18mm diameter photocathodes. The effective signal collection area for each PMT is about $0.6m^2$. The mount is computer controlled via stepping motors and its position verified by encoders to an accuracy of 0.04°. In normal operation, the source is tracked in right ascension, while the declination axis shifts at preset intervals to give each of the two apertures its turn on the source.

The PMTs are Hammamatsu model R1450, chosen for high quantum efficiency (~30% at 420 nm), small physical size, and good time resolution (~ .7ns). The gain is about 5×10^6, so that single photoelectron signals produce pulses of 2–3 mV amplitude and 3ns duration across a 50Ω load. Anode signals feed amplifiers with a voltage gain of 30 and 0.5 GHz bandwidth, driving 30m of 50Ω low-dispersion coaxial cable (RG-8/U) to the control room, where pulse heights from single photoelectron events range from 30 to 60mV.

Figure 3 shows the electronics. Discriminators convert PMT signals to standard NIM logic pulses of width 7ns with ~1ns rise and fall times. One set of pulses is combined in a pulse adder driving a discriminator that sets the trigger level. A second discriminator set at a lower threshold provides a small prescaled sample of primarily random triggers. Other discriminator pulses are routed to time digitizers (TDCs) with a least count of 0.1ns, pattern units to record the hit pattern, and scalers that monitor the PMT singles rates. A separate pulse drawn from the last two dynodes feeds pulse height digitizers (ADCs).

A trigger pulse initiates the following actions: i) readout electronics is latched and disabled, ii) a slave register of the Cesium clock is frozen, iii) a

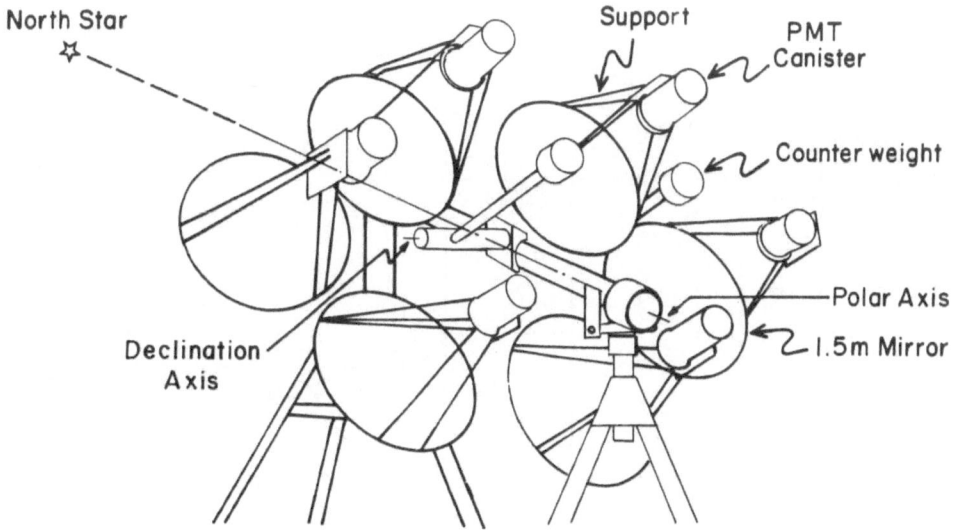

Figure 1 — Mechanical layout of the mirrors and equatorial mount.

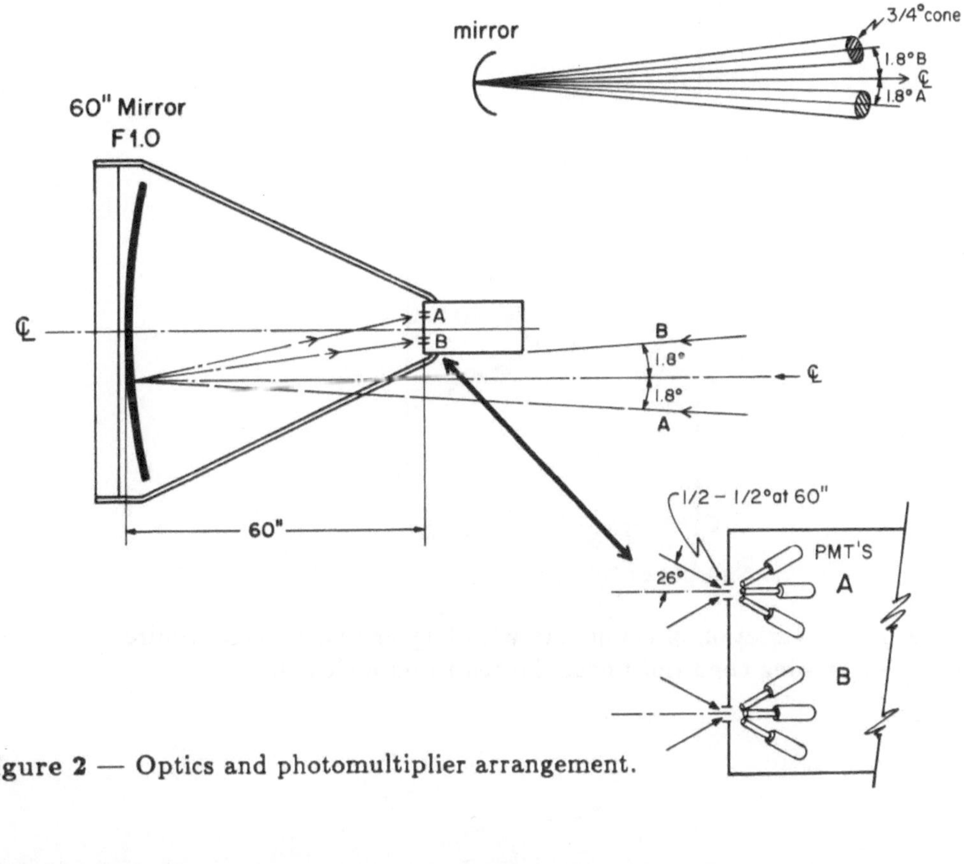

Figure 2 — Optics and photomultiplier arrangement.

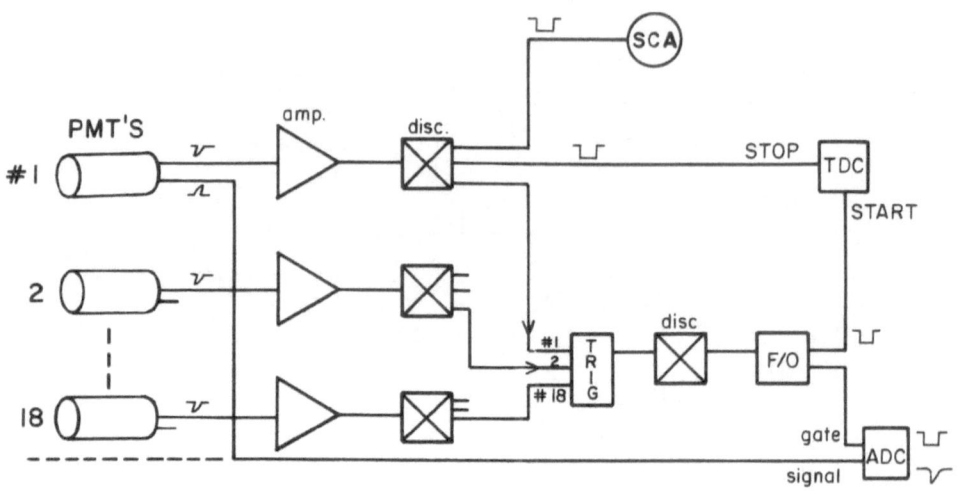

Figure 3 — Block diagram of the trigger logic and data digitizers.

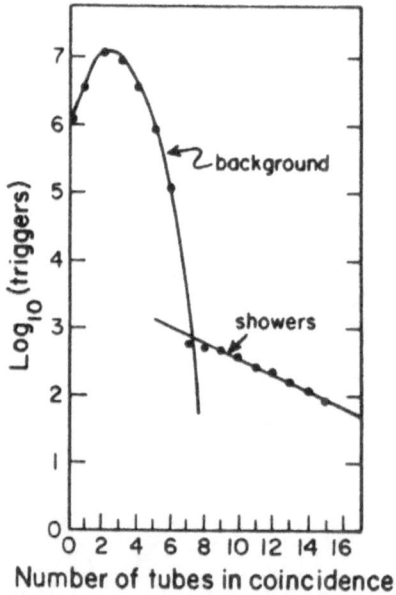

Figure 4 — Variation of count rate with trigger multiplicity requirement, for viewing conditions near the zenith on a clear night.

start timing pulse is sent to the TDCs, iv) a gate is sent to the ADCs and pattern units, and v) an LSI-11/73 computer is interrupted to initiate data transfer, which takes about 2ms per event. The record contains the pulse height and time for each tube, the hit pattern, the event time, and information on the position of the mount. Scalers are read and recorded once each second.

2. PERFORMANCE.

Figure 4 shows a typical example of the dependence of trigger rate on coincidence multiplicity. The steep slope at low multiplicity reflects the rapid decrease in random coincidences, while the flatter portion at high multiplicity is dominated by hadronic cascades. The operating point is set near the break in the slope of the curve, typically around 7 PMTs out of 18. With individual PMTs counting at 2–3 MHz, the trigger rate is about 1Hz per channel. At this rate the system is more than 98% live, and less than half the triggers are induced by random coincidences.

The time resolution of the hardware trigger is about 10ns. Digital filtering using TDC information yields a rate of about 0.5 to 0.7 Hz with a nearly perfect rejection of ambient light. Short-term increases in ambient light have a large effect on the trigger rate, but leave the rate after filtration unaffected. Typical showers give pulses that fall entirely within a 4–5ns window.

The observed shower rate, given our angular acceptance, corresponds to a threshold of 600 GeV for hadronic showers, which normally represent the bulk of the data after filtering. Shower simulations indicate that the yield of Cherenkov light from gamma-induced cascades is roughly three times that of hadronic showers of equal energy, so we estimate our threshold at 200 GeV for gammas.

The Cesium beam atomic clock is frequently checked against the Hawaii LORAN-C network, and several times each year it is compared to U.S. Naval Observatory standards by means of travelling atomic clocks. All checks to date have estimated it to be within $\pm 2\mu s$ of UTC. This timing is sufficiently accurate to determine absolute phase of the fastest known pulsar signals.

In the 14 months since the telescope was completed, it has operated an average of 80 hours per month out of a possible 100.

3. ACKNOWLEDGEMENTS.

We thank J. Pavlat of The CompuMotor Corporation of Petaluma, California for donating the computer-controlled stepping motors. The NASA Lunar Ranging Experiment maintains the atomic clock, and kindly allowed us to tap into its readout. The mount was designed fabricated at the Physical Sciences Laboratory of the University of Wisconsin under the careful supervision of T. Winch. The advice and assistance of many members of the staff of the Hawaii Institute for Astronomy, especially W. Lu, has been indispensable.

THE ADELAIDE VERY HIGH ENERGY GAMMA-RAY ASTRONOMY PROGRAMME

R.W. Clay, S.D. Elton, A.G. Gregory, J.R. Patterson & R.J. Protheroe
Department of Physics, University of Adelaide
Adelaide, South Australia 5001, Australia

ABSTRACT. The University of Adelaide group is in the final stages of designing a new V.H.E. gamma-ray telescope. The telescope will be located at Woomera, South Australia, and should be operational in 1987. The main features of the proposed telescope are described. Before completion of the new telescope, V.H.E. observations have already started using three mirrors of the White Cliffs Solar Power Station. The operation and performance of this system as a gamma-ray telescope is described.

1. PROPOSED TELESCOPE

The new telescope will initially consist of three mirrors each of area 10 m^2 on a single alt-azimuth mount. The focal length will be 6 m. Further telescope units may be added later to form an array. To facilitate access to the photomultipliers the mirrors will be mounted side by side. Mechanical design of the alt-azimuth mount is complete and construction will begin during the next few weeks.

The photomultiplier configuration will be decided after detailed simulations of the telescope response have been carried out. A number of options are currently being considered. These include: (1) the use of a 5" tube located at each focus, giving a 1.2° field of view, the three mirrors being operated in coincidence; (2) 5" diameter apertures at each focus, and several tubes located behind the aperture in such a way that each views the same 1.2° region of sky but using only a portion of the mirror area; (3) a camera comprising several small tubes placed in the focal plane. The first option follows the philosophy of trying to collect as much light as possible, as used in most V.H.E. telescopes to date, e.g. the new Durham telescope (Chadwick et al.,1985). In the second option the tubes might be operated at the single photon level, the approach adopted by the Hawaii, Wisconsin, Purdue, Athens collaboration at Haleakala. The third option follows the approach adopted at Mt. Hopkins (Cawley et al., 1985) and may allow the possibility of vetoing a large fraction of cosmic ray events (Hillas 1985) through imaging. Initially, the telescope would be operated in one of the non-imaging modes but we will consider upgrading to an imaging system in the future.

231

K. E. Turver (ed.), Very High Energy Gamma Ray Astronomy, 231–234.
© *1987 by D. Reidel Publishing Company.*

2. OBSERVATIONS AT WHITE CLIFFS SOLAR POWER STATION

2.1 Use of the Solar Mirrors

V.H.E. gamma-ray observations are currently being made using three mirrors of the White Cliffs Solar Power Station. The equipment is located at latitude 30° 51' South, longitude 143° 5' East, and elevation 160 m. The solar array incorporates fourteen mirrors, each 5 m diameter and 1.8 m focal length and was deveoloped under the direction of Professor S. Kaneff of the Australian National University.

The mirror units, having been designed for solar energy collection, are not ideal for gamma-ray astronomy. Specifically: there is no provision for the mirrors to point at azimuths between South and West; tracking is by a solar sensor; spot size is such that 60% of the solar energy falls in a circle of diameter 125 mm (4° at 1.8 m). The last two of these problems have been overcome as described below.

The two axes were fitted with shaft encoders and the drive motors put under computer control. Calibration of the shaft encoders was performed in the daytime during the normal operation of the power station while the mirrors were automatically tracking the Sun. The shaft encoders were read by the computer at regular intervals and the position of the Sun was calculated for these times at the location of the solar array. The calibration was checked by tracking the full Moon and observing its image projected on a screen placed in front of the photomultiplier.

Because the field of view is greater than 4° the night sky background is at least a factor of sixteen higher than in a system designed for gamma-ray astronomy (i.e. field of view about 1°). To eliminate spurious events due to night sky noise, it is then necessary to operate mirror units in coincidence and to set the discriminator thresholds at a sufficiently high level to reduce the chance background coincidence rate to an acceptable level. In each mirror the rate of cosmic ray background events is also more than sixteen times higher than for a 1° field of view. It was therefore decided to use the three mirrors with the longest baselines as an array and to use sub-nanosecond timing and shower arrival direction analysis to achieve an angular resolution of about 1°. We were thus able simultaneously to eliminate all but a few night sky noise events and reduce the cosmic ray background considerably. Because of the high discriminator levels however the energy threshold of the telescope is in excess of 5 TeV, higher than desirable for conducting periodicity searches.

2.2 The Fast Timing System

The three mirrors used formed a triangle with sides of about 43 m, 50 m, and 56 m. Surveying was performed to an accuracy better than 0.1 m. Each mirror unit was fitted at night with a Mullard XP2040 fast 5" photomultiplier. For a fixed pulse shape, and timing from one discriminator level, the pulse arrival time recorded will depend on the pulse amplitude. Signals from the photomultipliers were therefore input to double-discriminator units located at the mirrors. These units have been designed to minimize timing jitter (Clay and Ciampa, 1986) by splitting the signal and timing from when a low discriminator level is exceeded but only accepting events when a higher discriminator level is exceeded. Thus, timing occurs from near the start of the pulse (lower level) and is less sensitive to pulse amplitude, while night sky noise events are reduced by requiring a coincidence with the output from the upper level. An event is defined by signals exceeding the upper discriminator levels in all three systems during a 170 ns interval. A coincidence unit provides the common start pulse for three time-to-digital converters (TDC's). Pulses from the lower levels (suitably delayed) provide individual stop pulses for the TDC's. The timing precision of the TDC's is 0.77 ns. The time of each event was recorded to 1 ms. Additional details of the experimental arrangement are given by Clay et al. (1986).

2.3 Angular Resolution and Energy Threshold

A Monte Carlo simulation of the array response was performed to estimate the angular resolution of the telescope when pointed towards the South at an elevation of 60°. This direction is appropriate for many of our observations of potential V.H.E. gamma-ray sources. The location of an air shower axis was selected randomly and a primary energy was sampled from a differential power-law with index -2, appropriate to sources such as Cygnus X-3. Using an appropriate lateral distribution of Cerenkov photon density it was possible to estimate the pulse heights for each mirror unit. Knowing the impulse response for the photomultipliers it was possible to calculate the times at which the discriminator levels would be exceeded. The residual timing jitter together with TDC rounding errors were then added to the arrival times of a plane front incident from 60° elevation and due South. Two thousand events were generated in this way and the apparent arrival direction of each event was obtained from the simulated times. The resulting angular resolution function is plotted in Fig. 1 and is such that most events appear to arrive from within 1° of the true direction.

An estimate of the energy threshold of the telescope was obtained from the observed rate of cosmic ray events in the full field of view. When pointing at elevation 60° the event rate is 0.75 s^{-1}. Taking into account the variation of the effective area of the telescope system with primary energy based on the form of the lateral distribution of Cerenkov light within the mirrors' fields of view, and the cosmic ray energy spectrum, we obtain an effective energy threshold of around 5 TeV (see Clay et al., 1986, for the derivation).

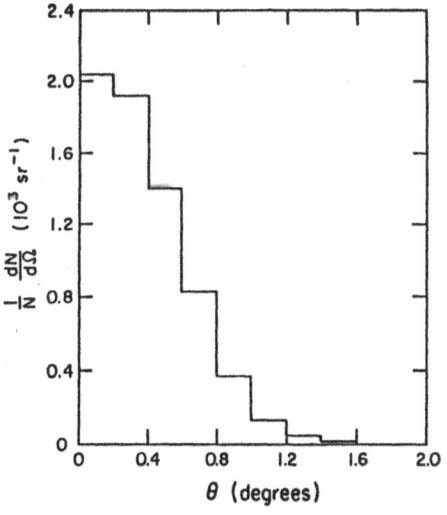

Figure 1. Angular resolution of the fast timing system at White Cliffs based on Monte Carlo simulations (see text).

2.4 Observations

Observations started during the March 1986 new moon period. Data were also taken during the April and June new moon and analysis is proceeding. Fig. 2 shows the arrival directions of air showers recorded during our April 1986 observation in the direction of X1417-624 and demonstrates the quality of our data. As noted above, individual arrival directions are obtained to within about 1°. The concentration of events within about 3° of the pointing direction corresponds to the detector's field of view. The scarcity of events outside this region shows the effectiveness of the coincidence system in eliminating most night sky noise events.

ACKNOWLEDGMENTS

We thank Professor J.R. Prescott for his generous and continued encouragement. We are most grateful to the New South Wales Energy Authority for the opportunity to use mirrors of the White Cliffs Solar Power Station during hours of darkness and to Professor S. Kaneff for his advice and encouragement. This work has been supported by grants from the Australian Research Grants Scheme and the University of Adelaide.

REFERENCES

Cawley, M.F., Fegan, D.J., Gibbs, K., Gorham, P.W., Hillas, A.M., Lamb, R.C., Liebing, D.F., MacKeown, P.K., Porter, N.A., Stenger, V.J., and Weekes, T.C., 1985: Proc. 19th Int. Cosmic Ray Conf. (La Jolla), 3, 453.

Chadwick, P.M., Dipper, N.A., Dowthwaite, J.C., Kirkman, I.W., McComb, T.J.L., Orford, K.J., and Turver, K.E., 1985: Proc. 19th Int. Cosmic Ray Conf. (La Jolla), 3, 406.

Clay, R.W., and Ciampa, D., 1986, Australian J. Phys., 39, 93.

Clay, R.W., Elton, S.D., Gregory, A.G., Patterson, J.R., and Protheroe, R.J., 1986, Proc. Astron. Soc. Australia, in press.

Hillas, A.M., 1985: Proc. 19th Int. Cosmic Ray Conf. (La Jolla), 3, 445.

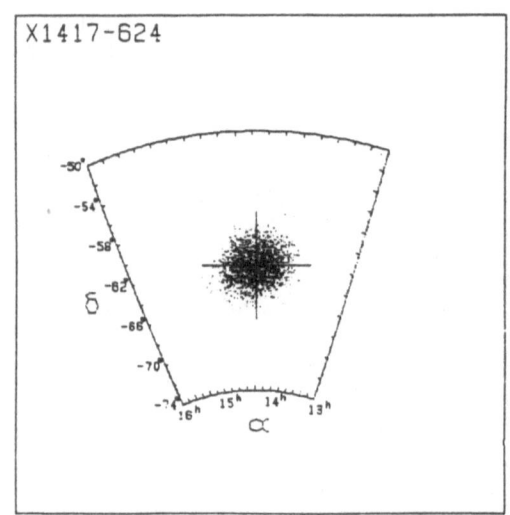

Figure 2. Arrival directions of events recorded during our April 1986 observations of X1417-624 (from Clay et al., 1986).

HERCULES - A NEW INSTRUMENT FOR TEV ASTRONOMY

T. C. Weekes
Harvard-Smithsonian CfA, Amado, AZ, 85645-0097, U.S.A.

R. C. Lamb
Iowa State U., Ames, IA, 50011, U.S.A.

A. M. Hillas
University of Leeds, eeds, UK

ABSTRACT. An improved photon detector, operating in the range of 10^{11} to 10^{14} eV, is described. The Whipple Observatory's atmospheric Cherenkov detector will be upgraded with the addition of another "10 m" class reflector and two high resolution Cherenkov cameras (pixel spacing 0.25°). The new apparatus, symbolized by the acronym HERCULES (High Energy Radiation Cameras Using Light Emitting Showers), will improve the sensitivity of the present detector by more than a factor of ten.

1. INTRODUCTION

The energy region around 10^{12} eV is a particularly fruitful one to study since (i) the energies are sufficiently high to be interesting; (ii) the source spectra of several sources, e.g., the Crab Nebula, are expected to steepen above these energies; (iii) gamma-ray photons can be readily detected using the atmospheric Cherenkov technique. The recent detection of a large number of sources at these energies (Cygnus X-3,[1] Hercules X-1,[2-4] 4U0115+63,[5-6] the Crab pulsar,[7-8] etc.) is evidence of the rich harvest that awaits the full exploitation of these wavelengths. Since many of the detected sources show strong evidence for variability, the ground-based atmospheric Cherenkov technique offers the additional advantage compared with satellite experiments that it can be used for long term monitoring of transient sources.

Despite its obvious advantages, these ground-based techniques have not been developed to their full potential; the total investment in all such experiments on five continents since the early sixties amounts to only a few million dollars, a small percentage of the cost of GRO, DUMAND or a major experiment in high energy physics. For this reason the improvement in sensitivity over the past two decades

K. E. Turver (ed.), Very High Energy Gamma Ray Astronomy, 235–242.
© *1987 by D. Reidel Publishing Company.*

has been slow and the full power of the technique has not been developed.

In the past three years the Whipple Observatory collaboration (consisting of the Smithsonian Astrophysical Observatory, Iowa State University, the University of Hawaii, and University College, Dublin) has developed a new technique using a fast large camera to record the Cherenkov light image from each shower. Recent simulations and experience with an early version of the camera indicate that this technique when fully developed is capable of isolating gamma-ray showers from the much more numerous proton shower background.

The most efficient Cherenkov detector consists of an array of cameras which cover an area roughly equal to the lateral dimension of the shower at ground level; the cameras should have apertures large enough to detect gamma-ray photons in the sub-TeV energy region and they should be located on a high, dark mountain plateau at a low latitude. The general principles involved in such a detector have been described by Weekes[9]. The best features of all the advances that have been made in the atmospheric Cherenkov technique in the past two decades are utilized.

In this paper major steps towards achieving the ultimate detector are described. The Whipple Observatory's atmospheric Cherenkov detector will be upgraded with the addition of another "10 m" class reflector and two high resolution Cherenkov cameras (pixel spacing 0.25°). The new apparatus, symbolized by the acronym HERCULES (High Energy Radiation Cameras Using Light Emitting Showers), will improve the sensitivity of the present single mirror detector by more than a factor of ten. In the remainder of this paper the atmospheric Cherenkov imaging technique is reviewed, with particular emphasis on those image differences which distinguish gamma-ray air showers from the proton shower background, and the sensitivity calculation for HERCULES is presented.

2. THE ATMOSPHERIC CHERENKOV IMAGING TECHNIQUE

The Cherenkov light from proton and gamma-ray initiated air showers are superficially similar; traditionally atmospheric Cherenkov detectors have been designed for the most efficient detection of gamma rays but have not sought to actively discriminate between the two types of showers. Recent simulations[10] have shown that the differences between the two kinds of showers are more pronounced than was originally supposed.

The Cherenkov light images of air showers were recorded using image intensifiers as early as 1960[11]. Some years later, Turver and Weekes[12] proposed an imaging system in which a nest of phototubes was coupled to a large optical reflector; by 1982 practical imaging systems were in operation at the Whipple Observatory[13] and the Crimean Astrophysical Observatory[14]. Similar systems are now under development at the University of California at Riverside, the University of Tokyo and the Yerevan Physics Institute.

The Whipple Observatory Imaging System has been developed as part of a collaboration involving Iowa State University, the University of Hawaii, University College, Dublin and the Smithsonian Astrophysical Observatory. It is based on the 10 m Optical Reflector with a 37-element camera in the focal plane consisting of 5 cm diameter phototubes; the pixel scale is 0.5° and the full field of view is 3.5°. The camera is activated when any two of the inner 19 tubes is triggered with a resolving time of 12 nsec. The trigger rate is 3–5 Hz and the energy threshold is 2×10^{11} eV. The camera has been in operation since 1983 and more than 30 suspected sources have been studied.[4,6,15]

Differences in the recorded Cherenkov light images from the randomly distributed background proton showers and the gamma-ray showers from a discrete source which are parallel to the detector axis arise from two causes:

a. the differences in arrival directions which cause the gamma-ray images to have a characteristic radial pattern about the detector axis (proton showers coming from the direction of the source would have the same pattern),

b. inherent differences in the images caused by the different development of electromagnetic showers and hadronic showers. These differences include the more rapid growth of the electromagnetic cascade, the presence of penetrating (and hence local) particles in the hadron shower and the greater angular divergence of particles in the hadronic shower due to the meson emission processes.

These processes can give rise to differences in the following observable parameters:

1. total light intensity
2. image size and shape
3. image displacement from pointing direction
4. image axis orientation
5. shower light-pulse width in time
6. spectral distribution of light at detector.

All Cherenkov detectors explicit property (1); in the system proposed below we will concentrate on exploiting (2–4) where the differences are well-established by simulations. Differences (5) and (6) are not so clear-cut and will not initially be used.

Fortunately, in the energy range of interest, it is possible to completely simulate the Cherenkov light image from both kinds of shower using Monte Carlo techniques. The most realistic simulations of the response of the Whipple Observatory Camera have been performed at the University of Leeds (U.K.).[10] The results of these simulations are dramatically illustrated in figure 1. The Cherenkov light images from a 10^{12} eV gamma ray are shown as recorded by five vertically directed cameras that are identical to the Whipple

Observatory Camera. The five cameras are positioned at the inter-
section and ends of a cross of arm length 80 m with the shower axis
coincident with the optic axis of the central detector. Two
conclusions are evident from this simulation; (i) by simultaneously
recording the image in two cameras the arrival direction can be fixed
to high precision (angular resolution < 0.2°); (ii) the precision of
the measurement is limited somewhat by the pixel scale. Although
this is for only a single shower, simulations of some hundreds of
showers at a variety of energies show that this is typical and that
shower-to-shower fluctuations are small. Note also that the recorded
light-signals at different orientations are a good measure of the

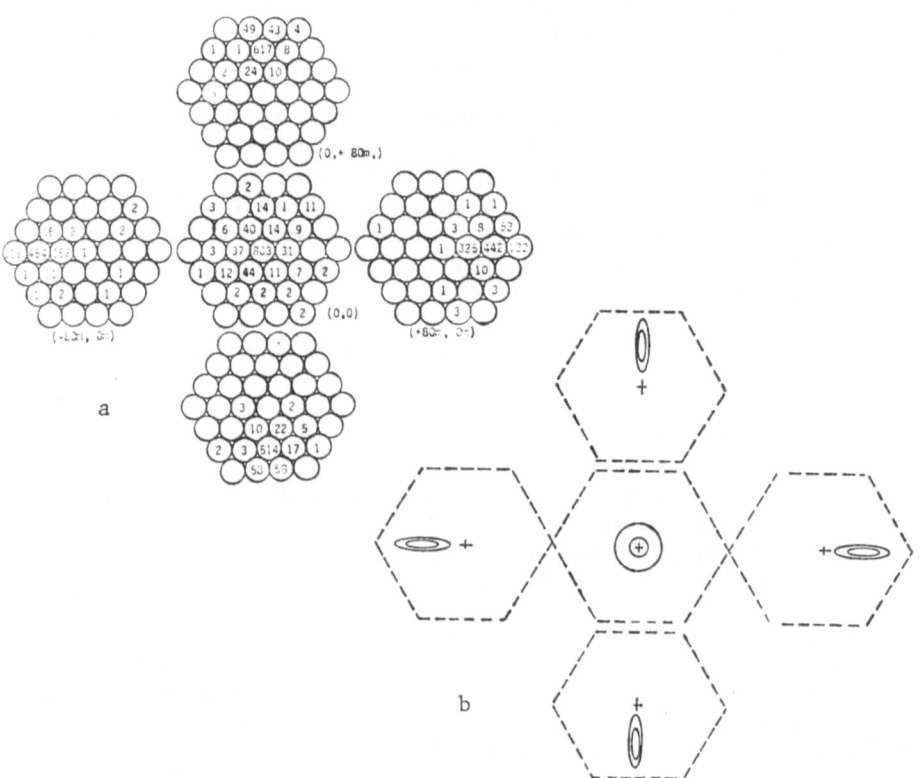

Figure 1. (a) The images recorded in a 37-element, 0.5° pixel
spacing, camera which is located on axis (central image) and at 80 m
distance along orthogonal axes. The shower axis and detector axis
are vertical and the shower energy is 1 TeV. (b) Contours deduced
from the images shower in (a). (Simulations by A. M. Hillas[10].)

primary particle energy. The simulations also show (as does
experience with the early version of the camera) that the images of
gamma ray showers are significantly smaller than images of proton
showers and they tend to have larger impact parameters and hence
locations further from the center of the field of view.

3. HERCULES

We propose to develop the atmospheric Cherenkov imaging concept by
(a) increasing the angular resolution of the 10 m optical reflector;
(b) adding another high resolution camera at a spacing 120 m from the
existing camera. All of the existing systems (optics, electronics,
computer, etc.) will be incorporated into the expanded system. Also,
the extensive software developed for the reduction of the existing
camera images will be directly applicable to the stereoscopic system.

3.1. Angular Resolution

The existing camera of 37 phototubes (full field of view 3.5°) will
be replaced with an array of 193 phototubes, with the inner 169 tubes
having a separation of 0.25°. The full field-of-view of the new
camera is 4.75°. The measured point spread function of the 10 m
reflector is 0.15° FWHM, significantly less than the new camera's
pixel size, and hence will not mask the meaningful scale of
variations in the shower images.

3.2. Additional Cameras

The increased sensitivity of two widely separated detectors was first
suggested by Grindlay and exploited by him and his co-workers in the
detection of a number of sources[16]. We propose to add a new camera
modeled on our existing camera at the distance of 120 m.

3.3. Sensitivity

The greatest improvement in flux sensitivity with the proposed
experiment will be in the energy region from 10^{11} to 10^{12} eV. To
fully evaluate the minimum flux sensitivity requires a complete Monte
Carlo simulation of the response of the system to both gamma rays and
background cosmic rays. This has not yet been undertaken but will be
the first task to be undertaken in the proposed experiment; it is
possible to make a conservative estimate of the gain to be achieved
by extrapolation from the existing camera. The extension of the
energy coverage to both lower and high energies will also be an
important factor in assessing the new system.
    As a standard for comparison we consider a conventional atmo-
spheric Cherenkov detector with an energy threshold of 10^{12} eV; with
a system of three 1.5 m reflectors, operated in coincidence with a
field of view (FWHM) of 1.7°, the background counting rate is ~ 1/sec.

For a collection area of $1.5 \times 10^4 \text{m}^2$ and an observation time of 10 hours, the minimum flux sensitivity (3σ level) is given by:

$$\frac{3 \ (1 \times 3.6 \times 10^4)^{1/2}}{1.5 \times 10^8 \times 3.6 \times 10^4} = 1.0 \times 10^{-10} \text{cm}^{-2} \text{s}^{-1}$$

This presumes that there is no background discrimination and that the signal is not modulated.

With a single imaging system[10] the background can be reduced by 97% by selection of events with gamma-ray characteristics; the collection area for gamma rays is $\pi(117)^2 \sim 4.3 \times 10^4 \text{m}^2$. (This larger collection area, by a factor of three, compared with that derived above results from the larger field of view.) With two cameras the combined collection area will be greater. However, because they overlap, the collection area is less than twice that of an isolated camera (approximately 1.5 times). As each shower within this area will be viewed by at least two cameras with high angular resolution we estimate that the background can be reduced to 1%. The flux

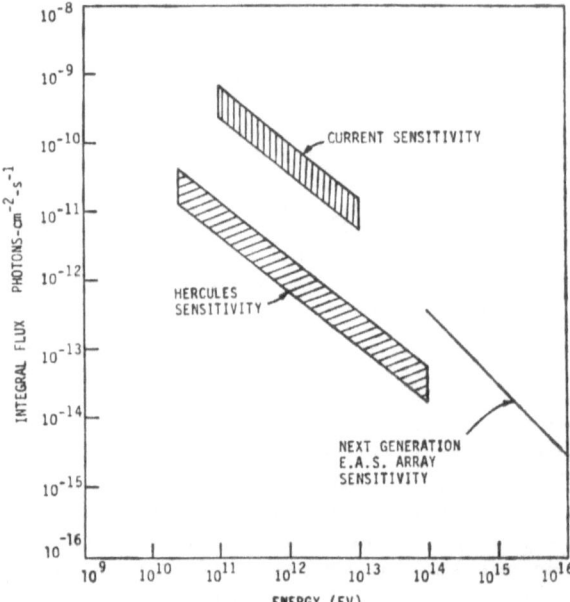

Figure 2. The flux sensitivity of atmospheric Cherenkov detectors for a steady source observed for 10 hours. The sensitivity range of the current detectors is contrasted with that of HERCULES. Also shown is the anticipated sensitivity of the next generation of EAS detectors.

sensitivity can thus be reduced by a factor $(0.01)^{1/2}/3\times1.5 = 0.022$, giving a minimum flux sensitivity of 2.2×10^{-12} photons/cm^2-sec. For a periodic signal (of known frequency) and duty cycle of 10% (or a longer exposure time on a steady source, say 100 hours), the flux sensitivity drops by a factor of three, to 7×10^{-13} photons/cm^{-2}-sec.

With the increased sampling due to the smaller pixels and the absence of dead spaces between phototubes, the cameras will have greater sensitivity over the entire range 10^{11} to 10^{13} as shown in figure 2. We will also extend our energy coverage into the decades below and above our primary range.

4. SUMMARY

The scientific pay-off very high energy astronomy has increased dramatically in the past several years. An instrument like HERCULES can provide answers to very fundamental questions of nature. For example, where do cosmic rays originate, what types of sources govern the high energy budget of our own galaxy, and under what conditions do spinning magnetized neutron stars (with or without binary companions) generate intense beams of high energy radiation? With the factor of 10 improvement one may expect the unexpected as well. (The "eyesight" of a frontier science to see where it is going is generally bad. Many times the biggest and most important developments are in totally unanticipated areas.)

We have confidence that HERCULES is straightforward to build and that it will achieve the predicted improvement in sensitivity. That confidence comes from two sources: our experience with the present 37-element camera (0.5° pixel spacing) and with detailed Monte Carlo simulations of gamma ray and cosmic ray air showers. Our present camera has demonstrated that imaging Cherenkov light from individual air showers is technically feasible. The Monte Carlo studies show that large gains in background reduction can be achieved with finer angular resolution cameras and an additional "10 m" class reflector. These improvements will proceed on a time scale which is determined by funding. HERCULES could be ready by 1989.

ACKNOWLEDGEMENTS

We are indebted to the members of the Whipple Observatory collaboration, particularly Michael Cawley and David Lewis. This work was supported by the U.S. Department of Energy.

REFERENCES

([1]) See for example references in T. C. Weekes: "New Particles 1985", eds. V. Barger, D. Cline, and F. Halzen, p. 288 (1986).
([2]) J. C. Dowthwaite et al.: Nature, 309, 691 (1984).
([3]) R. M. Baltrusaitis et al.: Ap. J. Letter, 293, L69 (1985).
([4]) P. W. Gorham et al.: Ap. J. in press (1986).
([5]) P. M. Chadwick et al.: Astron. Ap. 151, L1 (1985).

(6) R. C. Lamb et al.: Proc. NATO Workshop on High Energy Gamma Ray
     Astronomy, Durham, in press (1987).
(7) J. C. Dowthwaite et al.: Ap. J. Letter, 286, L35 (1985).
(8) T. Tumer et al.: 19th ICRC 1, 139 (1985).
(9) T. C. Weekes: 19th ICRC 3, 422 (1985).
(10) A. M. Hillas: 19th ICRC 3, 445 (1985).
(11) D. A. Hill and N. A. Porter: Q.J.R.A.S. 4, 275 (1963).
(12) K. E. Turver and T. C. Weekes: Nuovo Cimento 45B, 99 (1978).
(13) D. J. Fegan et al.: Nucl. Inst. & Meth. 211, 179 (1983).
(14) A. A. Stepanian et al.: Proc. Ooty Workshop, 43 (1982).
(15) M. F. Cawley et al.: 19th ICRC 1, 131 (1985).
(16) J. E. Grindlay et al.: Ap. J. Letter 197, L9 (1975).

CHARACTERISTICS OF CERENKOV IMAGES PRODUCED BY PRIMARY PHOTONS AND HADRONS AND BY LOCAL MUONS

A. M. Hillas
Physics Department, University of Leeds
Leeds LS2 9JT, England

J. R. Patterson
Department of Physics, University of Adelaide
Adelaide, SA 5000, Australia

ABSTRACT. The appearance of Cerenkov light sources in high-altitude air showers (and background signals due to local muons) has been calculated and typical differences between the images of showers induced by gamma rays and by background nucleons are illustrated by a few examples of high-resolution images. It seems that much higher resolution than is currently employed in detectors would be advantageous. The Cerenkov angle plays little part in the shape of the image: one almost has a simple geometrical perspective view of the shower in space. This perspective view determines what is seen when a shower is viewed from more than one detector position. The lateral distribution of light is then relevant: this is erratic in TeV nucleonic showers, where individual penetrating particles produce local light peaks.

1. OUTLINE

In search of characteristics of Cerenkov images of air showers that might distinguish primary photons from background hadrons, we have computed detailed forms of such Cerenkov images, as seen in a single focusing telescope, though only a few examples are displayed. For a shower falling within about 150m, the specifically Cerenkov features of the emission (fixed angle), are hardly apparent: one is virtually seeing simply a glowing shower, and its length, width and orientation and perspective effects are apparent. An exception is that individual nearby penetrating particles look unduly prominent. (Such local muon images can have uses in calibration.)

It may be advantageous to view showers from more than one location, in which case the lateral distribution of light is important. Hadronic showers usually have a much more irregular light distribution (to some extent on a scale of a few metres) - largely due to muons - whereas photon showers give a more regular light pool (Poissonian local fluctuations). We note that one often has to use an aperture stop in the image plane to reduce sky noise, and the lateral distribution should then be

K. E. Turver (ed.), Very High Energy Gamma Ray Astronomy, 243–248.
© 1987 by D. Reidel Publishing Company.

much curtailed. Showers from heavy nuclei would give much less light
per GeV than do proton showers (which are less efficient than gamma
showers), and so are not discussed.

2. CALCULATED CERENKOV SHOWER IMAGES

2.1. Differences between images of gamma- and nucleon- showers

It has been shown (Hillas, 1985) that even the crude imaging capabilities of the Mt. Hopkins 10m mirror (pixel spacing 0.5 degree) one expects
to be able to reject all but a very few percent of nucleonic images on
the basis of their shape and orientation, whilst accepting most gamma
images related to a point source. Briefly, the nucleon images tended to
be (a) broader - due to the emission angles of pions in nucleon collisions spreading the shower - (b) longer - since the TeV nucleonic
shower penetrates deeper into the atmosphere - and (c) only occasion-
ally, by accident, aligned (major axis of image) with the point source,
since they belong to an isotropic population. However, the selection of
gamma showers on the basis of image shape has not yet been so successful, so very high resolution images have been calculated, to check
whether instruments with better angular resolution would be worth considering. In the pictures shown here, every photoelectron in an image
obtained with a 10m mirror is included.

2.2. Examples of high-resolution calculated images

Figure 1 displays six images as would be seen with a focusing mirror of
10m diameter, at the altitude of Mt. Hopkins (2300m), pointing at a
source 30° from the zenith. Six different showers are shown - three
gamma-rays of 320 GeV and three 1 TeV nucleon showers - all coming from
the direction of the source, but falling at different distances from the
mirror. The orientation of the image in each case reflects the azimuthal
direction of the shower axis as seen from the mirror. The outer circle
on each image has a diameter of 4° and the inner 1.5°. Note the expected
perspective effect: the image gets longer and further from the source
when the impact parameter on the ground is larger. Beyond 140m, the
total amount of light begins to fall off rapidly, so very much more distant images are not expected. In the nucleon images, light emitted by
muons has been ringed: some characteristic arcs due to single muons are
seen (and one abnormal multiple arc).
     In general, nucleon showers have a more ragged appearance, as expec-
ted on the basis of the previous work referred to. And, in practice,
most nucleon showers will not come from precisely the direction of the
source (in the centre of the field), as these do, but will be aligned
randomly. The gamma images are aligned with the source to high accuracy,
so it would be permissible to reject all images whose axes missed the
source by even a very small amount. (Not all gamma images are narrow and
well defined, but one could afford to lose those.)

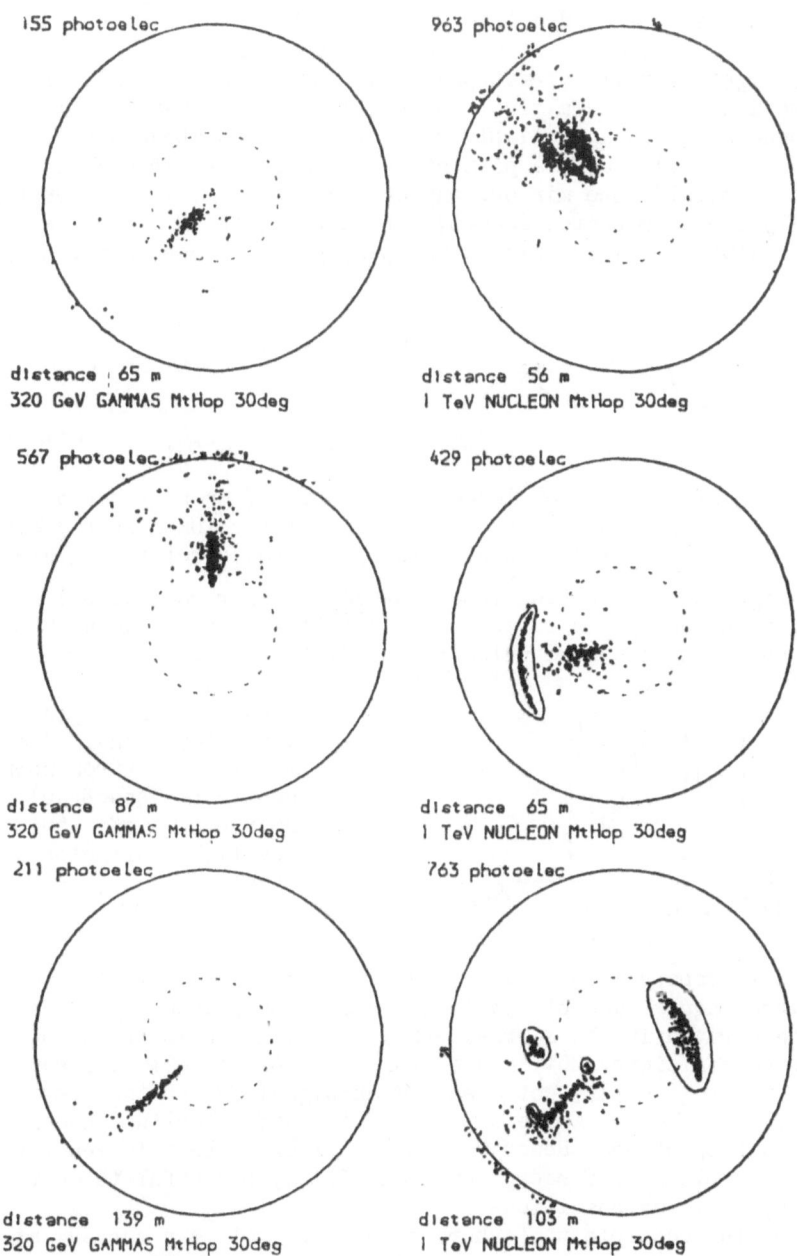

Figure 1. Simulated Cerenkov images of photon (L) and proton (R) showers - axes at various distances from detector (Mt. Hopkins 10m mirror),
showing every photoelectron. 4° field centred on source (inner dashed
circle 1.5° diameter). Light from muons is ringed. 30°from zenith.

2.3. Cerenkov images due to local muons

A particle track approaching close to a mirror produces a bright image:
the light intensity around a track reaching the ground falls as
(1/distance) out to at least several tens of metres. If a muon strikes
a mirror, light from all directions along its Cerenkov cone will be picked up: the image will be a circle of radius 1.3° (at sea level) - or less
if the muon energy is appreciably below 10 GeV (but above the Cerenkov
threshold, 4.3 GeV). Only a part of the cone of rays is picked up from a
muon passing outside the mirror, and one sees shorter arcs, though with
practically the same photon density per unit length of image, as illustrated in Figure 2. These images may perhaps be of use in calibrating

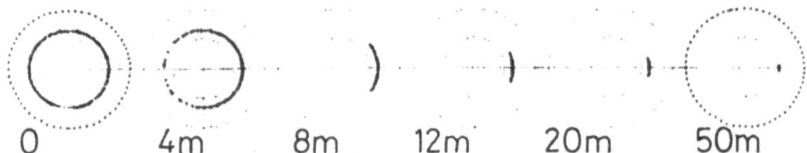

Figure 2. Cerenkov images due to 20 GeV muon at various distances from the centre of a mirror. (For 10m mirror diameter.
Distance scales with diameter.) Dotted circle: 4° diameter.

Cerenkov light detectors (for intensity and angular resolution) using a
scintillator telescope to select unaccompanied muons at known distance.
Figure 3 shows the expected pulse height distribution when a penetrating

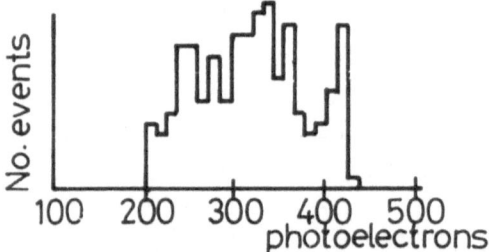

Figure 3. Histogram of 201
simulated signals from vertical muons (from normal
cosmic ray spectrum) passing
near centre of a 5m Cerenkov light collector at sea
level.

particle telescope selects tracks near the centre of a 5m-diameter mirror of large angular acceptance (e.g. 4° diameter) assuming no Cerenkov
light is generated in the mirror. (S-11 photomultiplier response with
peak quantum efficiency 20%, and mirror reflectivity 60% assumed.
Signal ∝ mirror diameter, not area. Mean signal 320 photoelectrons for
5m mirror.) Only muons above 4.3 GeV contribute. Inclined muons would
give a harder spectrum. However, there will often be a low-energy nucleon shower visible in the background, and it may be useful to do the calibration under cloudy conditions.
    It is also possible that short muon arcs, not very curved, might at
times be mistaken for narrow images of gamma showers.

3. AMOUNT OF LIGHT RECEIVED AT DIFFERENT AXIAL DISTANCES

Figure 4(a) shows the distribution of Cerenkov light on the ground (sea

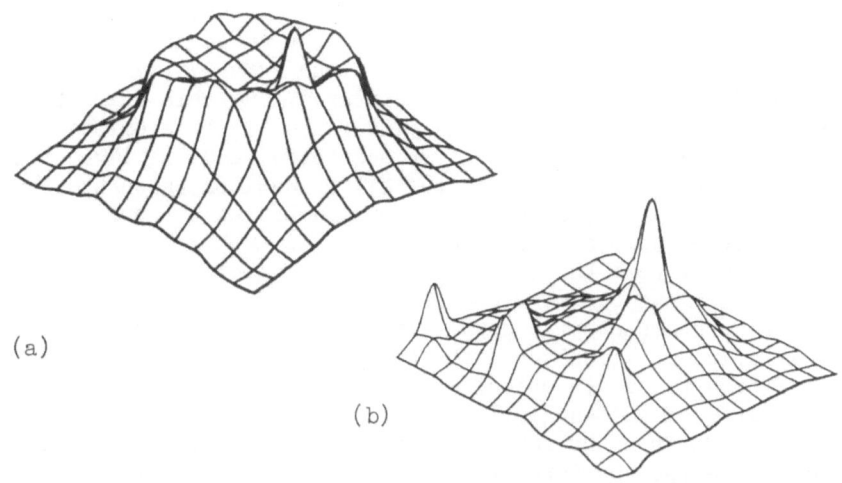

(a)

(b)

Figure 4. Typical examples of distribution of Cerenkov photon density on the ground: (a) for a 320 GeV photon shower, (b) a 1 TeV nucleon shower - both incident vertically at sea level. A 600m×600m area of ground is covered, with the shower axis in the centre, with a 50m grid. (Linear scales)

level) from a typical vertical 320 GeV gamma-ray shower. The distribution (at sea level) usually varies rather little from shower to shower, and shows an intensity peak near 140m from the axis, related to the characteristic Cerenkov angle. Nucleon showers, on the other hand, are much more variable, and have scattered peaks in intensity generally associated with local muons. Figure 4(b) shows a typical example for a 1 TeV proton shower. Figure 5 shows the lateral distribution of light from charged particles in different energy bands, in a gamma-ray shower,

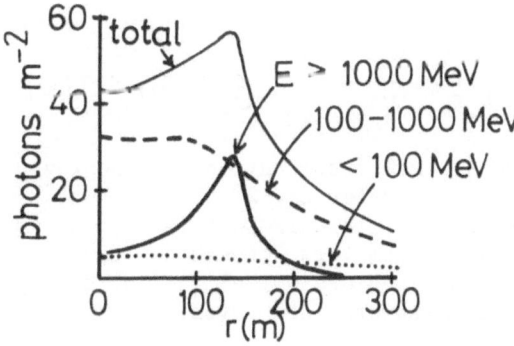

Figure 5. Lateral distribution at sea level of Cerenkov photons emitted by particles in different energy bands. (Vertical 1 TeV photon shower.)

and it can be seen that the striking ridge (Cerenkov ring) near 140m is
due to electrons above 1 GeV (noted also by Sinha: private communication). These retain the direction of the shower axis, and light emitted
at altitudes between 10km (where the Cerenkov angle is 0.8°) and 16km
(where it is 0.5°) reaches the ground at the axial distance of 140m.
Electrons of lower energy are more scattered, and this characteristic
Cerenkov cone is washed out. Nearly half the light is emitted by particles below 100 MeV: these are widely scattered, and the light reaches
the ground up to kilometres from the axis. Nucleon showers often develop mainly below 10km, and the high energy particles are more deflected by
nuclear interactions, so the Cerenkov ring is much less in evidence. At
angles well away from the zenith, the shower develops at greater altitudes, and the ring becomes more prominent.

If an aperture stop is used - light being collected only from
within a radius of 1°, say, of the source (in the case of gamma-ray
showers from a fixed direction) - one only collects light from a conical
volume of sky, of half-angle 1° (say), so in the case of a shower whose
axis falls 100m from the detector, only the part above 100m/sin 1° = 6km
is seen - thus the aperture stop will make the amount of light collected
fall off more rapidly with distance: see figure 6. This becomes important if one wishes to take stereo views from two (or more) separated
detectors: one may then find it best to centre the aperture not on the
source, but to aim at the centroid of a shower falling on the centroid
of the detector array: that is, a point about 9km up the axis from the
centroid of the array.

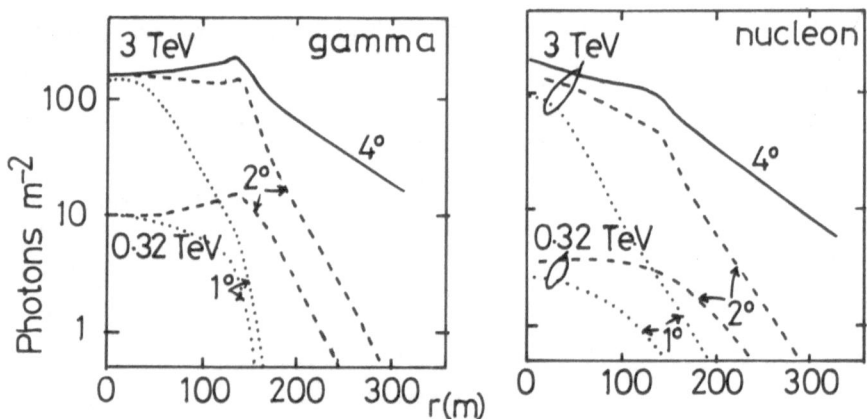

Figure 6. Average lateral distribution of Cerenkov photons emitted within cones of diameter 4°, 2° and 1° about the shower
axis. (For gamma showers, this is what is seen using aperture
stops of these angular diameters, though not for the background
nucleon showers, whose axes are not aligned with the viewing
direction.) Vertical showers, sea level.

REFERENCE

Hillas A M, 1985. *19th Int. Conf. on Cosmic Rays, La Jolla*, 3: 445-8.

OPTIMIZING THE DESIGN OF HIGH ENERGY GAMMA RAY TELESCOPES

A M Hillas
Department of Physics, Leeds University, Leeds LS2 9JT UK

J R Patterson
Department of Physics, University of Adelaide, Adelaide,
South Australia 5001

Abstract

From Cerenkov image simulations for different mirror arrays and
using integral spectra with powers -1.25 (photons) and -1.65
(nucleons), we seek to optimize the signal to noise $S/\sqrt{B}$ or Figure of
Merit $Q = S^2/B$. S is the number of gamma showers and B the number
of background events (nuclear plus sky noise). We find sky noise is
the overiding factor which affects the design. Optimum conditions
arise wth an aperture diameter of ~ 1.4° near 30° zenith angle and
trigger level such that sky noise is ~ 5% B. We find mirror
co-incidences, combined with image concentration and ratio tests, and
UV filters give some improvement. Fast timing does not near 1.4°,
although it can help for larger apertures. However, image selection
based on the width, length and pointing of images towards the source
with moderate resolution (≤ 0.5 degrees) should improve Q by a factor
of ~ 10 relative to counting in a 1.4° aperture.

1 Introduction

Cerenkov light signals from photon and nucleon – initiated
showers in the atmosphere have been simulated for a number of
mirror arrays on the ground (linear and triangular) at random
distances and orientations with respect to the cores. Gamma ray
showers only from the source direction (vertical or at 30° zenith
angle) were considered and nucleons initiated from random offsets
within 4° of the source direction. We use these to compare possible
observing strategies and mirror designs.

2 Definition : Figure of Merit Q

If S = number of gamma showers from the source,
and B = number of background events, nuclear + sky noise,
we need to maximise: signal/noise = $S/\sqrt{B}$
or: $Q = S^2/B$

Q factor improvements can be multiplied only if the strategies are quite
independent

K. E. Turver (ed.), Very High Energy Gamma Ray Astronomy, 249–253.
© *1987 by D. Reidel Publishing Company.*

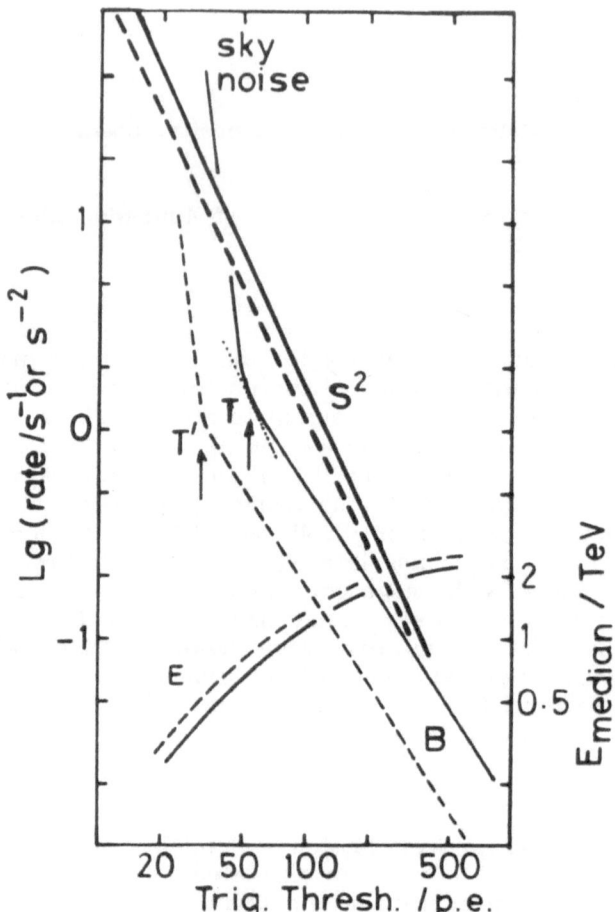

Figure 1: S² and B vs Trigger level for a single mirror (full-line) and two mirrors in co-incidence (dashed) spaced 120m apart. The aperture is 1.5° and the two mirrors are tilted in towards the radiant region, depth 300 gcm⁻². The vertical scale may be multiplied by 0.008 s⁻² (for S², assuming 10⁻⁶ photon m⁻² s⁻¹ above 150 GeV) or by 122 s⁻¹ (for B). With time S² becomes greater than B (as shown), a condition required for source detection. Also indicated are the median energies for gamma showers. (10 m mirror, 60% reflectivity)

3 How can we maximize Q?

Various possible strategies are briefly summarized.

Strategy 1 : Reduce Trigger Threshold T

$$\text{If} \quad S \propto T^{-1.1}$$
$$\text{and} \quad B = b_1 T^{-1.5} + b_2 T^{-8},$$
$$\text{(nucleons) + (sky noise)}$$
$$Q \propto T^{-0.7} \text{ above the break in slope,}$$

for a single mirror at 30° zenith angle. At the zenith $Q \propto T^{-0.8}$. Q is maximized when sky noise ~ 2 - 10% B, depending on power p (~ 8). Here and below we assume that the height of the pile up pulses varies approximately as the illumination to 0.7 power.

Figure 1 shows the behaviour of S^2 and B when the trigger threshold is varied. The optimum setting is where the slopes match before the sudden increase in sky noise. The calculations are for sea-level and 30° zenith angle. The photocathode is S11.

In practice a higher threshold may be used to guard against instabilities.

The source energy spectrum assumes an integral power -1.25 for photons and we use a nucleon power law of -1.65. If -1.0 is used instead, for photons, $Q \propto T^{-0.4}$.

Strategy 2 : Increase Aperture (Field of View)
But larger apertures increase sky noise: a compromise is needed. Allowing for sky noise, Q maximizes at ~ 1.4° diameter near 30° zenith angle.
With no sky noise, ~ 2.4° would be optimum, determined by the angular dstribution of bright gamma images (see accompanying paper). The loss in Q factor is ~ 0.3.

Strategy 3 : Increase Mirror Area A
Larger mirrors are better, even with the increased illumination because the threshold can be adjusted so that the noise rate increase due to increased illumination is offset by the higher threshold. More showers are collected because the light intensities are increased proportional to area.

$$Q \propto A^{0.6}$$

Strategy 4 : Use Blue or UV Filter
This is worthwhile when the moon is up.

Under best noise conditions, for S11 tube with borosilicate glass and UV filter

$$Q_{Filter} \sim 1.6 \, Q_{without}$$

For improved UV tubes, a further improvement may be expected. It depends on noise and detector characteristics. We used the night sky irradiance spectrum given in the *RCA Electro-Optics Handbook*.

Strategy 5 : Lower Altitude

At Mt Hopkins (~ 2300m) $Q_{MH} \sim 0.6 \, Q_{SL}$ at sea level for a single mirror.

Strategy 6 : Use N Mirrors

(a)   Independent (> 200m apart) : $Q_N \propto N$
      But $Q_N < Q_1$ for smaller apertures if close together, because more nucleons are picked up.

(b)   In Co-incidence with reduced thresholds,
      Two mirrors at the same place: $Q_2 \sim 3.2 \, Q_1$
      Two mirrors spaced 120 m apart: $Q_2 \sim 2.5 \, Q_1$

      The result depends sensitively on the slopes in figure 1 which depends on the mirror spacing. Noise exponents between 6 and 12 affect the result slightly. A 10ns time resolution is assumed.

(c)   A Ratio Test (for two mirrors R = (PH1 − PH2)/(PH1 + PH2)
      can test for flatness of showers and reject local muons.
      (PH = pulse ht). See Tumer et al, 1985). With same threshold, and 120m spacing,

$$Q \text{ Ratio} \sim 1.5 \, Q.$$

(d)   A Concentration Test (C = PH (1½° aperture)/PH (4° aperture))
      can test for smaller size of gamma images.
      For two mirrors and rejecting long images in either: $Q_C \sim 3Q_1$
      where Q_1 refers to counting in a single 1.4° aperture.

Strategy 7 : Fast Timing

(a)   Pulse-shape discrimination is difficult because pulses are very narrow. Some structured hadron showers may be rejected:
      $Q' \sim 1.4Q$

(b)   To Improve Arrival Directions
      Fast timing can locate the centroid of shower emission in the atmosphere, not necessarily the source direction. Thus showers will be uniformly spread inside a field of approximately 1.6°, so there is no advantage in trying to reduce the aperture by fast timing.

Strategy 8 : <u>Image Selection</u> based on the width, length and orientation of images (pointing at the source), is the most hopeful way of proceding, but requires experimental confirmation.

(i)   <u>Mt Hopkins Camera</u> (with 0.5° Pixel spacing, see Cawley et al 1985). We expect Q to be increased by a factor of ~ 10 relative to simple counting of pulses in 1.4° aperture. The factor also allows for lower thresholds because of reduced sky noise per tube.

(ii)   <u>Sector Selection</u> based on shaped pixels attached via Winston concentrators (Welford and Winston, 1978) is likely to be a more efficient use of photo-multipliers.

(iii)   <u>High Resolution and Multiple Images</u>
        Very small pixels enable the source alignment of image axis to be better defined and a much higher Q is expected. Half of the nucleon showers could be rejected if the sign of the distance between the centre of the image and centroid could be reliably determined. Multiple images enable more efficient rejection of nucleons.

<u>References</u>

Cawley M F, Fegan D J, Gibbs K, Gorham P W, Hillas A M, Lamb R C, Liebing D F, MacKeown P K, Porter N A, Stenger V J and Weekes T C, Proc 19th Int Cosmic Ray Conf (La Jolla) **2** 453 1985

Tumer O T, Wheaton W A, Godfrey C P and Lamb R C, Proc 19th Int Cosmic Ray Conf (La Jolla) **1** 139 1985

Welford W T and Winston R, *The Optics of Non-Imaging Concentrators*, Academic Press, 1978

The Angular Resolution of the Haverah Park Array for Ultra High Energy γ-Ray Astronomy

A Lambert, J Lloyd-Evans[#], J C Perrett and N J T Smith

Department of Physics, University of Leeds, Leeds LS2, UK

[#] Goddard Space Flight Centre, Maryland, USA

Abstract

Preliminary measurements of the angular resolution of the Haverah Park (GREX) array for PeV γ-ray astronomy are described. A resolution of about 1° has been achieved.

Introduction

An important parameter which must be established for any γ-ray telescope is its angular resolution. In the case of a satellite-borne telescope the angular resolution can be determined using accelerator calibrations or by observation of a strong source, eg for the COS B satellite both techniques were used. Using the Vela pulsar, the resolution of the γ-ray instrument was determined as a function of telescope orientation and photon energy. At 500 MeV the point spread function had an rms spread of about 3° (Hermsen 1980). It is unlikely that sufficient photons will be observed from a single source at ultra high energies (> 10^{14}eV) in the near future for this technique to be applicable and other techniques are required to establish the angular resolution to be used in searches for emission from known sources (eg X-ray binaries) and possible unknown sources which emit more strongly in the γ-ray waveband than elsewhere.

The new Haverah Park γ-ray experiment (GREX) has now been operational for 5 months and about 2 million events, due mainly to charged cosmic rays, have been recorded. The design of the array was based on preliminary studies (Lambert and Lloyd-Evans 1985) of the timing resolution of adjacent 1 m² scintillation detectors in air showers. The detector construction finally adopted consisted of a 1 m² x 10 cm slab of NE102A scintillator, viewed from below by a 2.5 ns risetime Phillips 2312B photomultiplier, in a black housing. The timing measurements are made using LeCroy 4208 TDC's. Details of the array have been given by Brooke et al (1985). The angular resolution is dependent on the size and core position of the showers. The preliminary analysis described below suggests that the angular resolution is better than 1° in certain cases.

K. E. Turver (ed.), Very High Energy Gamma Ray Astronomy, 255–260.
© *1987 by D. Reidel Publishing Company.*

2 Timing resolution of an individual detector

The GREX array is shown in figure 1. At two points of the
array pairs of detectors are placed side-by-side (15, 16 and 25, 32).
The distribution of time differences, recorded between pulses
> 20 m⁻², is shown for the detector pair (15, 16) in figure 2a. That
the spread of times ($\sigma(\Delta t)$ = 1.7 ($\pm$ 0.1) ns) is in part due to the
separation of the detectors (~ 1m) is clear from figure 2b where the
differences have been corrected to take account of the zenith and
azimuth angle of each event. The time spread is reduced to 0.86 $\pm$
0.06 ns. Other rms values have been obtained as a function of
detector density threshold and core distance and are shown in figure
3. The lines marked $\alpha = \pm 1.5^\circ$ and $\alpha = \pm 1.0^\circ$ indicate the trend
expected for these particular angular resolutions.

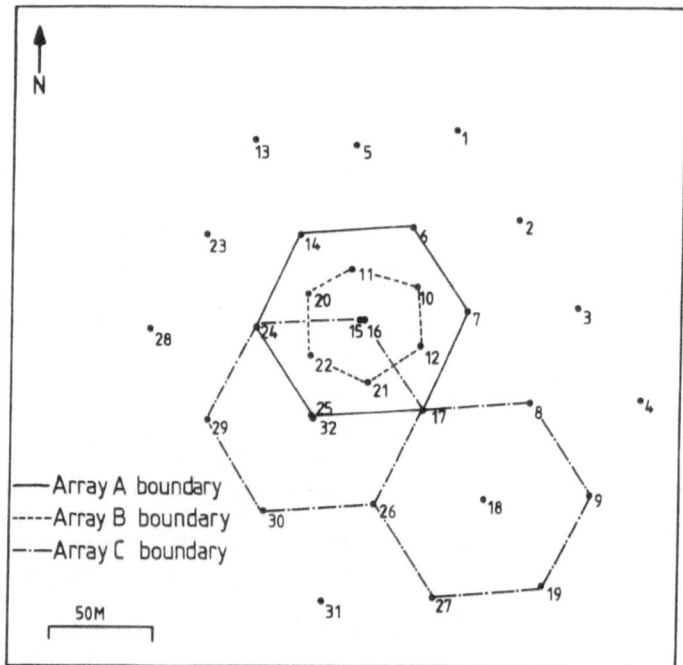

Figure 1:   The GREX array and boundaries of sub-arrays used in
the angular resolution analysis.

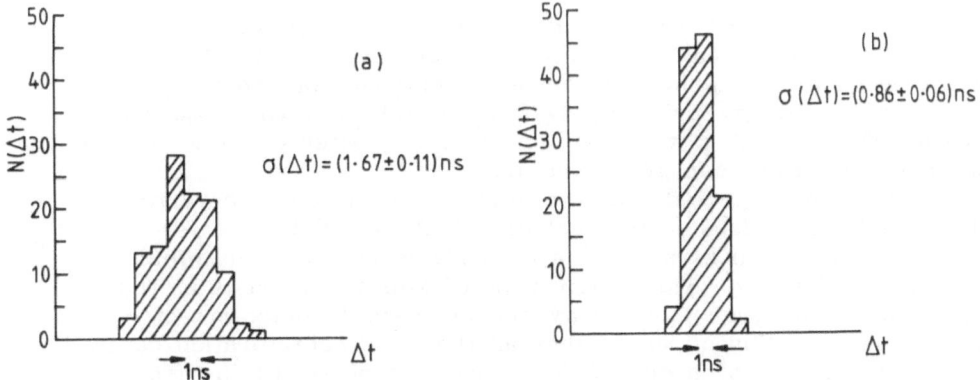

Figure 2:   Distribution of time differences for side-by-side detectors
            a) uncorrected
            b) corrected for disc traversal time

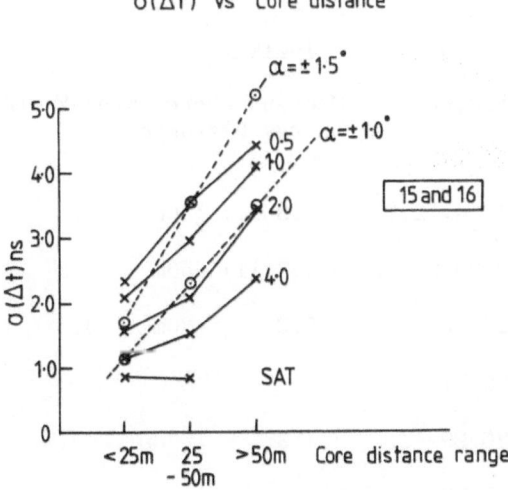

Figure 3:   RMS timing fluctuations in the triggering of side-by-side
detectors as a function of core distance and particle density.

3 Angular Resolution achieved with the GREX array

The detectors of the GREX array are not distributed uniformly over the available area. The central region (area ~ 8000m^2) has detectors on a 30 m grid and has been designed for the detection of relatively low energy primaries (> 2 x 10^{14}eV). For the remaining area a spacing of 50 m was adopted to give greater coverage for the less frequent, more energetic, events.

It is to be expected that the angular resolution of an array which encloses a close packed set of detectors will be greater than that achieved by an array with a comparable baseline but which is packed less densely. The former type of sub-array exists at the centre of the Haverah Park array and encloses 15 detectors on a 30 m grid within a radius of 50 m, whereas the same configuration centred on detector 18 encloses only 7 detectors on a 50 m spacing (see figure 1). The angular resolution has been estimated for both of these types of array by selecting two independent sub-groups of detectors from within each array (see Table 1). The two estimates of the shower arrival direction which are then obtained may be used to compile distributions of the angular deviations for that array. In figure 4 the distributions of the differences in right ascension ($\Delta\alpha$), declination ($\Delta\delta$), and the space angle (ψ) between the two directions are shown for the close packed array, A. Table 1 lists the results for this and the other detector combinations also investigated.

Table 1

Arrays		Median Energy PeV	Core Distance	$\sigma(\Delta\alpha)^0$	$\sigma(\Delta\delta)^0$	$\langle\psi\rangle^0$	$\sigma(\psi)^0$
A	15,7,11,14,12,22,25)	1.3	i) <30m	2.6	1.3	1.4	1.2
	16,6,10,20,21,17,24)	1.3	ii) <50m	3.3	1.6	1.7	1.6
B	15,11,12,22)	0.8	<30m	4.3	1.8	2.3	1.6
	16,10,20,21)						
C	16,25,26,29)	1.3	<30m	4.5	1.4	1.6	1.3
	17,24,30,32)						

Scintillator density threshold = 1 m^{-2}

Zenith Angle, Θ < 30^0

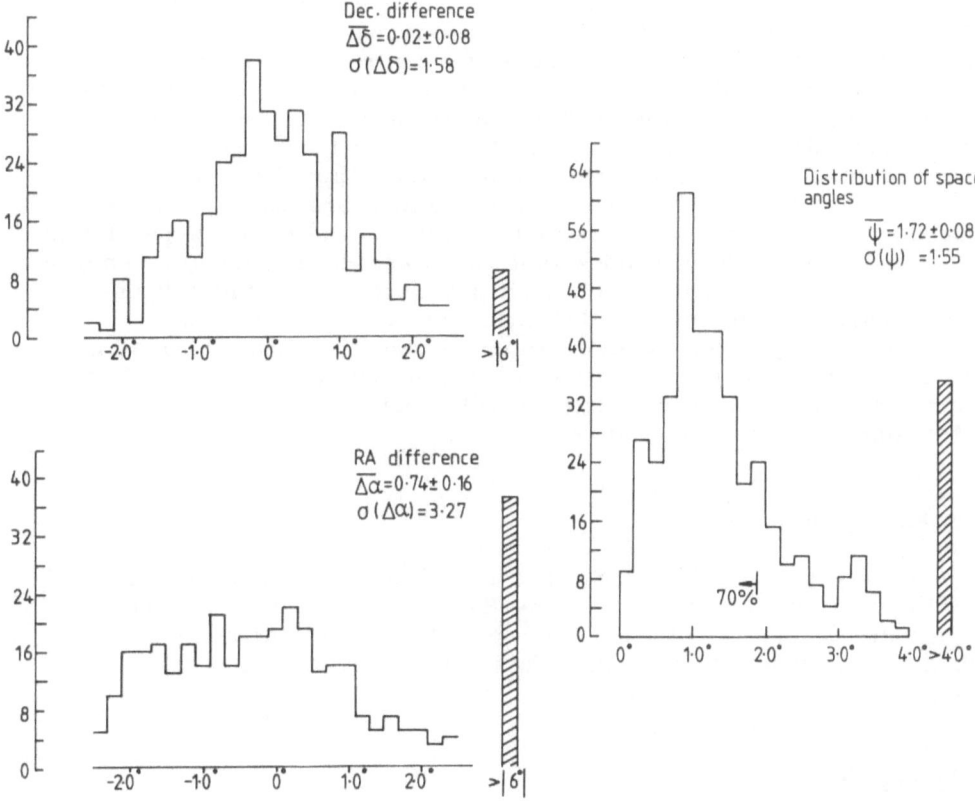

Figure 4:   Distributions of deviations in right ascension,
declination and space angle for array A selection (i).

    For the analysis summarised in Table 1 an unweighted plane fit
has been made to each detector set. Assuming that the errors in
each sub-array combine in quadrature and that the precision will
improve as the square root of the number of detectors used, we
estimate the mean angular difference for each of the arrays A, B and
C as ((i) 0.7° and ii) 0.85°)), 1.1° and 0.7° respectively. This

resolution is probably better than that achieved by Samorski and Stamm (1983) ($\sigma(\theta) \sim 1^{\circ}$) and at Ooty (Tonwar 1985) where an array of 24 detectors comparable to set A (15 detectors) gave $\sigma(\Delta\alpha) \approx \sigma(\Delta\delta) \approx 2.99^{\circ}$ for events within 30m of the array centre when a similar method of analysis was used. We expect to improve on the figures in table 1 when we have made further calibration refinements, used a weighted fitting procedure and corrected for the shower front curvature detected in some of the larger events studied thus far. Indeed, an improvement is expected since Arrays A and C are seen to perform similarly when a plane fit is used (see Table 1), although C has fewer detectors than A. This implies that the presence of curvature and/or equal weights assigned to each arrival time is preventing full use of the information contained in the extra detectors of Array A.

The distributions shown in figure 4 can be determined readily for each sub-array as a function of energy and zenith angle. These approximate point-spread functions will greatly aid the search for ultra high energy γ-ray sources.

4 Acknowledgements

We are grateful to the Science and Engineering Research Council for their continuing support of the Haverah Park project, to Paul Ogden and Mansukh Patel for their efforts in the construction of GREX and to Bob Reid and Alan Watson for their advice on the angular resolution analysis.

5 References

Brooke G et al, 1985, Proc 19th Int Cosmic Ray Conf (La Jolla) **3** 426

Lambert A and Lloyd-Evans J, 1985, Proc 19th Int Cosmic Ray Conf (La Jolla) **3** 449

Hermsen W, 1980, PhD thesis, Leiden

Tonwar S, La Jolla γ-ray Workshop, 1985, p40

Samorski M and Stamm W, 1983, <u>Ap J Lett</u> **268** L17

THE UCD/FLWO EXTENSIVE AIR SHOWER ARRAY AT MT. HOPKINS ARIZONA

G.H. Gillanders and D.J. Fegan
Physics Department, University College Dublin,
P.K. McKeown
Physics Department, University of Hong Kong, and
T.C. Weekes
Harvard-Smithsonian Center for Astrophysics

ABSTRACT: A new Extensive Air Shower array has been colocated with the 10m. Optical Cerenkov reflector at the F.L. Whipple Observatory, Mt. Hopkins, Arizona. The array is described and some preliminary calibration and response measurements are presented.

1. Introduction.

The positive identification of a number of celestial point sources of 10^{15} eV gamma rays by extensive air shower arrays working at or close to sea level (1) has motivated a number of groups to design and build small arrays at mountain altitudes. The main advantage of a mountain observational site is the lower operational threshold of the array. Also, the effects of fluctuations in cascade development tend to minimize close to shower maximum, giving a reasonably narrow spread in shower size distribution. By detecting showers close to the maximum of shower development, sampling errors are minimized as are errors in the estimation of both shower size and arrival direction. It has also been suggested (2) that gamma ray shower sizes may be 1.5 to 2.0 times greater than proton or heavy ion initiated showers of the same energy, close to shower maximum.

In order to exploit some of these advantages the UCD/FLWO array has been colocated with the 10m. Optical Cerenkov gamma ray facility at the F.L. Whipple Observatory, Mt. Hopkins Arizona, at an altitude of 2.3 Km above sea level. This arrangement offers the unique possibility of simultaneous operation of both the optical Cerenkov and particle array facilities, particularly during periods of the bright moon where narrow pass band optical filters may be employed with the 10m. imaging camera in order to raise its operating threshold to 5×10^{13} eV (3). The atmospheric depth at Mt. Hopkins is 780 gm cm^{-2} and the latitude is 31.7N. Cygnus X-3 and Her X-1 are both very favourably located with minimum zenith angles of 9.2 and 3.7 degrees respectively. Although the primary objective of the array will be to determine the energy spectrum and lightcurve of Cygnus X-3, the array will be operated continuously with all sky coverage with a view to looking for possible high energy gamma ray transients.

K. E. Turver (ed.), Very High Energy Gamma Ray Astronomy, 261–264.
© 1987 by D. Reidel Publishing Company.

2.   The scintillation counter array.

The EAS array consists of thirteen $1m^2$ scintillation counters
distributed around the 10m. Optical reflector at Mt. Hopkins, Arizona.
The array topology is shown in Figure 1. The inner four detectors
(D_1 to D_4) consist of slabs of NE102A plastic scintillator of one
inch thickness. Each slab is viewed by an RCA 4522 phototube located
60cm below the plastic. This inner group constitutes the master
trigger for the system. Detector D_1 is located at the center of an
equilateral triangle whose vertices are occupied by D_2-D_4, each of
which is 20m from D_1. Detector D_1 is located 3.25m above the
horizonal plane containing the other three. The outer eight detectors,
of similar design and construction, are currently used solely for
timing purposes. Three of these detectors D_5-D_7 are also located
20m. from D_1, in order to extend the timing arms of the array in the
area between D_2-D_4. Another three detectors D_8-D_{10} are located
at the midpoints of the sides of the equilateral triangle formed by
D_2-D_4. The final triad of detectors D_{11}-D_{13} constitute a close
knit group adjoining D_1. This group will be used mainly in
experiments which will operate in conjunction with the 10m. Optical
reflector. This group will also facilitate accurate measurements of
the thickness of shower fronts. Signals from each phototube are passed
along 55m. of RG 8/U signal cable to the adjoining control room at the
gamma ray facility.

3.   Signal processing and data acquisition.

The signal processing is performed exclusively by LeCroy
NIM/CAMAC fast modular instrumentation units. The data acquisition is
accomplished by an IBM PC which communicates with the CAMAC controller
along a GPIB bus, via a National Instruments driver. The data
acquisition software and interface drivers are written in FORTRAN. The
maximum system throughput rate is 0.6Hz. For the purpose of accurate
absolute timing, WWVB is monitored and injected into the system.

Figure 2 depicts the essential elements associated with the
generation of a coincidence master trigger pulse. Here the signals
from D_1-D_4 are processed to produce a 3 fold coincidence, with the
inclusion of the central detector D_1 mandatory. A pair of
discriminators are used for each channel D_1-D_4. One discriminator
of each pair is used for timing purposes and is set at 1/3 of minimum
ionization loss in each detector. This discriminator output feeds both
the TDC unit and the 429A Logic module which generates a common signal
for the TDC. The other discriminator of each pair is set at the
minimum ionization level and generates inputs for scalers and the
coincidence unit. A resolving time interval of 110ns is allowed for
registration of a 3 fold coincidence, before the TDC's and ADC's are
cleared. In cases where a coincidence forms, the TDC sends a LAM which
filters through to the IBM PC as a service request. This initiates the
reading of the CAMAC modules by the PC. In this way the amplitudes of
the signals from the PMT's together with the relative time of the
shower front striking each detector are recorded. The timing
resolution of the LeCroy 4208 TDC's is + 1ns.

For each master trigger we also interrogate and record the status

of the IBM PC clock and the contents of two CAMAC scalers which are fed
from a stable and accurate 1MHz crystal oscillator. Since absolute
time is injected from WWVB every minute we have absolute event timing
to 0.5ms and relative event timing to a microsecond.

The amplitudes and times of arrival of the signals in the other
scintillation counters are recorded in a similar fashion for each
master trigger pulse.

4.   Threshold energy, calibration and response of the array.

The array was first brought online in late 1985 with a minimum
configuration of D_1-D_4 operational. The system has been gradually
upgraded since that time and it is expected to be fully operational by
late 1986. The operational characteristics of each scintillator were
determined using a small muon telescope in order to establish Landau
distributions of the ionization loss of vertically incident single
muons. The operational singles rates for each scintillator has been
set at 250 Hz. Forming a 3 fold coincidence out of the inner four
scintillators operating at this single particle rate results in a
master trigger rate of 36m with a corresponding energy threshold
for triggering of 6 x 10^{13} eV.

With a multidetector array of the kind described here it is vital
that the timing characteristics associated with each channel remain
constant over the interval of time spanned by the observations. If
there is inherent creep in any channel, of magnitude greater than
0.5ns., then the direction finding capabilities of the array become
fuzzy and ambiguous. We check the relative timing response of each
scintillator by forming a coincidence between the scintillator under
test and a small 6" x 6" x 0.5" paddle scintillator, measuring the
threshold crossing time difference for the two components of the
telescope in response to the passage of some thousands of single
particles. The paddle is then moved to the next scintillator to be
tested and the measurements are repeated. This timing calibration
procedure is performed on a regular basis and in this way any longterm
tendancy for any timing channel to exhibit creep becomes obvious. The
maximum variation for any channel between December 1985 and April 1986
is <0.3ns, within the errors of measurement of the sample.

We wish to acknowledge the continuing support of the National
Board of Science and Technology, Ireland, the Scholarly Studies program
of the Smithsonian Institution and the University of Hong Kong. We
also wish to thank Dr. D.B. Swinson (University of New Mexico) for a
loan of some scintillators.

REFERENCES

1.   A.A. Watson. Rapporteur paper. Proc. 19th. Int. Conf. on Cosmic
       Rays, La Jolla (1985). To be published.
2.   K. Kasahara, S. Torri and T. Yuda. Proc. 19th Int. Conf. on
       Cosmic Rays, La Jolla, Vol.9, 473-476 (1985).
3.   M.F. Cawley, D.J. Fegan, K. Gibbs, P.W. Gorham, R.C. Lamb, D.F.
       Liebing, N.A. Porter, V.J. Stenger, T.C. Weekes and R.J.
       Williams. Proc. 19th Int. Conf. on Cosmic Rays, La Jolla, Vol.1,
       89-90 (1985).

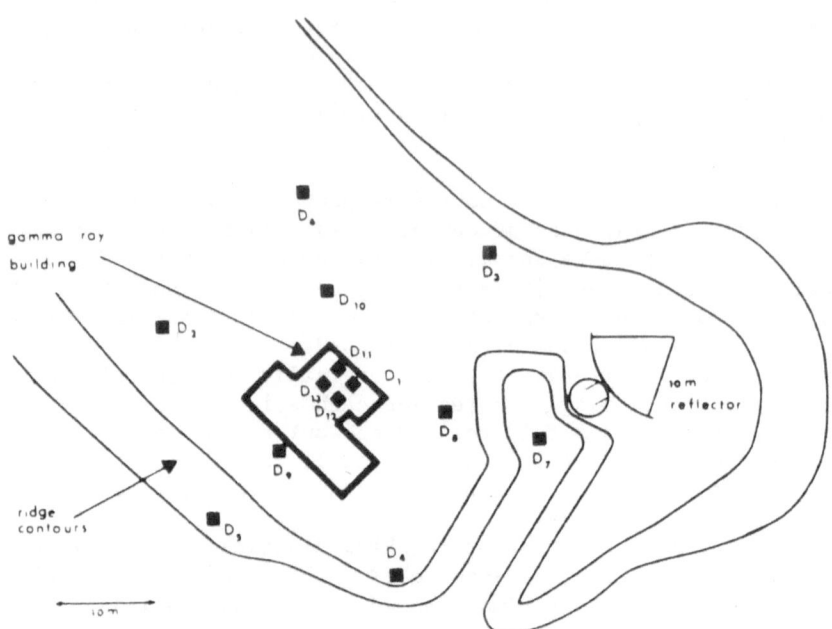

Figure 1.   The Geometry of the Array.

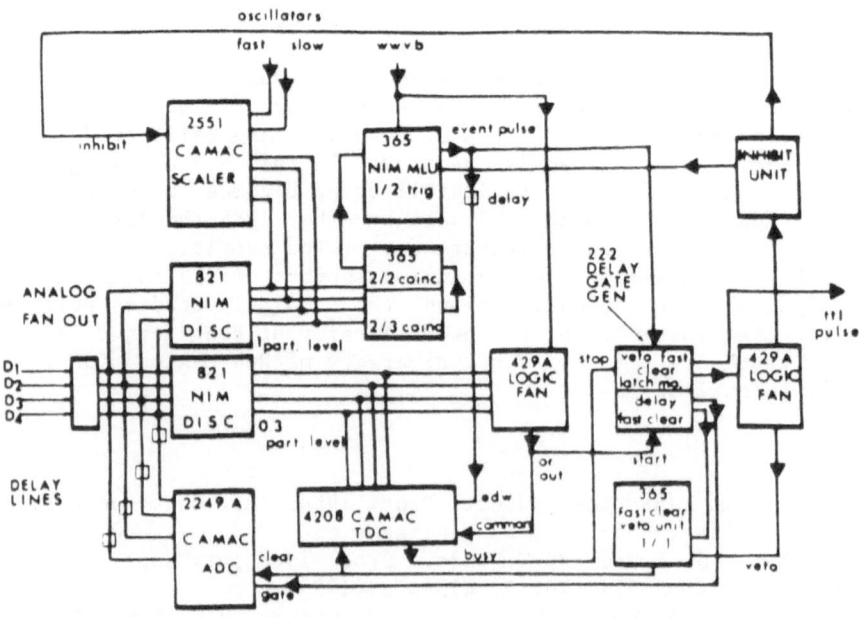

Figure 2.   The Signal Processing Electronics.

UHE GAMMA RAY ASTRONOMY WITH THE EAS-TOP ARRAY AT THE GRAN SASSO

M.Aglietta, C.Castagnoli, A.Castellina, B.D'Ettorre Piazzoli
W.Fulgione, P.Galeotti, G.Mannocchi, C.Morello, L.Periale,
G.Trinchero, P.Vallania, S.Vernetto
Istituto di Cosmogeofisica CNR - Torino - Italia
G.Badino, L.Bergamasco, G.Cini, M.Dardo, G.Navarra, P.Picchi
O.Saavedra
Istituto di Fisica Generale - INFN - Università di Torino
Italia

ABSTRACT. The EAS-TOP is an air shower array situated on the vertical
of the Underground Gran Sasso Laboratory, in Italy. Its performances
in Ultra High Energy (UHE) gamma-ray astronomy are described both in
the search for point sources ($E_o > 5.10^{13}$ eV) and in the identification of primary gamma-rays.

1. INTRODUCTION

The results obtained in the last few years in VHE and UHE gamma-ray
astronomy have opened a new window in astrophysics by establishing the
existence of sources of gamma-rays with energy $E > 100$ GeV.
Nevertheless, the experimental situation is still unclear.
For example:
1) difficulties arise when different experiments are compared, due to
the fact that the time and energy behaviour of the sources is a complex one;
2) the criteria for recognizing the nature of the primaries of the
detected air showers are still uncertain and sometimes contradictory.
For a review of the subject see ref. (1,2,3, general) and (4, Cygnus
X-3).
The EAS-TOP air shower array will be situated near the top of the Gran
Sasso mountain (42° 27' 09" lat. North, 13° 34' 28" long. East) at an
altitude of 2000 m a.s.l. It will run in coincidence with the muon detectors placed in the "Gran Sasso Underground Laboratory", at a depth
of 32 00 hg cm^{-2} (5).

K. E. Turver (ed.), Very High Energy Gamma Ray Astronomy, 265–269.

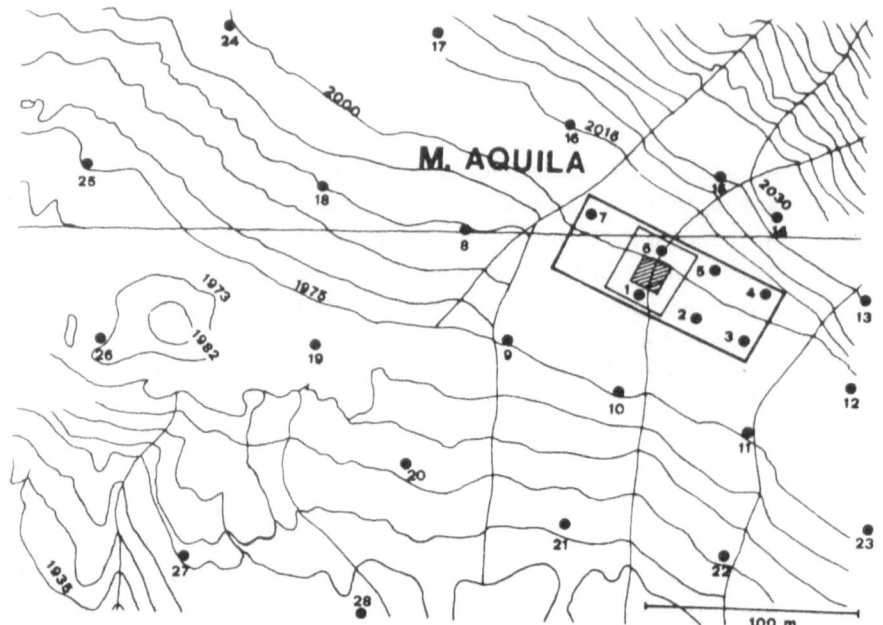

Fig. 1 - Location of the detectors on the ground (● scintillators, ▨ μ-h detector); the μ-h detector is located at the minimum distance from the underground laboratory.

The EAS-TOP will operate in the field of UHE gamma-ray astronomy to face some of the quoted problems. It will be built by :
1) a detector of the electromagnetic component of EAS, with large sensitive area and good timing resolution in order to guarantee a low threshold in primary energy and the required angular resolution (by the time of flight technique);
2) a central thick tracking detector for the study of the muon and hadron content of the extensive air showers.

2. THE ARRAY

2.1. THE ELECTROMAGNETIC DETECTOR (EMD).

It consists of 28 units. Each unit has a sensitive area of 10 m^2 and is composed of 16 liquid scintillators, housed in thermally insulated and stabilized boxes.

TABLE I - Performances of the EAS-TOP detector.

E_o (eV)	Ae (m^2)	$F(E > E_o)$ (y^{-1})	$\delta\theta$ (deg)	$Smin$ (3 s.d./ yr obs) ($cm^{-2} s^{-1}$)
5.10^{13}	10^4	$3.6 \cdot 10^7$	1.6	$1.6 \cdot 10^{-12}$
10^{14}	2.10^4	$2.2 \cdot 10^7$	1.3	5.10^{-13}
10^{15}	10^5	2.10^6	<1	10^{-14}
10^{16}	10^5	$1.8 \cdot 10^4$	<1	10^{-15}

The measured characteristics of the scintillators ($80 \times 80 \times 20$ cm^3), viewed by a fast P.M. Philips XP2041 are the following:
time resolution $\delta t = 1.7$ ns
amplitude resolution $\delta H / \bar{H} = 25\%$.
Fig. 1 shows how the scintillator units are scattered on the ground. The distances between the modules in the central part (~ 25 m) allow a density of some particles per unit at primary energy $E_o = 5.10^{13}$ eV (for vertical e.m. showers). The total enclosed area permits reasonable rates of showers up to primary energies $E_o > 10^{16}$ eV (*). In·Tab. 1 the effective area (Ae), the angular resolution ($\delta\theta$ obtained following ref. (6)), and the minimum continuous fluxes detectable from a point source are reported as a function of the primary energy (E_o).
The stability of a similar, smaller array devoted to UHE gamma-ray astronomy for long time operation and in the search for "burst" events is discussed in (7,8). It is shown there that the fluctuations are very near to the statistical ones.
By simulating the response of the array to air showers originated by gamma-rays of primary energy $E_o = 10^{15}$ eV, we obtain the following resolutions in determining the EAS parameters (for shower core located in a wide region near the center of the array):

core position : $(\delta r^2)^{1/2} = 1.8$ m
age : $\delta s < 0.1$
size : $\delta Ne / Ne = 5\%$
primary energy $\delta E_o / E_o < 50\%$.

(*) The expected rates are estimated assuming a reference gamma-ray source with energy spectrum: $S(>E)=30/E(eV)$ ($cm^{-2} s^{-1}$). This spectrum fits the positive observations from Cygnus X-3.

2.2 THE MUON-HADRON DETECTOR (MHD).

A central detector, with a sensitive area of 200 m^2, is devoted to
record the hadrons and the muons contained in the air shower. It consists of 12 layers of absorber (iron+concrete, total thickness t =
1000 gr cm^{-2}) and streamer tubes.For showers with their axis hitting
the detector, the MHD measures the absorption curve and the individual cascades in the core, thus operating as a hadron calorimeter.
Experimental evidence of μ-poor showers characterized by an absorption curve in the calorimeter compatible with pure e.m. cascades has
been reported in (9). From our reference source and for a primary
energy E_o = 5.10^{13} eV, we expect some showers per year with the core
striking the calorimeter, against $\sim$ 700 showers from cosmic ray nu-
clei.Notice that a primary proton of such energy produces a few hadrons of energy E $>$ 100 GeV at our observation level (10).
For showers with the core in the peripheral part of the array, the
MHD operates as a muon track detector (E_μ $>$ 3 GeV). At primary energy
E_o = $3 .10^{14}$ eV and core distance r = 50 m the number of muons detected by the array is N $\sim$ 20 if the shower is produced by a proton,
N_μ $<$ 1 if the shower is produced by a gamma-ray. In these conditions
we expect $\sim$ 50 events from the gamma-ray source against $\sim$600 cosmic
ray events.

3. CONCLUSIONS

The EAS-TOP array will therefore:
a) operate over a wide energy range at a sensitivity level 4-5 times
lower than the reported fluxes from Cygnus X-3;
b) allow selection of UHE gamma-ray primaries by correlating different EAS parameters (N_e - s - N_μ - N_h);
c) give systematic information on the accuracy in the determination
of the EAS arrival direction either by means of combined measurements
of delays and by comparing the air shower data with the directions
of the high energy muons recorded underground, which are practically

unscattered : $(\delta\theta^2)^{1/2}$ $<$ 1°.

REFERENCES

1 - Stepanjan A.A. (1984) Adv. Space Res.,3, 123.
2 - Porter N.A. (1983) 18th ICRC, Bangalore, Rapp. Paper.
3 - Watson A.A. (1985) 19th ICRC, La Jolla, Rapp. Paper.
4 - Vladimirskii B.M. et al. (1985) Sov. Phys. Usp., 28, 153.
5 - Aglietta M. et al. (1986) Il Nuovo Cimento (in the press).
6 - Linsley J. (1985) Proc. Workshop "Techniques in Ultra High Ener-
    gy Gamma-Ray Astronomy", La Jolla, 138.
7 - Morello C. and Navarra G. (1981) Nuclear Instrument and Methods,
    187, 533.
8 - Morello C. et al. (1986) Proc. Workshop "H.E. - U.H.E. Behaviour
    of Accreting X-Ray Sources " Vulcano.
9 - Stamenov J.N. et al. (1983) Proc. 18th ICRC, Bangalore, 6, 54.
10 - Grieder P.K.F. (1977) Rivista del Nuovo Cimento, 7, 1.

UHE GAMMA RAY OBSERVATIONS WITH THE K.G.F. AIR SHOWER ARRAY

P.N. Bhat, S.G. Khairatkar, M.R. Rajeev,
M.V.S. Rao, S. Sinha, K. Sivaprasad,
B.V. Sreekantan, S.C. Tonwar, P.R. Vishwanath
and K. Viswanathan

Tata Institute of Fundamental Research,
Bombay 400 005,
India.

ABSTRACT
    The KGF air shower array, with 7 muon detectors and 61 particle scintillators, is ideally suited for UHE gamma ray studies. Using the fact that the thickness of the shower disc varies with distance from the core, we have been able to obtain 0.5° accuracy in angular resolution. A preliminary look at the Muon data shows that the Cosmic Ray background can be considerably lessened using the muon information.

THE EXPERIMENTAL SET UP

    For a proper study of UHE gamma ray point sources, the array should have the following features: (i) high angular accuracy, (ii) large area muon detectors to establish the nature of the radiation, and (iii) large collection area. An array with all these features has been operating at Kolar Gold Fields. The array consists of 61 plastic scintillation detectors, each of area 1 m^2 and 7 muon detectors of threshold energy 1 GeV, each having an area of 30 m^2 (Bhat et al,1985) . The scintillators are arranged in a hexagonal pattern, with a spacing of 20 m between neighbouring detectors. The detectors extend upto 80 m from the center, covering an area of 1.66 x $10^4 m^2$. Each scintillator is instrumented for pulse height as well as fast timing measurement. The threshold particle density for timing measurement is kept at 0.3 to minimise rise time effects. The spread in arrival times at single particle level is 3.5 ns. The air shower trigger is provided by a coincidence of any three neighbouring detectors forming an equilateral triangle of side 20 m, in which the particle density exceeded 1.5. The trigger rate is 1 Hz. A 5 MHz Oscilloquartz crystal, with a stability of 1 part in 10 per day, provided real time information with an uncertainty of about 1 ms. Six of the muon detectors are located at the vertices of a hexagon of side 60 m and the seventh at its

K. E. Turver (ed.), Very High Energy Gamma Ray Astronomy, 271–274.
© *1987 by D. Reidel Publishing Company.*

center, which coincides with the center of the array.   It
consists of two layers of 48 proportional counters each under
600 g/cm^2 of concrete, separated by 4 radiation lengths of
brick. An on-line LSI-11 microprocessor recorded the events
and also continuously monitored all the detectors.

METHOD OF ANALYSIS

     We have used a two step process for the determination of
the arrival direction. At first, we use constant weight for
the uncertainty in the arrival time of the shower at the
detector. This initial determination of the arrival
direction is used for getting the other shower parameters
like the size , the age and the core position. At this
stage, we invoke the fact that the longitudinal thickness of
the shower front inreases rapidly with increasing distance
from the core(Linsley, 1985). Therefore, the weights depend
upon the density at the scintillator. We use Linsley's
formulation for the width of the arrival distribution.

     In order to obtain the zenith and the azimuth angles θ
and ϕ from the observed time delays, the relative time
delays T(offset)s of various detectors, for a vertically
incident shower are required. For this, T^{ij} , of the peaks of
the distributions of the differences in the time delays
between neighbouring pairs of detectors, after correcting the
differences in Z coordinates are used, to determine T(offset)s
, using the relation T^{ij} = T(offset)^i - T(offset)^j . For each
detector, 3 to 6 neighbouring pairs are available depending
on its location in the array. This method automatically
takes into account all the delays introduced in the path of
the signals.

     For individual showers, θ and ϕ are obtained by
minimising the equation of the shower front. We use these
initial values of θ and ϕ along with the NKG function to
determine the shower size, the age of the shower and the
position of the core. This gives the expected density in the
scintillator. The width of the arrival distribution σ is
(1.6/n**0.5)*(1+R/30)**1.5 where R is the distance from the
core and n is the expected density at distance R. W,the
weight factor is equal to 1/(σ**2+K) where K, the
instrumental resolution is equal to 1 nanosecond.

     The horizontality of the X-Y plane has been checked
using the distribution of the projected angles of the showers
in two perpendicular planes, the E-W and the N-S planes.
This distribution should be symmetric about the zenith and
the ratio, R, of the number of events from one hemisphere to
the number from the other, should be equal to unity. The

ratio of number of showers from north to south is 0.994 ± 0.011 and the ratio of number of showers from east to west is 1.011 ± 0.011. The ϕ distribution for the events is shown in Fig.1a and it is quite uniform with mean ϕ of 180.8±0.6 .

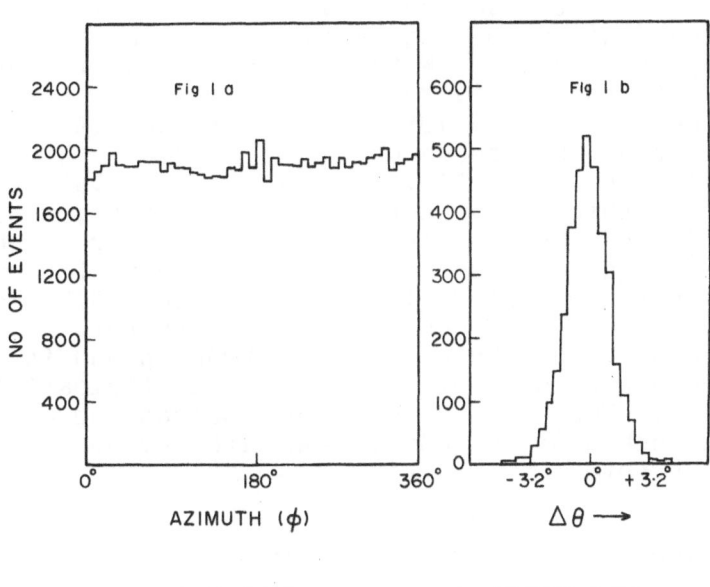

FIGURE 1(a)   FIGURE 1(b)

The error in the arrival direction of the showers is obtained in the following way. First, events with more than 2 n detectors with timing information were selected, where n is the minimum number of timing detectors needed for final analysis. These detectors were divided into two groups, each with a minimum of n detectors. Events were analysed using each group of detectors separately, thus obtaining two independent estimates of the zenith and azimuth angles for each event. Fig. 1b shows the $\Delta\theta$ distribution. The σ of this distribution is 1°. Since this is the resolution for n detectors, the angular resolution when 2n detectors are used is 0.5° . We note that this distribution is obtained only with 32 detectors in operation. When we use all the detectors, we expect the angular resolution to increase further. We note that this improvement over our earlier estimations (Bhat et al) is essentially because of using the more correct disc thickness.

MUONS

From random triggers , we found that half a tube per event per M-station can be ascribed to radioactivity. The requirement for a muon track is that both the top and the bottom layers should record at least one hit and the projected angle of the shower should be the maximum angle subtended by the muon and should be the minimum angle subtended by the muon. With such a requirement, the probaility of radioactivity imitating a muon is very small. For all showers above $5.10**4$, the probability that any single M-station has shown no muons is 43% . When we make cuts on size (>8 . $10**4$), age (0.4 to 1.6) and core position (within 50 meters from the centre), we are left with only 5% of the showers. In these showers, when one further asks that no muon be present in any M-station only 6 % of the showers qualify. Thus, out of about 300 showers in this group, 19 have no muons in any Mu detector. The mean number of muons in a shower in this shower group is $\sim$ 13 and thus the average number of M-stations participating in a shower is 2.6. Therefore, 6 % of the nucleon induced showers have to be looked at for comparison with any potential gamma ray induced events.

ACKNOWLEDGEMENTS

We thank B.K.Chatterjee, A.V.John, R.Mahalingam, B.K.Nagesh, N.S.Prasad, Ramesh Babu Raj, Shobha Rao , V.Ramu, C.V.Raisinghani, P.Reddy, A.J.Stanislaus, S.Swaminathan, Suresh Upadhyaya, M.Venkateshwarlu, B.L.Venkateshamurthy and R.P.Verma for their assistance in building and operation of the array and analysis of the data. It is a pleasure to acknowledge the kind co-operation extended by P.D.Gupta, Chairman and Managing Director of Bharat Gold mines and his staff.

REFERECNES

P.N.Bhat et al, TECHNIQUES IN ULTRA HIGH ENERGY GAMMA RAY ASTRONOMY, Editors: R.J.Protheroe and S.A.Stephens (University of Adelaide),1985,1

J.Linsley, Proceedings of the XIX ICCR, 1985, 3, 461

UHE GAMMA-RAY ASTRONOMY WITH THE BUCKLAND PARK ARRAY, A REVIEW

D. Ciampa, R.W. Clay, C.L. Corani, P.G. Edwards,
J.R. Patterson, R.J. Protheroe
Physics Department,
University of Adelaide, North Terrace, Adelaide. 5000
South Australia.

ABSTRACT

Data obtained by the Buckland Park EAS array have previously
been analysed to search for evidence for UHE gamma-rays.
Observations of Vela X-1 and LMC X-4 have been reported and a
tentative observation of an excess of events from the direction of
Centaurus A has been discussed. Additionally, we have reported upper
limits to the UHE gamma-ray flux from other possible point sources
and examined limits to the total excess flux of UHE gamma-rays from
the southern galactic plane. These 1979-81 data were obtained using
an earlier array configuration and since then the array has been
progressively developed for UHE gamma-ray work by reducing its
size/energy threshold and improving its angular resolution.
The development of the array and the status of the present
system are discussed in this paper together with a brief review of
past results and a presentation of some newer results on Centaurus A
and 4U1822-37.1. Future prospects will be examined.

THE ARRAY

Detector Configuration

The earliest Buckland Park array (see fig.1) consisted of
detectors at the corners of a square of side 90 m with an inner 30 m
fast timing square, eight detectors in all were used. This array had
a sea level size threshold in the region of 10^5 to 2×10^5 particles, a
fast timing angular resolution parameter of $1.25°$ and employed a
shower size analysis which was usually limited to fixed lateral
distribution functions. Three further detectors were added to give a
larger (~ 4x) collecting area for large showers (Crouch et al (1981))
and this was the array configuration employed to give the 1979-81
data set (Gerhardy and Clay(1983)). The angular resolution had been
unchanged but a shower lateral distribution was now routinely fitted
with a variable age parameter (Clay et al (1981)).

K. E. Turver (ed.), Very High Energy Gamma Ray Astronomy, 275–279.
© *1987 by D. Reidel Publishing Company.*

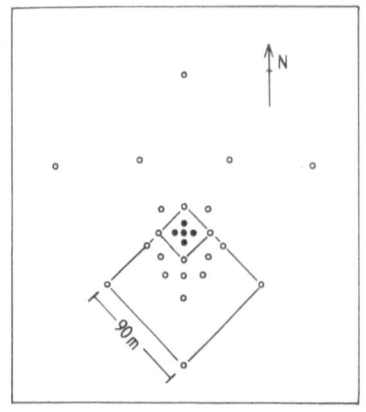

Fig. 1 Plan of the Buckland
Park Array. Open circles are
1 m^2 detectors, filled circles
are 2.25m^2. The inner group
of detectors is instrumented
for fast timing.

        Since 1981, further detectors were added (Clay et al (1985))
within the array (see fig.1) to lower the effective size threshold
to ~ 3x10^4 particles and more fast timing channels have been added,
initially from five to eleven in order to give directions for the
smallest showers (with only marginal resolution impovement for the
larger showers, see Ciampa et al (1986)). In 1986, the number of
fast timing channels has been increased to 24 with another 10
channels in preparation. The purpose of the more recent developments
has been to improve the overall angular resolution. These detectors
are still concentrated near the centre of the array where we expect
the shower front thickness to be, in general, the least.
        Individual detector timing uncertainties are typically found to
be ~2.5 ns. We expect to improve these substantially when sampling
large particle densities following the replacement of our present
discriminators with "double discriminators" designed specifically to
give good fast timing (Clay and Ciampa (1986)). At the present, our
timing resolution with the 1m^2 and 2.5m^2 detectors is fixed by the
electronic resolution in the detector, the light gathering time, and
the shower front thickness, each of which are ~2ns.

Electronics

        The array electronics are largely conventional. Most of the
fast discriminators and logic were designed and built in Adelaide,
the CAMAC analog to digital and time to digital converters are mainly
commercial LRS modules. Data acquisition is by a Data General Nova
4S computer and magnetic tape storage is employed. The data are
analysed in Adelaide. The data rate of ~ 1/10s makes data analysis
a bottleneck.
        The electronic logic records a coincidence when two out of 16
inner detectors trigger at about the 6 particle level. The fast
detectors trigger at a much lower level with a rate of ~ 1kHz for
the 1m^2 detectors. The only non-conventional technique we use is to
assume that, due to atmospheric collimation, showers on an average
come preferentially from the zenith. The long-term-mean timing

differences between all pairs of our timing detectors are used as the
time differences for vertical showers and directions for individual
showers are derived from measured time differences relative to these
'vertical' differences. We do not attempt to routinely measure
absolute delays in our individual timing channels.

Observations

 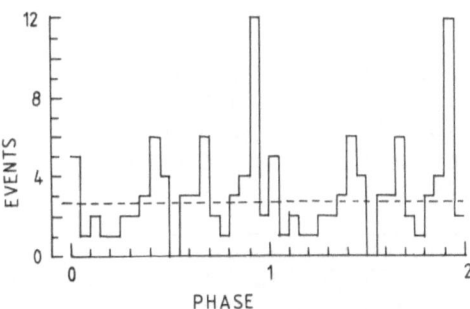

Fig. 2 Phase diagram for events
from the direction of Vela X-1.
The cyclic background is due to the
closeness of the period to 9 days.

Fig. 3 Phase diagram for events
from the direction of LMC X-4.

The earlier UHE gamma-ray observations employed a data set from
the period 1979-81. From these data was obtained evidence for
emission from the direction of Vela X-1 (Protheroe, Clay and Gerhardy
(1984)) and LMC X-4 (Protheroe and Clay (1985)) on the basis of
periodic analyses of data obtained from those source directions
(figures 2 and 3). The Vela X-1 data used a shower age cut (in our
case at S = 1.3) as had the original Samorski and Stamm (1983) data
but the LMC X-4 data were not cut since this source is only observed
from Adelaide at large zenith angles where array foreshortening makes
age determination less precise.
   Evidence was presented for a possible excess of events from the
direction of the prominent active radio galaxy Centaurus A (Clay,
Gerhardy and Liebing (1984)). This direction contained the largest
statistical excess in the overall directional distribution in the
sky. Since such a source contains no characteristic period for its
unambiguous identification, it is not possible to view the
observation with great confidence although the distance of the source
makes a possible positive effect very interesting from the point of
view of UHE gamma-ray interactions with the microwave background.
Absorption is expected to produce a deficit of events with energies
in the region of 10^{15} eV (see Protheroe (1986)). Table 1 shows the
distribution of observed shower sizes, compared to that expected,
from the overall direction of Cen A. As a rough rule, a factor of
10^{10} converts from vertical shower size to energy. Clearly, poor
statistics preclude a convincing conclusion but the lowest energy bin
certainly lacks any excess.

Table 1

The size distribution of showers from the direction of Cen. A.

Vertical Equivalent Size Range (particles)		Expected Number of Events	Observed Number of Events
6×10^4	2.5×10^5	262	263
2.5×10^5	10^6	544	624
10^6	4×10^6	209	235
4×10^6	1.5×10^7	44	40

The Buckland Park data set has also been searched for a number of other possible southern sources (Protheroe and Clay(1984)) and has been used to put a limit of ~ 40 x the Cygnus X-3 flux on the total excess gamma ray emission from the galactic plane as an admixture to the overall cosmic ray backgound (Clay, Protheroe and Gerhardy (1984)).

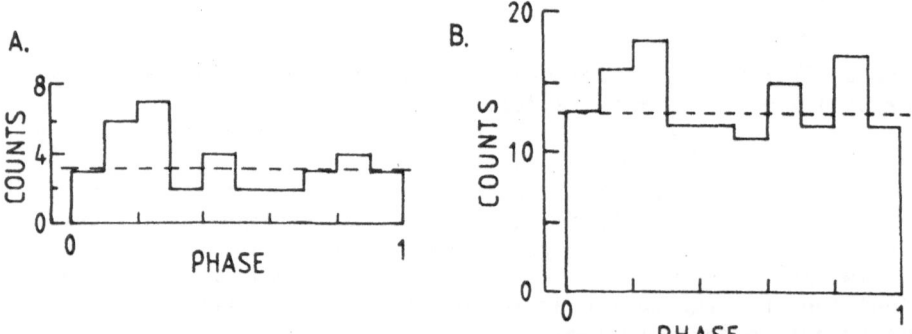

Fig. 4 Phase diagram for events from the direction of 4U1822-37.1, an object believed to be similar to Cygnus X-3. A. within 1° of the source. B. within 2°.

We have recently begun an examination of the direction of source 4U 1822-37.1 (Mason et al (1982)) with our earlier data set. This source is often compared to Cygnus X-3 and exhibits many similar characteristics. Unfortunately its period is not well known and we are unable to fold our data over three years in terms of the source period and still retain better than 0.1 cycle phase resolution. Current results are shown in figure 4.

FUTURE PROSPECTS

The Buckland Park array is being developed with the addition of further fast timing detectors. There are now 24 such detectors in operation and this will be increased to 34 in the near future.

We are currently operating a simple four detector array in Adelaide with 0.25 m^2 detectors on a 15 m baseline. This array will be upgraded and developed as a pilot array for a proposed southerly experiment to be operated first in Tasmania and then possibly at the Australian base at Mawson, Antarctica.

REFERENCES

Ciampa et al (1986) *Proc. Astron. Soc. Aust* (In the press)
Clay et al (1981) *Il Nuovo Cimento,* 4C, 668
Clay et al (1985) *Proc 19th Int. Cosmic Ray Conf., La Jolla (NASA pub. 2376)* 3, 414
Clay, R.W., Protheroe, R.J., and Gerhardy, P.R. (1984) *Nature,* 309, 687
Clay, R.W., Gerhardy, P.R., and Liebing, D.F. (1984) *Aust. J. Phys,* 37, 91
Crouch, P.C. et al (1981) *Nucl. Instrum. Meth.,* 179, 467
Gerhardy, P.R., and Clay, R.W. (1983) *J. Phys G: Nucl. Phys,* 9 1279
Mason, K.O. et al (1982) *M.N.R.A.S.,* 200, 793
Protheroe, R.J., and Clay, R.W. (1984) *Proc. Astron. Soc. Aust,* 5 586
Protheroe, R.J. and Clay, R.W., (1985) *Nature,* 315, 205
Protheroe R.J., Clay, R.W., and Gerhardy, P.R. (1984) *Ap. J. Lett,* 280, L47
Samorski, M. and Stamm, W. (1983) *Ap. J.,* 268, L17.

Acknowledgement

This work is supported by the Australian Research Grant Scheme.

THE KIEL EAS EXPERIMENT FOR DETECTING UHE GAMMA RAYS

O. C. Allkofer, M. Samorski, W. Stamm
Institut fuer Reine und Angewandte Kernphysik
University of Kiel
Olshausenstr. 40
D - 2300 Kiel
Federal Republic of Germany

ABSTRACT. A new air shower experiment specially designed for UHE gamma-ray astronomy is now under construction. It comprises 37 scintillation detectors of 1 m^2 each. It is planned to install it at an altitude of 2200 m at the Canary Island La Palma. Observations should start in summer 1987.

1.   INTRODUCTION

Our first plan to build an air shower array specially designed for UHE gamma-ray astronomy goes back to 1983 when the results about the UHE gamma-ray emission from Cygnus X-3 were published. A detailed proposal has been presented at the 19th ICRC at La Jolla (1). Here we give information about the first stage of the new Kiel EAS-array HEGRA (High Energy Gamma Ray Array) which is now under construction.

2.   DETECTOR AND ARRANGEMENT

The scintillation detectors have a size of 1 m^2 and are composed of 4 blocks of 50 x 50 x 5 cm^3 (NE 102A).
They are inside of white painted (TiO$_2$, Reflector Paint NE 560) aluminium boxes and are viewed by two phototubes (11 cm diameter) from below. One phototube is used for fast timing and the other one for particle density measurements. As a protection against sun and rain each detector box is inside a stable white hut.
The arrangement of the detectors is shown in Figure 1.
Detector separation is 15 metres. The full dots represent the 37 detectors of the initial stage of the experiment which is now under construction. All detectors of the inner 5 x 5 grid are equipped with fast phototubes. The open dots represent the planned extension at a later stage.

K. E. Turver (ed.), Very High Energy Gamma Ray Astronomy, 281–284.

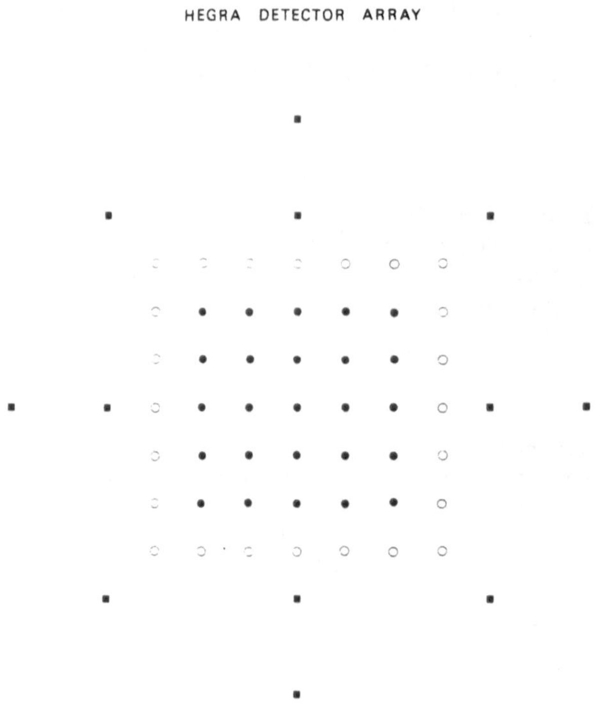

HEGRA DETECTOR ARRAY

Figure 1.   Arrangement of the HEGRA scintillation detectors.
             Grid spacing is 15 metres. The full dots represent
             the detectors of the initial stage.

3. PARTICLE DENSITY MEASUREMENT

For particle density measurements all scintillation detectors are
equipped with VALVO 54 AVP phototubes. To cover the wide range,
from one particle to several 10^4 particles in a detector, the signal is picked up from the anode and from two dynodes at lower gain.
After amplification the signals are transmitted by cables to charge
sensitive analog to digital converters (ADC's, Le Croy CAMAC-Modules
2282 B). Calibration is based on the one-particle peak of the pulse
hight distribution picked up at the anode.
From the overlaps of the measuring ranges the calibration can be extended to the highest particle densities. The pulse height corresponding to one particle will be kept constant by varying the high
voltage of the phototube automatically during run time.

4.   FAST TIMING MEASUREMENTS

For fast timing measurements an inner grid of 5 x 5 detectors is
equipped with large area RCA 4522 phototubes. Viewing from below
prevents an obliteration of the fast timing pattern by those
shower particles hitting the phototube itself. The anode pulses
are transmitted by cables to discriminators for noise suppresion
and pulse shaping. The layout of the fast timing channels is
shown in Figure 2.

HEGRA FAST TIMING LAYOUT

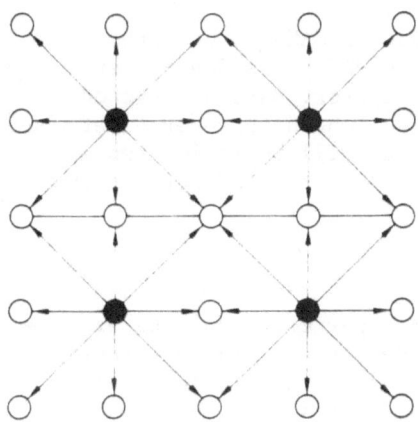

Figure 2.   Central detector array for fast timing measurements.
             The 4 detectors marked by full dots start an 8 channel
             TDC. The 8 surrounding detectors provide the stop sig-
             nals.

Four 8-fold time to digital converters (TDS's Le Croy CAMAC-Modules 2228 A) will be used for 25 detectors.
Because time measurements are based on the first particle of the
shower front the discriminator threshold must be low (leading
edge timing). For the 4 scintillation detectors giving the start
signal we will use the double discriminator technique to reduce
triggering by noise. Calibration and monitoring of the fast
timing system will be done by light pulses emitted from LED's in

front of each phototube. Short pulses of about 100 Volts peak height and a risetime of less than 2 ns are distributed by cables of well-known lengths to the LED's.

5. UNIVERSAL TIME AND ATMOSPHERIC PRESSURE

In order to look for correlations between known pulsar periods and the arrival time of air showers, the UTC (Coordinated Universal Time) of each event must be known to submilliseconds accuracy. We will use a Rubidium time standard from Rohde & Schwarz. Synchronisation to UTC will be done by time transportation, using a borrowed Caesium standard. Atmospheric pressure will be recorded during the run time.

6. THE RECORDING ELECTRONICS

The recording system is based on CAMAC. The adresses of two CAMAC-Crates are converted by the CAMAC-Crate Controller to the bus adresses of the PDP 11/73 computer. Data are stored on 5 1/4 inch Winchester-type magnetic disks which can operate at mountain altitude. All electronic equipment will be inside two air-conditioned 20 feet ISO-containers located near the centre of the array.

7. SITE OF THE EXPERIMENT

The best site for the HEGRA experiment would be at a geographical latitude of about 30°. Besides Cygnus X-3 additional objects of astrophysical interest would be in the field of view some already known from gamma-ray observations in the VHE range. A site at mountain altitude would lower the energy threshold of the array. It is planned to install the HEGRA experiment at top of the Roque de los Muchachos (2200 m a.s.1) at a latitude of 28.8° N. Observations of UHE gamma-ray sources should start in summer of 1987.

HEGRA is supported by the Deutsche Forschungsgemeinschaft DFG under Grant Al 61/31-3 and the University of Kiel.

REFERENCE

(1) O. C. Allkofer, M. Samorski and W. Stamm, 19th ICRC, La Jolla, Techniques in Ultra High Energy Gamma Ray Astronomy, eds. R.J. Protheroe and S.A. Stephens, University of Adelaide 1985, p. 60.

MUONS IN PHOTON INDUCED EXTENSIVE AIR SHOWERS

M. W. Beard, P. R. Blake, A. D. Bullock, P. J. Edwards,
W. F. Nash and G. B. Stanley
Department of Physics,
University of Nottingham,
NOTTINGHAM NG7 2RD
U.K.

ABSTRACT. An experiment is now operational at the Haverah Park EAS
detector array to study the muon content of photon induced EAS. The
muon detector (~315 MeV threshold) consists of 40 m² of liquid
scintillator shielded by barytes/lead. The recording of the muon
response is triggered by the University of Leeds GREX array at a rate
of ~6 minute^{-1}. The aim is to identify EAS coming directly from
sources such as Cygnus X-3 and to compare the muon content with
background nucleus induced EAS.

1. INTRODUCTION

The muon content of γ-induced EAS is expected to be ~10% of that of
proton-induced EAS [eg. (1)]. Thus muon measurements are expected to
yield identification of ultra-high energy photons from sources such
as Cygnus X-3. However there are some indications that the abundance
of muons in EAS from "particles" arriving directly from Cygnus X-3 is
closer to that expected from proton induced EAS rather than photon
induced EAS [eg. (2)]. Clearly there is an important need for more
accurate muon density measurements.

2. MUON DETECTOR

This consists of a total of 40 m² of liquid scintillator shielded by
barytes/lead and supporting steel. The 40 m² is made up of 16 x
2.5 m² units. The scintillator liquid used is medicinal liquid
paraffin plus paraterphenyl and P.O.P.O.P. The liquid is 10 cm deep
and is contained in perspex trays (2.84 m x 1.07 m). The ends of
the trays are extended by plane mirror light guides (~30 cm long) and
the scintillator is viewed by two 12 cm diameter phototubes, one at
either end. Summing the responses from the two phototubes yields
excellent uniformity over the whole area of each unit and a very good
resolution of single muon responses. The sides of the detector are
well shielded so that the shielding is very efficient up to high

285

K. E. Turver (ed.), Very High Energy Gamma Ray Astronomy, 285–287.

zenith angles.

The detector is calibrated in terms of its mean response to single penetrating background cosmic ray muons. The dynamic range of the detector/recording system extends from 0 up to 30 m^{-2}. The muon detector is situated close to the centre of the Haverah Park GREX array (3).

3. RECORDING

The response from each phototube is fed into an ADC. The digitalisation is initiated if the response from any tube rises above that equivalent to 0.3 of a vertical muon response. The full digitisation process lasts ~100 μs but is terminated if no trigger pulse from the EAS array is received within 1.5 μs. The EAS trigger rate is ~6/minute. The muon data is output onto magnetic tape for merging with the GREX array data and further analysis.

4. PRECAUTIONS

A number of checks have been built into the recording system to ensure the reliability of the data. It is necessary to identify the deadtime due to digitisation of background events. After each local trigger there is a deadtime of ~2 μs whilst the ADC is being cleared in the absence of a EAS trigger. During this time the detector is inoperative. The local trigger rate is set at 333/s for each separate unit. From this it is calculated that the deadtime will be $\lesssim$ 1%. Thus it is anticipated that ~1% of data contain extraneous "zero" events. Clearly it is essential that such events are identified and to this end a method of "flagging" has been adopted. This hardware flag is interrogated by the system prior to recording; if it is set then an appropriate indicator is placed in the data stream.

In order to maintain a check on the accuracy of detector response, two gain stability monitoring systems have been installed. First a two level discriminating system keeps track on the performance of the phototubes. The lower discrimination threshold is set at the local trigger level, permitting levels of noise contamination to be assessed and the system deadtime to be accurately known. The second threshold is set ~1.5 muons per unit and measurements of the integral count rate above this level provide a measure of gain stability. The system continuously cycles around the 16 detector units, monitoring the integral count rates. Each element is tested sequentially for two minutes at each level, the time elapsing between successive checks of a particular subunit being approximately one hour. If the count rate in any individual unit strays to an unacceptable level, that unit is withdrawn from operation pending investigation and correction. Results of the monitoring process are stored on an independent recording system along with the local time so that the overall detector stability over a given time period may be checked in subsequent analysis.

In order to check the recording system reliability every hour the system generates a mock GREX trigger. At this time LEDs present at each phototube are flashed, supplying a false EAS signal. This signal is recorded with the genuine data. In this way the response of the system to a known energy input is continuously monitored. In merging the Nottingham and Leeds data these events are used to check on the integrity of the data in hourly increments.

5. BURST CONTAMINATION

Anomalous responses arising from electromagnetic bursts in the absorber or scintillator are rejected by comparing the responses from neighbouring units. Any unit response which is significantly higher than the mean of the neighbouring units can be ignored.

6. OPERATION

At the time of the NATO Workshop at Durham the muon detector had been operational with high efficiency for some months and data on more than 100,000 analysed EAS had been accumulated. Lateral density distributions for background events have been derived. The mean EAS core distance of recorded events is ~60 m and the mean density ~0.2 muons m^{-2}. No analysis of photon induced EAS had yet been undertaken.

REFERENCES

(1)   S. Karakula and J. Wdowczyk, Acta Phys. Polon. **24**, 231 (1963).

(2)   M. Samorski and W. Stamm, Proc. 18th ICRC **11**, 244 (1983).

(3)   G. Brooke et al, Proc. of the Workshop on Techniques in U.H.E. γ-ray Astronomy, La Jolla, p.13 (1985).

CALCULATED TIME STRUCTURE OF AIR SHOWER FRONTS

A. M. Hillas
Physics Department
University of Leeds
Leeds LS2 9JT
England

ABSTRACT. The time structure of an air shower front is important in the accurate determination of arrival direction of showers, as is necessary to locate point sources. Experimental determinations of this structure are not yet consistent, and do not refer to gamma-ray-induced showers. Monte-Carlo simulations of showers in the PeV range show an early peak with a very long tail to the pulse obtained in a scintillator. One consequence is that statistical fluctuations in triggering time are very bad when only a single particle is detected, but get very rapidly better with 3 or more particles. (The jitter improves much more rapidly than $1/\sqrt{N}$.) Time structure will not distinguish gamma showers from hadronic showers.

1. THE PRACTICAL IMPORTANCE OF SHOWER FRONT TIME STRUCTURE

It has become important to be able to determine the arrival direction of gamma-ray-induced showers with high angular accuracy, in order to detect weak point sources of 10^{15} eV photons. This is attempted by timing the arrival of the "shower plane" at several detectors, spaced some tens of metres apart. The direction of motion of the primary photon is the normal to this plane. This plane, always moving forward at speed c, is an idealization: the shower particles straggle behind it, more so at larger distances, r, from the shower axis. Two questions prompted a calculation of the expected shower front structure. (a) How should one attempt to reconstruct the plane from observations of the passage of particles which always lie behind it? and (b) Does the observable time structure of the shower front look different for hadronic showers so that they might be distinguished from gamma showers and rejected?

2. MONTE-CARLO SIMULATIONS

A full three-dimensional Monte-Carlo simulation of shower development follows particles down to an energy of 0.05 MeV. By using thin sampling (Hillas, 1981) it is possible to average over many showers of primary energy 10^{14} or 10^{15} eV, as here, (or even 10^{19} eV) without greatly leng-

K. E. Turver (ed.), Very High Energy Gamma Ray Astronomy, 289–293.

thening the computing time. (In the present work, only a fraction of
particles having energy less than 1/200 of the primary energy were followed, the fraction decreasing with decreasing energy, and the particles
that were followed were given a "weight" greater than 1, to compensate
precisely for the incomplete sample.)

   The showers considered here were incident vertically, and observed
near sea level. The signal produced in a scintillator (a few cm thick)
was calculated (normalised to the signal of an average vertical muon).

3. RESULTS

3.1. Distribution of delay times

The delay time distribution has an early peak and a very long tail.

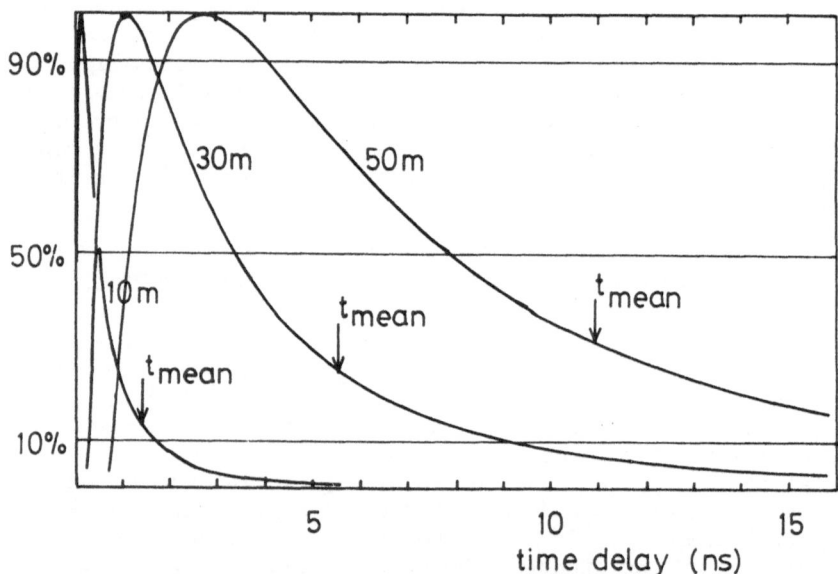

Figure 1. Shower front thickness: time distribution of scintillator signal at three distances from axis of PeV vertical
gamma-ray-induced shower at sea level.

Figure 1 shows the time distribution, measured from the passage of the
ideal plane front, at three axial distances. The tail extends well beyond the graph: the standard deviation (r.m.s. spread) of the time is
generally appreciably larger than the mean. Hence the r.m.s. spread
will be hard to determine in practice: it will be greatly affected by
the instrumental cut-off time. According to this distribution, one
would expect often to see much-delayed late pulses following the main
pulse. The distributions shown here were averaged over 10^{14} and 10^{15} eV
showers. No change was detected over this energy range.

Over at least the range 2 - 70 m from the shower axis, the distribution of scintillator signal in time may be represented by the
expression

$$f(t) = K.(1.57 + u) \; t^{-1} \exp(-\tfrac{1}{2}u^2) \; ,$$

where $u = 0.6 + w \ln(50 \, t / r^{1.5})$

$w = 0.244 + 0.4 \ln(3 + r) \; .$

Here, r is the distance in metres from the shower axis, and t the
delay time (behind the plane front) in nanoseconds. K is a normalising
constant. The formula has been fitted to the simulation results for
vertical photon-induced showers observed at sea level - in the energy
range 10^{14} - 10^{15} eV, as stated earlier.

3.2. Fluctuations in time at which recording apparatus is triggered

If the time of passage of the shower front is recorded by measuring the
time when the scintillator signal passes a very low triggering level,
well below the one-particle level, *the fluctuation in triggering delay
varies very rapidly with particle density.*
The effect was studied by drawing N samples from the time distributions described above, and finding in each case when the trigger level
was exceeded. For this purpose (though the precise figures are not critical), each scintillator particle was assumed to give a waveform as
illustrated in figure 2: a constant current pulse of duration 10 ns
was integrated to give a voltage ramp of 10 ns risetime, and the trigger
level was set at 0.1 of the height of this pulse. (Triggering thus
occurred 1 ns after the start of the pulse of the first particle, unless
another particle arrived in that 1 ns interval to speed up the voltage
rise.)

Figure 2. Simplified voltage waveform, due to a single particle, assumed in simulations of triggering time delay. (A
faster rise would reduce the mean delay but not the fluctuation in triggering delay.)

Figure 3 shows the r.m.s. fluctuation in triggering delay (timing jitter)
for four axial distances, r.
With N=1 (1 particle detected), this just represents the r.m.s.
width of the distributions of figure 1, and is very long. If several
particles are detected, it is very likely that at least one of them is in
the early part of the distribution: the delay and its fluctuation become
small. It is seen that with 3 particles detected in the scintillator,
the r.m.s. fluctuation has typically been reduced by a factor 5 or more.
The jitter is certainly not proportional to $1/\sqrt{N}$, especially for small N.
The timing jitter is so much worse when N=1, that it is advisable to
ignore all signals with N<(2 or 3) when fitting a shower front.

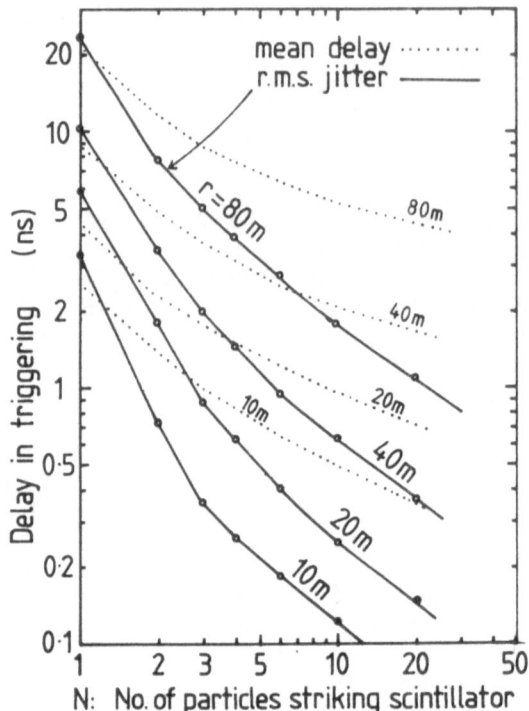

Figure 3. Effect of particle density on triggering delay
(delay relative to ideal plane shower front) at several axial
distances, r, in photon-induced PeV vertical showers at sea
level. Thick lines give r.m.s. jitter in triggering delay,
dotted lines give mean triggering delay. (The latter depends
on the rise time of the single particle response - given in
figure 2 - and would be somewhat shortened if this were faster.) Trigger level set at 0.1 particle.

3.3. Mean triggering delay: correction to obtain ideal plane front

The mean triggering delay is also shown in figure 3, by the dotted lines.
It should be practicable, from a good estimate of the distance of each
scintillator from the shower axis, and the number of particles striking
it, to subtract this expected delay from the observed triggering time,
and then to fit a plane to the corrected times, provided that at least
three detectors reached the 3-particle level.
   Although the waveform generated by a single particle has only a
small effect on the jitter in the triggering delay (assuming the fluctuation in rise time of individual particle pulses is small), it does affect
the mean delay. Thus if the pulse width is reduced from 10 ns to 4 ns,
delays are reduced - more for N=1 than for large N - and for precise
work, the graph of mean delays would be recalculated by convolution of

the waveform with the sampled particle delay times. In practice, the difference - when $N \geqslant 3$ - will not be very great.

3.4. Comparison with hadronic showers

Although muons contribute to the scintillator signal in hadronic showers, but hardly at all in gamma-ray showers, the time distributions for the two types of shower, in the PeV range, differ too little to be detected (within the range of axial distance discussed here), and the form of the scintillator pulse does not seem to offer any assistance to the rejection of background hadronic showers, when looking for gamma-ray showers from a point source. Full details will be published elsewhere.

4. CONCLUSIONS

The main conclusions are the recommendations

(a) to avoid using signals corresponding to less than about 3 particles in fitting a shower front, noting that the timing accuracy improves more like $1/N$ than $1/\sqrt{N}$ for small numbers of particles, N, and

(b) to use an N- and r- dependent correction to observed triggering times to reconstruct a plane shower front.

REFERENCE

Hillas A M, 1981. *17th Int. Conf. on Cosmic Rays, Paris*, 9: 193-6

THE NORMAL KERNEL DENSITY ESTIMATOR APPLIED TO γ-RAY LIGHT CURVES

O.C. De Jager[1], J.W.H. Swanepoel[2], B.C. Raubenheimer[1]
1 - PU-CSIR Cosmic Ray Research Unit, Potchefstroom University, 2520 Potchefstroom, South Africa
2 - Dept. of Statistics, Potchefstroom University, 2520 Potchefstroom, South Africa

The problems with the histogram estimator is well known: Rebinning makes a claimed signal questionable. In its place we propose using the reliable, consistent, objective and smooth kernel density estimators for γ-ray light curves. The important smoothing parameter can be reliably estimated from the data alone without making any a priori assumptions. We propose in particular the normal kernel which ensures good sensitivity, accuracy and little computer time.

1. Introduction

A gamma ray light curve is a periodic function which shows the source intensity as function of the phase θ. Due to the measuring and folding technique such a light curve can only be identified by estimating the population density function $f(\theta)$. Gamma ray astronomers all use the classic histogram to estimate $f(\theta)$. The most important aspect of any density estimator is the choice of the bin width or smoothing parameter h. In the past subjective choices of h for the histogram has caused many debates concerning the validity of a claimed signal. This is not the only problem that besets the user of the histogram - there also exists no optimal and objective method to chose the position of a bin. Due to these problems and the fact that information within a bin is lost, we proposed more reliable estimation techniques for γ-ray light curves (De Jager et al., 1986, hereafter Paper I). In the place of the histogram we propose the use of the reliable, objective and consistent 'kernel density estimators'. The latter phrase refers to a wide class of estimators, covered extensively in the statistical literature. In Paper I we discussed the 'naive' or moving average estimator (NE) and the truncated Fourier series estimator (FSE).

In this paper we give a brief review of these kernel techniques and substituted the naive kernel with the normal kernel, by which one gains in sensitivity (i.e. lower signal strengths can be extracted), accuracy and a considerable amount of computer time. Although the FSE is better than the naive- and normal kernel estimators for broad peaks, the losses by using the normal kernel for broad light curves may not

K. E. Turver (ed.), Very High Energy Gamma Ray Astronomy, 295–298.
© *1987 by D. Reidel Publishing Company.*

be too evident (see Figure 2 for such an example).   This kernel may also
be used for any possible form of light curve.

2. Kernel density estimators, their errors and the normal kernel

        Let $\theta=(\theta_1, \ldots, \theta_n)$ be the phases which are obtained by folding
the arrival times modulo the period of the observed object. Let there exist
an unknown underlying population density function $f(\theta)$ which generated
the sample θ in an independent an identically fashion. However, this may
never be the case for VHE γ-ray astronomy due to the possible instability
of natural VHE particle accelerators (Weekes, 1986).   A wide class of
estimators for $f(\theta)$ is the so-called 'kernel density estimators'
(Rosenblatt, 1956; Parzen, 1962):

$$\hat{f}_h(\theta;\theta) = \frac{1}{nh} \sum_{j=1}^{n} K \left(\frac{\theta-\theta_j}{h}\right) \tag{1}$$

where $K(x)$ refers to the kernel or weighting function for which there
exists a large amount of choices. The smoothing parameter is h while
$x=(\theta-\theta_j)/h$, where θ refers to the phase where the density is to be estimated and θ_j refers to a data point.   Remember that a light curve is
periodic so that data near phase 0 should contribute to the estimator for
$\theta\approx2\pi$ and vice versa.   For the naive kernel one has $K(x)=\frac{1}{2}I(-1\leq x\leq1)$, while
$K(x)=\exp(-\frac{1}{2}x^2)/\sqrt{(2\pi)}$ for the normal kernel. It is clear that the latter
kernel yields a smoother estimate than the former.
        The next step is to examine the error one makes by using (1).
A good measure of the error involved is the well known mean squared
error, MSE,   evaluated at θ:

$$\text{MSE}(\theta) = E\,(\hat{f}_h(\theta;\theta)-f(\theta))^2$$

$$= (E\hat{f}_h^2(\theta;\theta)-(E\hat{f}_h(\theta;\theta))^2) + (E\hat{f}_h(\theta;\theta)-f(\theta))^2 \tag{2a}$$

$$\approx cf(\theta)/nh + rh^4(f''(\theta))^2 \tag{2b}$$

Equation (2a) shows that the MSE consists of a variance component (first
term) and a bias squared component (second term). These terms can be
approximated by (2b) for large. n (c and r are constants depending on
the kernel used). As h increases, the variance decreases while the bias
increases and vice versa.   The variance is well known to all researchers,
but not the bias, which is difficult to work with when estimating a light
curve, since the second derivative $f''(\theta)$ of $f(\theta)$ must be known or estimated.   This problem for the kernel estimators is solved to the extend
where one obtains an objective data based choice of h which minimises
the sum of the variance and bias (see Section 3).   Since the kernel estimators are asymptotically normally distributed, one can use
$\pm qs\sqrt{\text{MSE}(\theta)}$ (with $q>1$) to construct a CONFIDENCE BAND for the estimator at a significance level of s standard deviations. Fortunately, $q\rightarrow1$
very fast with increasing n.   Usually $f(\theta)$ is unknown so that one ap-
proximates the error band of $\hat{f}_h(\theta;\theta)$ by using an estimator of the variance component of (2b):

$$f(\theta) \simeq \hat{f}_h(\theta;\underset{\sim}{\theta}) \pm s\sqrt{(c\hat{f}_h(\theta;\underset{\sim}{\theta})/nh)} \tag{3}$$

where c equals 0.28 and 0.5 for the normal and naive kernels respectively. In regions where $|f''(\theta)|$ is large, (3) will give an underestimation of the true error.

A global measure of the error is the MISE which is the MSE(θ) integrated over the whole phase range. The use of the MISE will become evident when we discuss the method to obtain a data based choice of the optimal h:

3. Reliable smoothing techniques for kernel estimators

We demand that two properties should hold for kernel density estimators: (a) The estimator $\hat{f}(\theta;\theta)$ should converge to the true density $f(\theta)$ with probability one as $n\to\infty$. This is the consistency' property and imposes the restriction that h should equal a constant times $n^{-1/5}$. This ensures a rate of convergence at a speed proportional to $n^{-4/5}$ (see (2b) for this choice of h). (b) With this restriction on h, one choses the constant for h such that the MISE is a minimum. When using the MISE as a criterion, no a priori weight is placed on any part of the light curve - thus ensuring objectivity. Let h^* be that value, then MISE(h^*) < MISE(h) for all $h \neq h^*$. However, the MISE is still a function of $f(\theta)$ and $f''(\theta)$, so that one has to obtain an unbiased estimator of the MISE without any knowledge of $f(\theta)$ and find the value of h which minimises the h-dependent terms of this estimator. Say this value is $\hat{h}$, then one will also have that $E(\hat{h})=h^*$. Another prerequisite is that the variance of $\hat{h}$ should be small. In Paper I we developed a method by which $\hat{h}$ can be estimated for the NE. That method is however applicable to ANY positive kernel density estimator. For the normal kernel the optimal h can be estimated by finding that h which minimises

$$\sum_{t=1}^{\infty} \left\{ \exp(-(ht)^2) \; \|\hat{\phi}(t)\|^2 - 2 \exp(-\tfrac{1}{2}(ht)^2) \left[\frac{n\|\hat{\phi}(t)\|^2 - 1}{n-1} \right] \right\} \tag{4}$$

where $\hat{\phi}(t)=\hat{\alpha}(t)+i\hat{\beta}(t)$ is the empirically trigonometric moments:

$$\hat{\alpha}(t) = \frac{1}{n} \sum_{j=1}^{n} \cos\theta_j \quad \text{and} \quad \hat{\beta}(t) = \frac{1}{n} \sum_{j=1}^{n} \sin\theta_j \tag{5}$$

The only difference between (4) and eq. (14) in Paper I is the substitution of (sin ht)/ht by $\exp(-\tfrac{1}{2}(ht)^2)$. The gain in computer time can now easily be seen: the latter term decreases to zero much faster than the first one. After obtaining $\hat{h}$, one can replace (1) by its Fourier representation which eases the computational technique considerably. For the normal kernel this will be

$$\hat{f}_h(\theta;\underset{\sim}{\theta}) = \frac{1}{2\pi} \left[1+2 \sum_{t=1}^{\infty} \exp(-\tfrac{1}{2}(ht)^2)(\hat{\alpha}(t)\cos t\theta + \hat{\beta}(t)\sin t\theta) \right] \tag{6}$$

The efficiency of the estimation of h^* for a single peaked light curve with duty cycle d and pulsed fraction p is determined by the fraction of times (power) for which $2h \leq d$ is obtained from (4). The case

2h>>d yields oversmoothed or flat estimates of f(θ). The criterion 2h≈d
implies a theoretically limiting signal strength or detection threshold of
1/√n. Figure 1 gives this power as function of the signal strength for
different values of d.

Using the normal kernel, we reanalysed the TeV data on
PSR1802-23 (Raubenheimer et al., 1986) to obtain the light curve in
Figure 2. Comparison with their Figure 3 shows a similar light curve
but this estimate is a continuous function with the same amplitude as that
of the histogram. Note that no rebinning was necessary to obtain Figure
2 - an obvious advantage over the histogram method. The true light
curve always lies within the confidence band at the specified confidence
level. The latter cannot be said for the histogram because confidence
intervals instead of bands are used. A confidence interval can only be
used to predict the behaviour of f(θ) in a point but nothing can be said
about the behaviour of f(θ) over the whole phasogram - however, the
latter can be predicted with confidence bands.

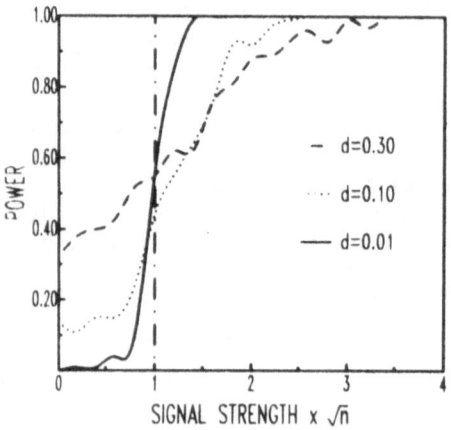
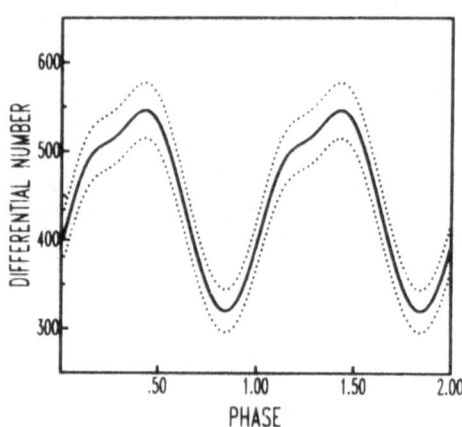

Fig. 1. The power as a function of
the signal strength for a sample
size of 500 and different duty cycles. The vertical line gives the
theoretical detection threshold.

Fig. 2. The TeV light curve for
PSR1802-23. Here ĥ=0,1 and n=447.
The confidence band is at significance level of ≈1σ. The phase is
arbitrary.

References

De Jager, O.C. et al., Astron. Astrophys., in press (1986) (Paper I)
Parzen, E., Ann. Math. Statist., **33**, 1065-1076 (1962)
Raubenheimer, B.C. et al., Ap. J. (Lett.), in press (1986)
Rosenblatt, M., Ann. Math. Statist., **27**, 832-837 (1956)
Weekes, T.C., Proc. of the VIth. Astrophysics Meeting, preprint, Les
Arcs, France (1986)

LIST OF PARTICIPANTS

Dr. P.R. Blake,
Department of Physics,
University Park,
Nottingham NG7 2RD

Professor K. Brecher,
Department of Astronomy,
Boston University,
725 Commonwealth Avenue,
Boston, MA 02215,
U.S.A.

Professor U. Camerini,
DESY-F1 85 Notkestrasse 2000,
Hamburg 52,
West Germany

Dr. M. Cawley,
Whipple Observatory,
P.O. Box 97,
Amado,
Arizona 85645 0097

Miss P.M. Chadwick,
Department of Physics,
University of Durham,
Durham DH1 3LE,
England

Dr. G. Chardin,
DPhPE/CEN Saclay,
91191 GIF/Yvette Cedex,
France

Dr. N.A. Dipper,
Department of Physics,
University of Durham,
Durham DH1 3LE,
England

Dr. David J. Fegan,
11 Knocklyon Heights,
Templeogue,
Dublin 16,
Ireland

Professor W.F. Fry,
1150 University Avenue,
Department of Physics,
University of Wisconsin,
Madison,
Wisconsin 53706,
U.S.A.

Professor J.A. Gaidos,
Department of Physics,
Purdue University,
W. Lafayette,
IN 47907,
U.S.A.

Dr. Philippe Goret,
CEN-Saclay DPhG/Sap,
91191 GIF sur Yvette,
France

Dr. W. Hermsen,
Lab. for Space Research Leiden,
Wassanaarseweg 78,
P.O. Box 9504,
2300 RA Leiden,
The Netherlands.

Dr. A.M. Hillas,
Department of Physics,
University of Leeds,
Leeds, LS2 9JT

Dr. J.V. Jelley,
29 Abbott Road,
Abingdon,
Oxon OX14 2DT

Dr. T. Kifune,
Inst. for Cosmic Ray Research,
University of Tokyo,
Midori, Tanashi,
Tokyo,
Japan.

Dr. W.R. Kropp,
Department of Physics,
University of California,
Irvine 92717,
U.S.A.

Professor R.C. Lamb,
Department of Physics,
Iowa State,
Ames,
IA 50010,
U.S.A.

Dr. R. Loveless,
3918 Yuma Drive,
Madison, Wisc. 53711,
U.S.A.

Dr. T.J.L. McComb,
Department of Physics,
University of Durham,
Durham DH1 3LE,
England.

Dr. Carlo Morello,
Instituto di Cosmogeofisica,
Corso Fiune 4,
Torino,
Italy.

Dr. R. Morse,
Department of Physics,
University of Wisconsin,
Madison WI 53706.
U.S.A.

Professor X. Moussas,
Department of Astrophysics,
National University,
GR15771 Athens,
Greece.

Professor W.F. Nash,
Department of Physics,
University of Nottingham,
Nottingham NG7 2RD

Dr. K.J. Orford,
Department of Physics,
University of Durham,
Durham, DH1 3LE,
England.

Professor T.R. Palfrey,
Department of Physics,
Purdue University,
W. Lafayette, IN 47907,
U.S.A.

Professor N.A. Porter,
Department of Physics,
University College,
Belfield,
Stillorgan Road,
Dublin 4,
Eire.

Dr. R.J. Protheroe,
Department of Physics,
University of Adelaide,
Adelaide, 5001,
Australia.

Dr. J.J. Quenby,
Blackett Laboratory,
Imperial College,
London SW7

Professor P.V. Ramana Murthy
Tata Inst. of Fundemental Res.,
Colaba,
Bombay 5,
India.

Dr. B.C. Raubenheimer,
Department of Physics,
Potchesfstroom University,
Potchesfstroom 2520,
South Africa.

Professor L. Resvanis,
Physics Laboratories,
University of Athens,
104 Solonos Street,
Athens 144,
Greece.

Professor K. Ruddick,
School of Physics,
University of Minnesota,
Minneapolis MN, 55455
U.S.A.

Dr. S. Samorski,
Institut fur Kernphysik,
University of Kiel,
Olshausestr. 40,
D-2300 Kiel,
West Germany.

Dr. P. Schlatter,
Laboratory for High Energy Physics,
University of Berne,
Sidlestrasse 5,
CH-3012 Berne,
Switzerland.

Dr. G.H. Sembroski,
Purdue University,
Department of Physics,
W. Lafayette, IN 47905,
U.S.A.

Professor Sir Graham Smith FRS,
Nuffield Radioastronomy Laboratories,
Jodrell Bank,
Macclesfield,
Cheshire, SK11 9DL

Professor G.R. Smith,
Department of Physics,
University of Manitoba,
Canada R3T 3TR

Dr. S. Stamm,
Institut fur Kernphysik,
Universitat Kiel,
Olshauenstr. 40,
D-2300 Kiel,
West Germany.

Professor J. Trümper,
MPI fur Extraterrestriche Physik,
8046 Garching BEI Munchen,
West Germany.

Dr. K.E. Turver,
Department of Physics,
University of Durham,
Durham DH1 3LE,
England.

Dr. P.R. Vishwanath,
Tata Institute of Fundamental Res.,
Calaba,
Bombay 5,
India.

Professor A.A. Watson,
Department of Physics,
University of Leeds,
Leeds 2.

Professor J. Wdowczyk,
Department of Physics,
University of Durham,
Durham DH1 3LE,
England.

Dr. T.C. Weekes,
Whipple Observatory,
Box 97,
Amado AZ 85640,
U.S.A.

Professor A.W. Wolfendale,
Department of Physics,
University of Durham,
Durham DH1 3LE,
Durham,
England.